MATHEMATICS FOR ECONOMICS

Student's Solutions Manual

Michael Hoy
John Livernois
Chris McKenna
Ray Rees
Thanasis Stengos

MATHEMATICS FOR ECONOMICS

Student's Solutions Manual

Michael Hoy
John Livernois
Chris McKenna
Ray Rees
Thanasis Stengos

Department of Economics
University of Guelph

Addison-Wesley Publishers

Don Mills, Ontario • Reading, Massachusetts
Menlo Park, California • New York • Wokingham, England
Amsterdam • Bonn • Sydney • Singapore • Tokyo • Madrid
San Juan • Paris • Seoul • Milan • Mexico City • Taipei

Publisher: Ron Doleman
Managing Editor: Linda Scott
Coordinating Editor: Madhu Ranadive
Page Layout/Desktopping: Nelson Gonzalez
Production Coordinators: Melanie van Rensburg/Wendy Moran
Manufacturing Coordinator: Sharon Latta Paterson
Cover Design: Anthony Leung
Printing & Binding: Maracle

ISBN 0-201-55367-8

Printed and bound in Canada

A B C D E -MP- 00 99 98 97 96

Contents

Note: For Answers to Chapter 17, see the back of the text. No further Solutions are required.

INTRODUCTION

This Student's Solutions Manual contains the full solutions to the odd-numbered questions in the text. Brief answers to these questions were also given at the end of the text.

Chapter 1 is a review of basic arithmetic and algebra, which may prove to be useful to those students who need to have a quick refresher course in basic math. Each section contains self-tests to help you identify gaps in your knowledge. The remaining chapters follow the chapter sequence in the text, and are organized in corresponding sections. Only odd-numbered solutions are provided. Chapter 17 has a number of exercises for which the answers in the text are sufficient. We have therefore omitted a "Chapter 17" from this Manual.

The figures are numbered according to chapter, section, and question number, so it is easy to associate a figure with its corresponding question. For example, Figure 13.1.3 is the figure for question 1 of Section 3 in Chapter 13. Figure 12.R.5 is the figure for question 5 of the Chapter Review in Chapter 12.

Updated information about the text and the solutions manual is available on our Worldwide Web site at

http:/www.css.uoguelph.ca/econ/pubfac/books.html.

Michael Hoy
John Livernois
Chris McKenna
Ray Rees
Thanasis Stengos

CHAPTER 1: REVIEW AND SELF-TESTS

In this chapter, we present a review of basic arithmetic and algebra along with a series of tests designed to help you identify your initial strengths and weaknesses. You should read this chapter during the first few days of your course. It is possible that the instructor will not be covering this basic material in class.

The chapter is in two sections. Section 1.1 reviews basic arithmetic and Section 1.2 reviews basic algebra. Each section is in turn composed of three parts: a pretest, a review, and an exit test. Use the pretest to identify your strengths. If you have difficulties, you should read the review part carefully. If you have no difficulties, you could just refresh your memory by reading the review quickly. The exit test helps you assess what you have learned from studying the review. If you remain unsure, you should discuss your problems with the instructor. Answers to all self-tests are on page 26.

1.1 Review of Basic Arithmetic

Pretest

1. What is $1/10$ of $3/4$?
a) $1/10$
a) $15/5$
b) $2/15$
b) $3/40$
c) None of the above

2. One number is 3 more than 2 times another. Their sum is 21. Find the numbers.
a) 7, 14
a) 2, 19
b) 6, 15
c) 10, 11
d) 8, 13

3. Evaluate $4\frac{1}{3} - 2\frac{5}{6}$.
a) $3\frac{1}{3}$
b) $1\frac{1}{2}$
c) $3\frac{1}{2}$
d) $2\frac{1}{2}$
e) None of the above

4. What is the product of $(\sqrt{3}+6)(\sqrt{3}-2)$?
a) $9+4\sqrt{3}$
b) -12
c) $-9+4\sqrt{3}$
d) $-9+6\sqrt{3}$
e) 9

5. What is the value of the expression $1/[1+1/(1+1/4)]$?
a) $9/5$
b) $5/9$
c) $1/2$
d) 3
e) 5

6. Which of the following has the smallest value?
a) $1/0.2$
b) $0.1/2$
c) $0.1/1$
d) $0.2/0.1$
e) $2/0.2$

7. Which is the smallest number?
a) $5(10^{-5})/3(10^{-5})$
b) $0.3/0.2$
c) $0.3/0.3(10^{-4})$
d) $5(10^{-3})/0.1$
e) $0.3/0.3(10^{-2})$

8. Evaluate $10^3 + 10^5$.
a) 10^8
b) 10^{15}
c) 20^8
d) 2^{10}
e) $101,000$

9. Evaluate $-10 - \{(2^3+27)/[3-2(8-10)]\}$.
a) 5
b) -15
c) 25
d) 10
e) 35

10. Which is the largest fraction?
a) $1/5$
b) $2/9$
c) $2/11$
d) $4/16$
e) $3/19$

11. Evaluate $[2^{(1-\sqrt{3})}]^{(1+\sqrt{3})}$.
a) 4
b) -4
c) 16
d) $1/8$
e) $1/4$

12. Evaluate $(2^{100}+2^{98})/(2^{100}-2^{98})$.
a) 2^{198}
b) 2^{99}
c) 64
d) 8
e) $5/3$

13. Evaluate $(2^{-4}+2^{-1})/2^{-3}$.
a) 9/27
b) 18
c) 1/2
d) 2^{-3}
e) 9/2

14. What is 1/10 of 3/8?
a) 1/8
b) 15/4
c) 15/2
d) 4/15
e) None of the above

15. Evaluate $3/6+2/12$.
a) 1/12
b) 5/6
c) 2/3
d) 8/9
e) 5/18

16. Simplify $6\sqrt{7}+4\sqrt{7}-\sqrt{5}+5\sqrt{7}$.
a) $10\sqrt{7}$
b) $15\sqrt{7}-\sqrt{5}$
c) $\sqrt{5}\sqrt{21}-\sqrt{5}$
d) $15\sqrt{16}$
e) 50

17. Evaluate $\sqrt{75}-3\sqrt{48}+\sqrt{147}$.
a) $3\sqrt{3}$
b) $7\sqrt{3}$
c) 0
d) 3
e) $\sqrt{3}$

18. The cheapest among the following prices is
a) 10 oz. for 16 cents
b) 2 oz. for 3 cents
c) 4 oz. for 7 cents
d) 20 oz. for 34 cents
e) 8 oz. for 13 cents

19. Evaluate $|4+(-3)|+|-2|$.
a) -2
b) -1
c) 1
d) 3
e) 9

20. Evaluate $|-42||7|$.
a) -294
b) -49
c) -35
d) 284
e) 294

ARITHMETIC REVIEW

Introduction to the Real Number System

Arithmetic is concerned with certain operations like addition, subtraction, multiplication, and division carried out on numbers and the relations between numbers expressed by such phrases as "greater than" or "less than."

We will illustrate how the general concept of a real number can serve as the basis of a mathematical theory. Let us suppose that we wish to measure the interval AB by means of the interval CD being the unit of measurement in Figure 1.1

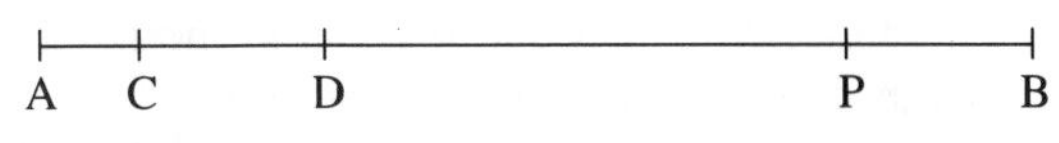

Figure 1.1

We apply the interval CD to AB by determining how many times CD fits into AB. Suppose this occurs n_0 times. If after doing this, there is a remainder PB, then we divide the interval CD into ten parts and measure the remainder with these tenths. Suppose that n_1 of the tenths go to the remainder. If after this there is still a remainder we divide our new measure (the tenth of CD) into ten parts again (that is, we divide CD into a hundred parts) and repeat the same operation. Either the process of measurement comes to an end or it continues. In either case, we reach the result that in the interval AB the whole interval CD is contained n_0 times, the tenths are contained n_1 times, the hundredths are contained n_2 times, and so on. In other words, we derive the ratio AB to CD with increasing accuracy: up to tenths, to hundredths, and so on. The ratio itself is represented by a decimal fraction with n_0 units, n_1 tenths, n_2 hundredths, and so on or

$$(\mathrm{AB}/\mathrm{CD}) = n_0 n_1 n_2 n_3 \ldots$$

The decimal fraction may be infinite, corresponding to the possibility of an indefinite increase in the precision of measurement.

The ratio of two intervals or of any two magnitudes in general is always representable by a decimal fraction, either finite or infinite. A real number may be formally defined as a finite or infinite decimal fraction.

Our definition will be complete if we say what we mean by the operations of addition and multiplication for decimal fractions. This is done in such a way that the operations defined on decimal fractions correspond to the operations on the magnitudes themselves. Therefore, when intervals are put together, their lengths are added; that is, the length of the interval AB + BC equals the sum of the lengths AB and BC. In defining the operations on real numbers there is a difficulty that these numbers are represented in general by infinite decimal fractions, while the well-known rules for these operations refer to finite decimal fractions. A rigorous definition of the operations for infinite decimals may be made in the following way. Suppose, for example, that we must add the two numbers a and b. We take the corresponding decimal fractions up to a given decimal place, say the millionth, and add them up. We thus obtain the sum $a + b$ with corresponding accuracy up to two millionths, since the errors in "a" and "b" may be added together. We are able to define the sum of two numbers with an arbitrary degree of accuracy and in that sense their sum is completely defined up to the chosen degree of accuracy.

We will define the following collections of numbers on which we will define arithmetic and algebraic operations.

The collection of **natural numbers** is denoted by

$$Z_+ = \{1, 2, 3, \ldots\}$$

and contains **all positive integers**. Adding zero to the above collection yields the set of **whole numbers** denoted by

$$Z_0 = \{0, 1, 2, 3, \ldots\}$$

whereas adding to the above all negative integers yields the collection of **all integers** denoted by

$$Z = \{\ldots, -3, -2, -1, 0, 1, 2, 3, \ldots\}$$

The next important collection of numbers is that of **rational** numbers.

Definition 1.1: A rational number is a number that can be expressed as the ratio of two integers, the divisor, of course, not being zero. The set or collection of all rational numbers is denoted by Q.

For example 3/5, 5, and 1/2 are all rational numbers, since they are expressed as the ratio a/b where "a" and "b" are integers. In this case for $3/5 = a/b$, $a = 3$ and $b = 5$. For $5 = a/b$, $a = 5$ and $b = 1$, and for $1/2 = a/b$, we have that $a = 1$ and $b = 2$. Rational numbers can be expressed in decimal form. In the above examples $1/2 = 0.5$ and $3/5 = 0.6$.

Every rational number can be expressed as a **terminating** or a **periodic** decimal. By a periodic decimal we mean a number that includes a repeating portion in its decimal part. For instance

$$\begin{aligned} 1/3 &= 0.333\dot{3} \\ 1/6 &= 0.166\dot{6} \\ 1/11 &= 0.090\dot{9}0\dot{9} \end{aligned}$$

are periodic decimals. In a sense all integers can be thought of as periodic decimal numbers with period 0. In that case

$$4 = 4.0000$$

and so on.

Any number that cannot be expressed as a terminating or periodic decimal is **irrational**. An irrational number therefore is a nonperiodic decimal. Examples of irrational numbers are

$$\begin{aligned} \sqrt{2} &= 1.414221356\ldots \\ \pi &= 3.141519265\ldots \end{aligned}$$

and

$$0.101001000100001\ldots$$

The last number has a pattern that is not periodic. The set of all irrational numbers is denoted by $\bar{Q}$.

Squaring a number means multiplying the number by itself. In this case

$$2^2 = 2 \times 2 = 4$$

$$3^2 = 3 \times 3 = 9$$

Finding the **square root** is the inverse operation to squaring a number. Numbers such as $\sqrt{2}$, $\sqrt{3}$, $\sqrt{5}$ and $\sqrt{8}$ are called **radicals.** Note that there are two square roots to any number, a positive and a negative one. For instance the square roots of 4 are 2 and -2, since $2^2 = (-2)^2 = 4$. Some square roots cannot be expressed by a periodic decimal such as

$$\begin{aligned} \sqrt{7} &= 2.6457513\ldots \\ \sqrt{3} &= 1.7320508\ldots \end{aligned}$$

These are all irrational numbers.

Real numbers

The set of real numbers $\mathbb{R}$ is the set of periodic and nonperiodic decimals. We can also define $\mathbb{R}$ as the union of the set of rational numbers Q and the set of irrational numbers $\bar{Q}$.

Examples of real numbers are 1, -2, 3, $\sqrt{2}$, $2/3$, $1/3$, $\sqrt{5}$ etc, where $\sqrt{2}$ and $\sqrt{5}$ are irrational. Each real number can be associated with a position on a number line known as the **real line**, denoted $\mathbb{R}$ as shown in Figure 1.2.

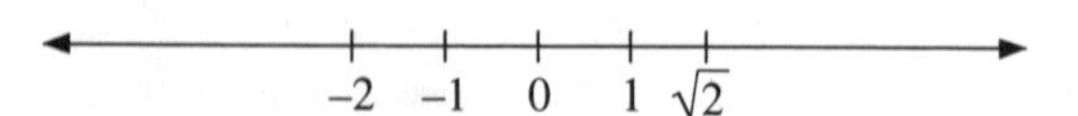

Figure 1.2

According to their positions on the real line we can determine whether one number is less than, equal to, or greater than another number. There are the following possibilities when comparing two real numbers a and b:

(i) $a < b$ (a is less than b)
(ii) $a > b$ (a is greater than b)
(iii) $a = b$ (a equals b)

Note that the two strict inequalities $<$ (less than) and $>$ (greater than) can be modified to be $\leq$ (less than or equal to) and $\geq$ (greater than or equal to). It follows that if

(i) $a < b$ and $b < c$, then $a < c$
(ii) $a > b$ and $b > c$, then $a > c$.

Example 1.1 The two inequalities $3 < 5$ and $5 < 7$ imply that $3 < 7$. Equivalently, $7 > 5$ and $5 > 3$ imply that $7 > 3$.

■

Absolute values

The absolute value of a number is represented by two vertical lines around the number and is equal to the number *without* its sign. In other words

$$|a| = \begin{cases} a \text{ if } a \geq 0 \\ -a \text{ if } a < 0 \end{cases}$$

Example 1.2

$$|5| = 5 \qquad |-3| = -(-3) = 3$$

■

The following rules apply:
a) $|-a| = |a|$
b) $|a| \geq 0$, equality holds only if $a = 0$.
c) $|a/b| = |a|/|b|$
d) $|ab| = |a||b|$
e) $|a|^2 = a^2$
Note that in general, $|a + b| \neq |a| + |b|$.

Example 1.3

$|4 - 10| = |-6| = 6$, but $|4| - |10| = 4 - 10 = 1 - 6$, and so $|4 - 10| \neq |4| - |10|$.

■

Manipulating radical numbers

We have defined a radical number to be the positive square root of a number. The following rules apply:

When multipying two radicals we have

$$\sqrt{a} \times \sqrt{b} = \sqrt{ab}$$

where $a, b \geq 0$.

Example 1.4 Simplify

$$\sqrt{100} = \sqrt{25} \times \sqrt{4} = 5 \times 2 = 10$$

■

A radical number is in its simplest form when it has the smallest possible number under the radical sign.

Example 1.5 Simplify

$$\sqrt{50} = \sqrt{25 \times 2} = \sqrt{25} \times \sqrt{2} = 5\sqrt{2}$$

A number like $5\sqrt{2}$ is known as a **mixed radical**. ■

Radicals can also be multiplied using the distributive rule that is expressed as $2x + 2y = 2(x + y)$, where x and y are any two numbers. In other words, the unequal parts x and y are added first and then they are multiplied by the common factor 2.

Example 1.6 Simplify

$$\sqrt{5}(\sqrt{2} - 2) = \sqrt{5}\sqrt{2} - 2\sqrt{5} = \sqrt{10} - 2\sqrt{5}$$

and

$$3\sqrt{7}(\sqrt{3} + 2\sqrt{5}) = (3\sqrt{7} \times \sqrt{3}) + (2 \times 3 \times \sqrt{7} \times \sqrt{5})$$
$$= 3\sqrt{21} + 6\sqrt{35}$$

■

Division of radicals is also performed by noting that

$$\sqrt{ab}/\sqrt{b} = (\sqrt{a}\sqrt{b})/\sqrt{b} = \sqrt{a}, \quad b \neq 0$$

Example 1.7 Simplify

$$\sqrt{14}/\sqrt{2} = \sqrt{14/2} = \sqrt{7}$$

and

$$[(24\sqrt{21})/6\sqrt{3}] = (24/6)\sqrt{21/3} = 4\sqrt{7}$$

■

The addition and subtraction of radicals takes place by using the distributive rule of addition, whereby we factor out the common part and, if possible, the other parts.

Example 1.8 Add the mixed radicals $5\sqrt{2}$ and $7\sqrt{2}$.

$$5\sqrt{2} + 7\sqrt{2} = (5 + 7)\sqrt{2} = 12\sqrt{2}$$

■

Note that if we have two radicals given by $\sqrt{3}$ and $\sqrt{5}$ they cannot be added using the distributive rule because they are not alike. Of course they can be added as two real numbers expressed as decimals.

Powers with integral bases

In the context of multiplication, the terms to be multiplied are called **factors**. A repeated multiplication of equal factors can be expressed as a power. For example $2 \times 2 \times 2 \times 2 = 2^4$. In the expression 2^4, 2 constitutes the **base** and 4 constitutes the **exponent**. It reads "2 to the power of 4," "2 raised to the fourth power," or "2 to the fourth." We also say that $2^4 = 16$ is the fourth power of 2. Similarly 81 is the fourth power of 3, since $3^4 = 81$ and 1000 is the third power of 10, since $10^3 = 1000$ etc.

Example 1.9 Express the following numbers as powers of integral bases.

$$25 = 5 \times 5 = 5^2$$

and

$$27 = 3 \times 3 \times 3 = 3^3$$

and

$$256 = 4 \times 4 \times 4 \times 4 = 4^4 = (2^2)^4 = 2^8$$

■

A negative base must be enclosed in brackets. A power with a negative base gives a positive result when the exponent is even and a negative result when the exponent is odd.

Example 1.10 Expand

$$(-2)^2 = (-2) \times (-2) = 4$$

and

$$(-2)^3 = (-2) \times (-2) \times (-2) = -8$$

■

Example 1.11 If $x = -2$ and $y = 4$, evaluate x^3 and $3x^2y$

$$x^3 = (-2)^3 = (-2)(-2)(-2) = -8$$

$$3x^2y = 3(-2)^2 4 = 3(-2)(-2)4 = 48$$

■

Example 1.12 If $x = -2$ and $y = 4$, evaluate $(x^2 - y^2)^2$.
In this case

$$\begin{aligned}(x^2 - y^2)^2 &= [(-2)^2 - 4^2]^2 \\ &= (4 - 16)^2 = (-12)^2 = 144\end{aligned}$$

■

Suppose that we have two numbers expressed as 3^4 and 3^7. Then multiplication between these two numbers takes place as follows:

$$\begin{aligned}3^4 \times 3^7 &= 3 \times 3 \times 3 \times 3 \times 3 \times \\ &\quad 3 \times 3 \times 3 \times 3 \times 3 \times 3 \\ &= 3^{11}\end{aligned}$$

In general the exponent rule for multiplication, where the number y^a is multiplied by y^b, with y being the base and a and b being the exponents, is given by

$$(y^a)(y^b) = y^{a+b}$$

When we multiply two powers of the same base we add the exponents. On the other hand, when we divide two powers of the same base, we subtract the exponents. For instance

$$\begin{aligned}3^7/3^4 &= (3 \times 3 \times 3 \times 3 \times 3 \times 3 \times 3)/(3 \times 3 \times 3 \times 3) \\ &= 3 \times 3 \times 3 = 3^3\end{aligned}$$

Hence $3^7/3^4 = 3^{7-4} = 3^3$.
In general,

$$y^a/y^b = y^{a-b}, \quad \text{for } y \neq 0$$

In the case of a power of a power, we multiply the exponents. For instance

$$(2^3)^2 = (2^3) \times (2^3) = (2 \times 2 \times 2) \times (2 \times 2 \times 2) = 2^6$$

In general,

$$(y^a)^b = y^{ab}$$

If we have two powers with different bases such as 5^3 and 7^4, then the expression $(5^3 \times 7^4)^2$ becomes

$$\begin{aligned}(5^3 \times 7^4) \times (5^3 \times 7^4) &= 5^3 \times 5^3 \times 7^4 \times 7^4 \\ &= (5^3)^2 \times (7^4)^2 \\ &= 5^6 \times 7^8\end{aligned}$$

In general

$$(x^a y^b)^c = x^{ac} y^{bc}$$

Also the power of a quotient becomes

$$(x^a/y^b)^c = x^{ac}/y^{bc} \quad \text{for } y \neq 0$$

Example 1.13 Simplify $(2y^2)(3y^5)$, $(x^2y^3)(x^4y)$, and $(20/15)(x^5/x^3)$
We have

$$(2y^2)(3y^5) = (2)(3)(y^2)(y^5) = 6(y^7)$$

$$(x^2y^3)(x^4y) = x^2x^4y^3y = x^6y^4$$

and

$$(20/15)x^5/x^3 = (4/3)x^{5-3} = (4/3)x^2$$

■

Suppose we divide 2^2 by 2^2. Then we have

$$2^2/2^2 = 2^{2-2} = 2^0 = 1$$

In general x^a divided by x^a, for $x \neq 0$, becomes

$$x^a/x^a = x^{a-a} = x^0 = 1$$

Suppose we divide 2^2 by 2^5. Then we have

$$\begin{aligned}2^2/2^5 &= (2 \times 2)/(2 \times 2 \times 2 \times 2 \times 2) \\ &= 1/(2 \times 2 \times 2) = 1/2^3\end{aligned}$$

Therefore, $2^2/2^5 = 2^{2-5} = 2^{-3} = 1/2^3$.
Generally

$$x^a/x^b = x^{a-b} = 1/(x^{b-a}) \quad \text{for } x \neq 0, \; a < b$$

Furthermore

$$x^{-a} = 1/(x^a) \quad \text{for } x \neq 0$$

Example 1.14 Simplify -3^{-2}, $-3/4^{-1}$, $(-3)^{-2}$, and $(x^5y^2)/(x^3y^3)$
We have

$$-3^{-2} = -1/3^2 = -1/9$$

$$-3/4^{-1} = -3/(1/4) = -12$$

$$(-3)^{-2} = 1/(-3)^2 = 1/9$$

and

$$\frac{(x^5y^2)}{(x^3y^3)} = \frac{x^{5-3}}{y^{3-1}} = \frac{x^2}{y}$$

■

EXIT TEST

1. Evaluate $|43 - 62| - |-17 - 3|$.
a) -39
b) -19
c) -1
d) 1
e) 39

2. Evaluate $|-76|/|-4|$.
a) -20
b) -19
c) 13
d) 19
e) 22

3. Simplify $\sqrt{3}\sqrt{12}$.
a) 3
b) $\sqrt{15}$
c) $\sqrt{36}$
d) 6
e) 8

4. Simplify $(3\sqrt{5})(2\sqrt{5})$.
a) $5\sqrt{5}$
b) 25
c) 30
d) 6
e) $6\sqrt{5}$

5. Simplify $\sqrt{10}/\sqrt{2}$.
a) $\sqrt{8}$
b) $2\sqrt{2}$
c) $\sqrt{5}$
d) $2\sqrt{5}$
e) $5\sqrt{2}$

6. Simplify $\sqrt{48}/\sqrt{8}$.
a) $4\sqrt{3}$
b) $3\sqrt{2}$
c) $\sqrt{6}$
d) 6
e) 12

7. Simplify $\sqrt{7} + 3\sqrt{7}$.
a) $3\sqrt{7}$
b) $4\sqrt{7}$
c) $3\sqrt{14}$
d) $4\sqrt{14}$
e) $2\sqrt{21}$

8. Simplify $3\sqrt{32} + 2\sqrt{2}$.
a) $5\sqrt{2}$
b) $\sqrt{34}$
c) $14\sqrt{2}$
d) $5\sqrt{34}$
e) $6\sqrt{64}$

9. Simplify $6\sqrt{5} + 2\sqrt{45}$.
a) $12\sqrt{5}$
b) $8\sqrt{50}$
c) $40\sqrt{2}$
d) $12\sqrt{50}$
e) $8\sqrt{5}$

10. Evaluate $2^2 2^5 2^3$.
a) 2^{10}
b) 4^{10}
c) 8^{10}
d) 2^{30}
e) 8^{30}

11. Simplify $a^4 b^2 a^3 b$.
a) ab
b) $2a^7 b^2$
c) $2a^{12} b$
d) $a^7 b^3$
e) $a^7 b^2$

12. Simplify $m^8 n^3 m^2 n m^4 n^2$.
a) $3m^{16} n^6$
b) $m^{14} n^6$
c) $3m^{14} n^6$
d) $3m^{14} n^5$
e) m^2

13. Simplify $11^8/11^5$.
a) 1^3
b) 11^3
c) 11^{13}
d) 11^{40}
e) 88^5

14. Simplify $x^{10} y^8 / x^7 y^3$.
a) $x^2 y^5$
b) $x^3 y^4$
c) $x^3 y^5$
d) $x^2 y^4$
e) $x^5 y^3$

15. Simplify $c^{17}d^{12}e^4/c^{12}d^8e$.
a) $c^4d^5e^3$
b) $c^4d^4e^3$
c) $c^5d^8e^4$
d) $c^5d^4e^3$
e) $c^5d^4e^4$

16. Evaluate $(3^6)^2$.
a) 3^4
b) 3^8
c) 3^{12}
d) 9^6
e) 9^8

17. Simplify $(a^4b^3)^2$.
a) $(ab)^9$
b) a^8b^6
c) $(ab)^{24}$
d) a^6b^5
e) $2a^4b^3$

18. Simplify $(r^3p^6)^3$.
a) r^9p^{18}
b) $(rp)^{12}$
c) r^6p^9
d) $3r^3p^6$
e) $3r^9p^8$

19. Simplify $(m^6n^5q^3)^2$.
a) $2m^6n^5q^3$
b) m^4n^3q
c) $m^8n^7q^5$
d) $m^{12}n^{10}q^6$
e) $2m^{12}n^{10}q^6$

20. Simplify $10\sqrt{32}+6\sqrt{2}$.
a) $46\sqrt{2}$
b) 64
c) $26\sqrt{2}$
d) $80+6\sqrt{2}$
e) $20\sqrt{16}+6\sqrt{2}$

1.2 Review of Basic Algebra

Pretest

1. Find the value of x in the equation $y = (n/2)(x + b)$.
a) $(2y - b)/n$
b) $2n/(y - b)$
c) $2y - b$
d) $(2y/n) - b$
e) None of the above

2. What is the factorization of $x^2 + ax - 2x - 2a$?
a) $(x + 2)(x - a)$
b) $(x - 2)(x + a)$
c) $(x + 2)(x + a)$
d) $(x - 2)(x - a)$
e) None of the above

3. What is the value of x in the equation $\sqrt{(5x - 4)} = -1$?
a) 2
b) 5
c) No value exists
d) 1
e) -4

4. If $T = 2\pi\sqrt{(L/g)}$, then L is equal to
a) $T^2/2\pi g$
b) $T^2 g/2\pi$
c) $T^2 g/4\pi^2$
d) $T^2 g/4\pi$
e) $T^2/4\pi^2 g$

5. Simplify $1 - y/(x + 2y) + y/(x - 2y)$.
a) 0
b) 1
c) $1/[(x - 2y)(x + 2y)]$
d) $2x - y/[(x - 2y)(x + 2y)]$
e) $x^2/[(x - 2y)(x + 2y)]$

6. Given $[(a+x)+y]/(x+y) = (b+y)/y$, what is x/y?
a) a/b
b) b/a
c) $b/a - 1$
d) $a/b - 1$
e) 1

7. Which of the following statements are true if

$$\begin{aligned} x + y + z &= 10 \\ y &\geq 5 \\ 3 \leq z &\leq 4? \end{aligned}$$

a) $x < y$
b) $x > y$
c) $x + z \leq y$
d) $x < y$ and $x + z \leq y$
e) $x > y$ and $x + z \leq y$

8. What is $\sqrt{x\sqrt{x\sqrt{x}}}$?
a) $x^{7/8}$
b) $x^{7/4}$
c) $x^{1/16}$
d) $x^{3/4}$
e) $x^{1/8}$

9. If $z = x^a$ and $y = x^b$, then $z^b y^a$ is
a) $x^{(ab)^2}$
b) x^{ab}
c) x^0
d) x^{2ab}
e) x

10. If $4^{x-3} = (\sqrt{2})^x$, then the value of x is
a) 0
b) 5
c) 4
d) 2
e) 1/4

11. If $f(x) = 2x - 5$, then $f(x + h)$ is
a) $2x + h - 5$
b) $2h - 5$
c) $2x + 2h - 5$
d) $2x - 2h + 5$
e) $2x - 5$

12. If $x > 1/5$, then
a) x is greater than 1
b) x is greater than 5
c) $1/x$ is greater than 5
d) $1/x$ is less than 5
e) None of the above

13. If $x + 2y > 5$ and $x < 3$, then $y > 1$ is true
a) Never
b) Only if $x = 0$
c) Only if $x > 0$
d) Only if $x < 0$
e) Always

14. The quotient $(x^2 - 5x + 3)/(x + 2)$ is
a) $x - 7 + 17/(x + 2)$
b) $(x - 3) + 9(x + 2)$
c) $x - 7 - 11/(x + 2)$
d) $x - 3 - 3/(x + 2)$
e) $x + 3 - 3(x + 2)$

15. If x and y are two different real numbers and $xz = yz$, then the value of z is
a) $x - y$
b) 1
c) x/y
d) y/x
e) 0

16. Solving for x in $5/x = 2/(x-1) + 1/[x(x-1)]$, we obtain the value of x to be
a) -1
b) 0
c) 3
d) 2
e) 5

17. If $2x + y = 2$ and $x + 3y > 6$, then
a) $y \geq 2$
b) $y > 2$
c) $y < 2$
d) $y \leq 2$
e) $y = 2$

18. The expression $(x + y)^2 + (x - y)^2$ is equivalent to
a) $2x^2$
b) $2y^2$
c) $2(x^2 + y^2)$
d) $x^2 + 2y^2$
e) $2x^2 + y^2$

19. If $x + y = 1/a$ and $x - y = a$, then the value of $x^2 - y^2$ is
a) 4
b) 1
c) 0
d) a^2
e) $1/a^2$

20. If $3/(x - 1) = 2/(x + 1)$, then x is
a) -5
b) -1
c) 0
d) 1
e) 5

ALGEBRA REVIEW

Polynomials

A mathematical expression using numbers or variables combined to form a product or a quotient is called a **term**. Examples of terms are $6x, 4y^3, 2xy, 3$, etc. The number part of a term is called the numerical coefficient. For example, 6 is the numerical coefficient of the term $6x$. A term may also have a variable part. A **variable** is a symbol that may take any value of a particular set. For example, y is the variable component of the term $4y^3$. A **polynomial** is an algebraic expression formed by adding or subtracting terms whose variables have positive integral exponents. We then classify polynomials by the number of terms they contain:

$6x$: one term - monomial
$3x + 5y$: two terms - binomial
$3x + 5y - 5$: three terms - trinomial

The **degree** of a term is the sum of the exponents of its variables.

$3y^2$ is a term of degree 2
$5xy^2$ is a term of degree 3
$2x^2y^4$ is a term of degree 6

Terms that have the same variable factors such as $5xy$ and $8xy$ are known as *similar* terms. Using the distributive rule of collecting together the similar terms we see that

$6xy + 3xy = (6 + 3)xy = 9xy$
$5x + 2x + 3x = (2 + 5 + 3)x = 10x$

To add polynomials we collect together all the similar terms.

Example 1.15 Add together $2x^2+3x+1$ and $3x^2+x-3$.
We have

$$(2x^2 + 3x + 1) + (3x^2 + x - 3) = 5x^2 + 4x - 2$$

■

To subtract a polynomial from another polynomial, multiply each term of the first one by -1 and add them to the terms of the second polynomial.

Example 1.16 Subtract (x^2+2x-3) from $(4x^2-3x+1)$.
Multiply $x^2 + 2x - 3$ by (-1) to obtain

$$(-1)(x^2 + 2x - 3) = -x^2 - 2x + 3$$

Then add to $(4x^2 - 3x + 1)$ to obtain

$$\begin{aligned}(4x^2 - 3x + 1) \quad + \quad & (-x^2 - 2x + 3) \\ = \quad & 4x^2 - x^2 - 3x - 2x + 1 + 3 \\ = \quad & 3x^2 - 5x + 4\end{aligned}$$

■

Multiplying monomials involves multiplying the coefficient and then multiplying the variables. For example

$$(3x)(4x^2) = (3)(4)(x)(x^2) = 12x^3$$

$$(2x)(3y^2) = (2)(3)(x)(y^2) = 6xy^2.$$

Using the distributive rule, the product of a polynomial and a monomial is treated as successive products of monomials.

Example 1.17 Multiply $3x(5x^2 + 2y^2)$

$$3x(5x^2 + 2y^2) = (3x)(5x^2) + (3x)(2y^2) = 15x^3 + 6xy^2$$

■

We also use the distributive law to multiply two binomials.

Example 1.18 Multiply $(x + 2)(x^2 + 1)$

$$\begin{aligned}(x + 2)(x^2 + 1) &= (x)(x^2) + x + 2(x^2) + 2 \\ &= x^3 + 2x^2 + x + 2\end{aligned}$$

In other words, we multiply each term of one binomial by each term of the other binomial.

■

Example 1.19 Multiply $(2x - 3)(5x + 3)$

$$\begin{aligned}(2x - 3)(5x + 3) &= (2x)(5x) - 3(5x) + 3(2x) - 9 \\ &= 10x^2 - 15x + 6x - 9 \\ &= 10x^2 - 9x - 9\end{aligned}$$

■

The same procedure is used when we multiply binomials involving radicals.

Example 1.20 Multiply $(3\sqrt{2}-4\sqrt{3})(3\sqrt{3}+3\sqrt{2})$

$$\begin{aligned}(3\sqrt{2}-4\sqrt{3})(3\sqrt{3}+3\sqrt{2}) &= (3\sqrt{2})(3\sqrt{3}) \\ &\quad +(3\sqrt{2})(3\sqrt{2}) \\ &\quad -(4\sqrt{3})(3\sqrt{3}) \\ &\quad -(4\sqrt{3})(3\sqrt{2}) \\ &= 9\sqrt{6}+9\sqrt{4} \\ &\quad -12\sqrt{9}-12\sqrt{6} \\ &= -(18+3\sqrt{6})\end{aligned}$$

■

Special products of polynomials

Let us look at the following two special products, $(a+b)^2$ and $(a-b)^2$. To square a binomial add the square of the first term, the square of the second term, and twice the product of both terms. In the case of $(a-b)^2$ we can write it as $(a+(-b))^2$. Therefore

$$\begin{aligned}(a+b)^2 &= (a+b)(a+b) = a^2+ba+ab+b^2 \\ &= a^2+2ab+b^2 \\ (a-b)^2 &= (a+(-b))^2 = a^2+b^2+2(-b)a \\ &= a^2-2ab+b^2\end{aligned}$$

Another important special product is given by the product $(a+b)(a-b)$. This becomes

$$(a+b)(a-b) = a^2+ba-ab-b^2 = a^2-b^2$$

In this case, the product reduces to the difference of the squared terms.

Example 1.21 Simplify $(3x-2y)(3x+2y)$

$$(3x-2y)(3x+2y) = 9x^2-4y^2$$

■

Example 1.22 Simplify $(2\sqrt{3}+3\sqrt{2})(2\sqrt{3}-3\sqrt{2})$

$$\begin{aligned}(2\sqrt{3}+3\sqrt{2})(2\sqrt{3}-3\sqrt{2}) &= (2\sqrt{3})^2-(3\sqrt{2})^2 \\ &= 2^2 3 - 3^2 2 \\ &= 12-18 = -6\end{aligned}$$

■

In this case, when we multiply two binomial expressions involving radicals and we obtain a solution that reduces to a rational number then the radicals are known as **conjugate** radicals.

We can simplify radical expressions with a binomial denominator by multiplying the numerator and the denominator by the conjugate of the denominator.

Example 1.23 Simplify $3/(3+\sqrt{2})$

Multiply and divide $3/(3+\sqrt{2})$ by $(3-\sqrt{2})$, the conjugate of the denominator.

We get

$$\begin{aligned}[3/(3+\sqrt{2})][(3-\sqrt{2})/(3-\sqrt{2})] &= \frac{3(3-\sqrt{2})}{(9-4)} \\ &= (9-3\sqrt{2})/5\end{aligned}$$

■

The above results can be extended to higher powers of binomials. For instance

$$\begin{aligned}(a+b)^3 &= (a+b)^2(a+b) \\ &= (a^2+2ab+b^2)(a+b) \\ &= (a^2+2ab+b^2)a+(a^2+2ab+b^2)b \\ &= a^3+2a^2b+b^2a+a^2b+2ab^2+b^3 \\ &= a^3+3a^2b+3ab^2+b^3\end{aligned}$$

Similarly,

$$\begin{aligned}(a-b)^3 &= (a-b)^2(a-b) \\ &= (a^2-2ab+b^2)(a-b) \\ &= (a^2-2ab+b^2)a-(a^2-2ab+b^2)b \\ &= a^3-2a^2b+b^2a-a^2b+2ab^2+b^3 \\ &= a^3-3a^2b+3ab^2-b^3\end{aligned}$$

Common factors

Expressing a polynomial as a product of two or more polynomials is called factoring. In fact, factoring is the opposite operation of expanding a polynomial by means of the distributive property. For instance

$$6x(2x-1) = 12x^2-6x$$

This operation is an expansion. Its opposite works as

$$12x^2-6x = 6x(2x-1)$$

When a factor is contained in every term of an algebraic expression, it is called a common factor.

Example 1.24 Factor the following expressions

(i) $5xy+20y$, (ii) $6x^3-3x^2+12$

(i)

$$5xy+20y = 5y(x)+5y(4) = 5y(x+4)$$

(ii)

$$\begin{aligned}6x^3-3x^2+12x &= 3x(2x^2)-3x(x)+3x(4) \\ &= 3x(2x^2-x+4)\end{aligned}$$

■

Many polynomials such as $x^2 + 3x - 4$ can be written as the product of two polynomials of the form $(x + r)$ and $(x + s)$. In other words

$$\begin{aligned}(x+r)(x+s) &= x^2 + rx + sx + rs \\ &= x^2 + (r+s)x + rs \\ &= x^2 + bx + c\end{aligned}$$

where $b = s + r$ and $c = rs$.

For example, write $x^2 + 3x - 4$ as the product $(x + r)(x + s)$ we have to find r and s such that they satisfy the requirement that $b = s+r$ and $c = rs$ in x^2+bx+c. In our case $b = 3$ and $c = -4$. Therefore, $3 = s+r$ and $-4 = sr$. Choose $r = 4$ and $s = -1$. Then $r + s = 4 - 1 = 3$ and $rs = 4(-1) = -4$. Therefore

$$x^2 + 3x - 4 = (x + 4)(x - 1)$$

Example 1.25 Factor $x^2 - 3x - 10$

We have that $r + s = -3$ and $rs = -10$. Choosing $r = -5$ and $s = 2$ (or $s = -5$ and $r = 2$) we have that $-5 + 2 = -3$ and $(-5)2 = -10$. Therefore

$$x^2 - 3x - 10 = (r - 5)(s + 2)$$

■

Factoring $ax^2 + bx + c, a \neq 1$

Factoring trinomials of the above type can be simplified if we break up the middle term into two parts. Let us fix ideas by looking at the following example $6x^2 + 15x + 9$. In that case

$$\begin{aligned}6x^2 + 15x + 9 &= 6x^2 + 9x + 6x + 9 \\ &= 3x(2x+3) + 3(2x+3) \\ &= (3x+3)(2x+3)\end{aligned}$$

Note that we split $15x$ into $9x + 6x$. We could have split it up into $10x + 5x$ or $11x + 4x$ etc. However, we will see that $9x + 6x$ is the correct break up of $15x$ in the above trinomial. Let us analyze the general expansion

$$\begin{aligned}(px+r)(qx+s) &= pqx^2 + psx + rqx + rs \\ &= pqx^2 + (ps+qr)x + rs\end{aligned}$$

Denoting $pq = a$, $ps + qr = b$ and $rs = c$, we can write

$$pqx^2 + (ps + qr)x + rs$$

as $ax^2 + bx + c$. If we break the middle term bx into two terms, say mx and nx, then it is clear that

$$m + n = ps + qr = b, \qquad mn = psqr = ac$$

In the example, $6x^2 + 15x + 9$ where $a = 6$, $b = 15$, $c = 9$, we can see that $m = 9$ and $n = 6$ (since $mn = ac$). Therefore, $m + n = b$, since $9 + 6 = 15$ and $mn = ac = (6)(9) = 54$.

The general solution to $ax^2 + bx + c = 0$, gives the values of x that satisfy the quadratic equation above. They are known as the **roots** of the quadratic equation and are given by

$$\begin{aligned}x_1 &= \frac{-b + \sqrt{b^2 - 4ac}}{2a} \\ x_2 &= \frac{-b - \sqrt{b^2 - 4ac}}{2a}\end{aligned}$$

Example 1.26 Factor $6x^2 + 7x + 2$ and explicitly obtain its roots.

In the above case, $a = 6$, $b = 7$, and $c = 2$. We have that $m + n = 7$ and $mn = 12$. For $m = 3$ and $n = 4$, we have

$$\begin{aligned}6x^2 + 7x + 2 &= 6x^2 + 4x + 3x + 2 \\ &= 2x(3x+2) + 3x + 2 \\ &= (2x+1)(3x+2)\end{aligned}$$

The roots are given by

$$x_1 = \left[-7 + \sqrt{(49 - 48)}\right]/12$$

and

$$x_2 = \left[-7 - \sqrt{(49 - 48)}\right]/12$$

Hence, $x_1 = -1/2$ and $x_2 = -2/3$.

■

Factoring special quadratic polynomials

We can use the following identities to factor certain polynomials

$$a^2 + 2ab + b^2 = (a+b)(a+b) = (a+b)^2$$

$$a^2 - 2ab + b^2 = (a-b)(a-b) = (a-b)^2$$

Also

$$a^2 - b^2 = (a-b)(a+b)$$

Example 1.27 Factor the following polynomials (i) $9x^2 + 30x + 25$, (ii) $4x^2 - 12x + 9$, and (iii) $16x^2 - 49$

(i)

$$\begin{aligned}9x^2 + 30x + 25 &= (3x)^2 + 2(3x)5 + (5)^2 \\ &= (3x+5)(3x+5)\end{aligned}$$

(ii)

$$\begin{aligned}4x^2 - 12x + 9 &= (2x)^2 - 2(2x)3 + (3)^2 \\ &= (2x-3)^2\end{aligned}$$

(iii)

$$16x^2 - 49 = (4x)^2 - 7^2 = (4x-7)(4x+7)$$

■

Dividing a polynomial by a monomial

The rule for the division of exponents is

$$x^b/x^a = x^{b-a} \quad \text{for } x \neq 0$$

For example $x^5/x^2 = x^{5-2} = x^3$.

To divide a polynomial by a monomial, each term of the polynomial is divided by the monomial. In other words, we apply the distributive rule.

Example 1.28 Simplify $(10x^5 - 3x^2 + 3)/2x$

$$(10x^5 - 3x^2 + 3)/2x = (10x^5/2x) - (3x^2/2x) + 3/2x = 5x^4 - (3/2)x + 3/(2x)$$

■

In simplifying rational expressions, it may be necessary to factor both the numerator and the denominator, if possible, and then to divide by any common factors.

Example 1.29 Simplify $(3x^2)/(6x^2 - 12x)$

$$(3x^2)/(6x^2 - 12x) = (3x^2)/[3x(2x-4)] = x/(2x-4) \quad \text{for } x \neq 2, 0$$

Note that we need the restrictions $x \neq 2$ and $x \neq 0$ because division by zero is not defined.

■

Multiplying and dividing rational expressions

When we multiply rational expressions, we may want to first factor the numerator and denominator and then divide by any common factors. We then express the product as a rational expression.

Example 1.30 Multiply $(x^2-3x-10)/(x-5)$ by $1/(x^2+4x+4)$

Note that $x^2-3x-10$ can be factored as $(x+r)(x+s)$ where r and s satisfy $r+s=-3$ and $rs=-10$. We can choose $r=-5$ and $s=2$. Then $(x^2-3x-10) = (x-5)(x+2)$. Also (x^2+4x+4) can be factored as $(x+2)^2$. Therefore

$$\frac{(x^2-3x-10)}{(x-5)}\frac{1}{(x^2+4x+4)} = \frac{(x-5)(x+2)}{(x-5)(x+2)^2} = 1/(x+2)$$

■

Rational expressions are divided in the same way as we divide rational numbers.

Example 1.31 Divide $[(2x-2)/(x^2-x-20)]$ by $(x-5)/(x+4)$.

We first factor (x^2-x-20) into $(x-5)(x+4)$, since $4-5=-1$ and $4(-5)=-20$. Then we have that

$$\frac{[(2x-2)/(x^2-x-20)]}{[(x-5)/(x+4)]} = \frac{(2x-2)}{(x-5)(x+4)}\frac{(x+4)}{(x-5)} = (2x-2)/(x-5)^2$$

■

Adding and subtracting rational numbers

To add or subtract rational expressions with equal denominators we use the distributive property:

$$a/b + c/b = a(1/b) + c(1/b) = (1/b)(a+c) = (a+c)/b$$

Example 1.32 Simplify $1/x + 7/x - 3/x$

$$1/x + 7/x - 3/x = (1+7-3)/x = 5/x$$

■

In the case, where the denominators are not the same we have to find the **least common multiple** of the denominator and then transform the original expressions to new ones with a common denominator. The least common multiple (LCM) for two numbers 10 and 25 is found as follows. We take each number and expand it in terms of the factors that make it up that cannot be further reduced. Therefore

$$10 = (2)(5) \quad \text{and} \quad 25 = (5)(5)$$

The LCM for 10 and 25 must include all the separate irreducible factors that make up 10 and 25 and is given by $(2)(5)(5) = 50$.

The **least common denominator** (LCD) is the LCM of the denominators.

Example 1.33 Simplify $(3x-1)/3 + (x-2)/2 - (x-1)/4$

The LCD of 3, 2, and 4 is $(3)(2)(2) = 12$. Then

$$\begin{aligned}&(3x-1)/3 + (x-2)/2 - (x-1)/4\\ =\ &4(3x-1)/[(4)(3)] + 6(x-2)/[(6)(2)]\\ &-3(x-1)/[(3)(4)]\\ =\ &(12x-4+6x-12+3x-3)/12\\ =\ &(21x-17)/12\end{aligned}$$

In order to transform the first term of the sum $(3x-1)/3$ to a term with a denominator of 12, we multiply both numerator and denominator by 4 so that the term remains unchanged. We do the same with the other two terms $(x-2)/2$ and $(x-1)/4$, where $(x-2)/2$ has both its numerator and denominator multiplied by 6 and $(x-1)/4$ has its numerator and denominator multiplied by 3. ■

Equations in One Variable

An equation is a statement that two expressions are equal. The equation $3x - 2 = 4$ states that the value of the left-hand side, $3x - 2$, is equal to the value 4 that makes up the right-hand side. For this to be true, x has to equal to 2. Then $3(2) - 2 = 4$.

To solve an equation, we have to find the value of the variable that makes the statement true. This value is called a **root**. The set consisting of all possible roots of an equation is called the **solution set**.

Rules for solving equations

If $x = y$, then

$$\begin{aligned} x+a &= y+a && \text{Addition Rule (AR)} \\ x-a &= y-a && \text{Subtraction Rule (SR)} \\ (x)(a) &= (y)(a) && \text{Multiplication Rule (MR)} \\ x/a &= y/a, \quad a \neq 0 && \text{Division Rule (DR)} \end{aligned}$$

Example 1.34 Solve $3x - 2 = 2x + 3$ for all x that are real numbers, or for $x \in \Re$

$$\begin{aligned} 3x-2 &= 2x+3 \\ 3x-2-2x &= 2x+3-2x, \quad \text{using the SR} \\ x-2+2 &= 3+2, \quad \text{using the AR} \\ x &= 5 \end{aligned}$$

Hence the root is 5. ■

Example 1.35 Solve $5x - 3.2 = 3.2 + 1.5x$

We collect all the terms with the x's together and the ones without any x's together.

$$5x - 1.5x = 3.2 + 3.2$$

(We add to both sides of the equation $-1.5x$ and 3.2). Then

$$3.5x = 6.4$$

Dividing both sides by 3.5 yields the root $x = 6.4/3.5$. ■

Equations involving rational expressions

In this case, we have to find first the LCD and we then have to multiply both sides of the equation by it.

Example 1.36 Solve $(2/3)x-(1/4)(x-4) = (2x-3)/2$.

The LCD is $12((3)(2)(2) = 12)$. Then we multiply both sides of the equation by 12.

$$\begin{aligned} 12[(2/3)x - (1/4)(x-4)] &= 12[(2x-3)/2] \\ 8x - 3(x-4) &= 12x - 18 \\ 8x - 3x + 12 &= 12x - 18 \end{aligned}$$

Collecting all the terms with x's together and the ones without x's together we obtain

$$\begin{aligned} 12x - 8x + 3x &= 18 + 12 \\ 7x &= 30 \\ x &= 30/7 \end{aligned}$$

■

Inequalities in one variable

Inequalities are expressions that involve the symbols "greater than, >", "less than, <", "greater than or equal to, ≥" and "less than or equal to, ≤". The rules that govern inequalities are given below

If $a > b$ then

$$\begin{aligned} &a+c > b+c && \text{Addition Rule (AD)} \\ &a-c > b-c && \text{Subtraction Rule(SR)} \\ &ac > bc \text{ for } c > 0 && \text{Multiplication Rule (MR)} \\ &ac < bc \text{ for } c < 0 && \text{Multiplication Rule(MR)} \\ &a/c > b/c \text{ for } c > 0 && \text{Division Rule (DR)} \\ &a/c < b/c \text{ for } c < 0 && \text{Division Rule (DR)} \end{aligned}$$

The solution of inequalities typically involves a set of possible values that satisfy the inequality in question.

Example 1.37 Solve $3 + 4x > 3x + 4$

Add $(-3x)$ to both sides of the inequality

$$3 + 4x - 3x > 3x + 4 - 3x$$

so

$$3 + x > 4$$

Add (-3) to both sides of the inequality

$$3 + x - 3 > 4 - 3$$

so

$$x > 1$$

The solution set is $\{x \in \mathbb{R} : x > 1\}$ or in words all the real numbers that are greater than one, see Figure 1.3.

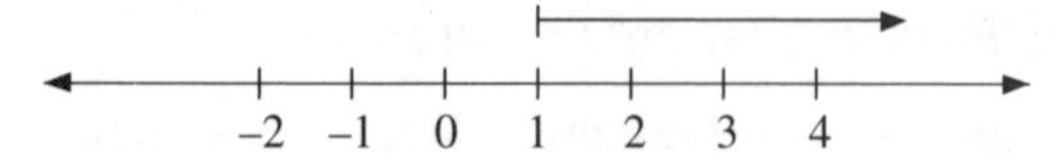

Figure 1.3

■

Graphs: Binary Relations

Let us analyze the following table that provides information on the distance travelled by a stone dropped from the top of a building in seconds.

Table 1.1

Time (seconds)	Distance (meters)
1	2.45
2	9.80
3	22.05
4	39.20
5	61.25

The above information can be displayed as a set of ordered pairs

$$\{(1, 2.45), (2, 9.8), (3, 22.05), (4, 39.20), (5, 61.25)\}$$

This set makes up what is known as a **binary relation**. The first component in each pair represents the time in seconds and the second entry the distance that it takes for the stone to travel. The pairs are ordered and the order matters. The set of all first components in each pair of the relation is called the **domain** of the relation and the second component is the **range** of the relation. In the example above the set $\{1, 2, 3, 4, 5\}$ is the domain and $\{2.45, 9.8, 39.2, 61.25\}$ is the range.

We can represent the above relation in terms of a graph (Figure 1.4). On the horizontal axis we put the time in seconds and on the vertical axis the distance that the object travels.

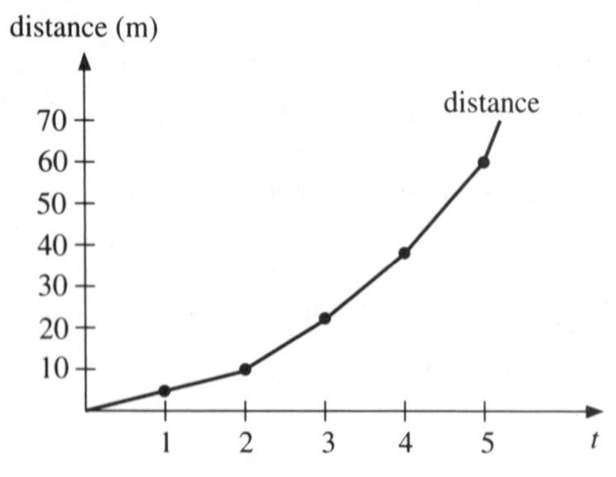

Figure 1.4

In fact the above relation follows the formula $d = 2.45t^2$, where d stands for distance and t for time. In graphing the relation we first identify the points corresponding to each pair and then draw the line between these points.

The example below gives the distance travelled by an airplane flying a constant speed of 300 km per hour.

Table 1.2

Time (hours)	Distance (kilometers)
1	300
2	600
3	900
4	1200
5	1500

We see from Table 1.2 that if the time is doubled the distance is doubled as well. If the time is tripled, the distance is tripled. In fact any change in time will bring about a proportionate change in distance. In this case, we write $d = mt$, where $m = 300$, since $m = d/t = 300/1 = 600/2 = \ldots = 1500/5$.

The graph is a straight line through the origin in Figure 1.5.

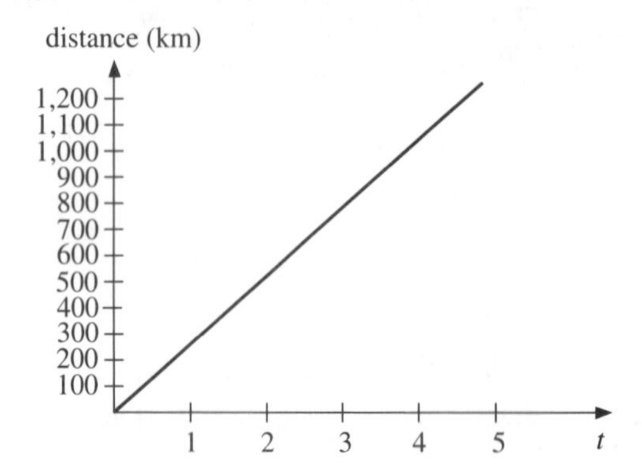

Figure 1.5

There are relations that can be expressed in terms of the equation $y = a + mx$. In this case, the graph is a straight line but the line starts at some point other than the origin.

For example suppose that it costs \$0.05 to print each newspaper plus a fixed cost of \$1000 to set up the press. In that case, the relation that describes the cost of printing a number of newspapers "n" is given by $C = 0.05n + 1000$. If $n = 20000$, then

$$\begin{aligned} C &= 0.05(20000) + 1000 \\ &= 1000 + 1000 \\ &= 2000 \end{aligned}$$

Therefore it costs \$2000 to print 20000 newspapers.

Slope of a line

The slope of a line represents a measure of "steepness" of the line. It is defined as the ratio of the vertical change (the rise) divided by the horizontal change (the run), see Figure 1.6.

$$\text{slope} = \frac{\text{rise}}{\text{run}} = \frac{\text{vertical change}}{\text{horizontal change}}$$

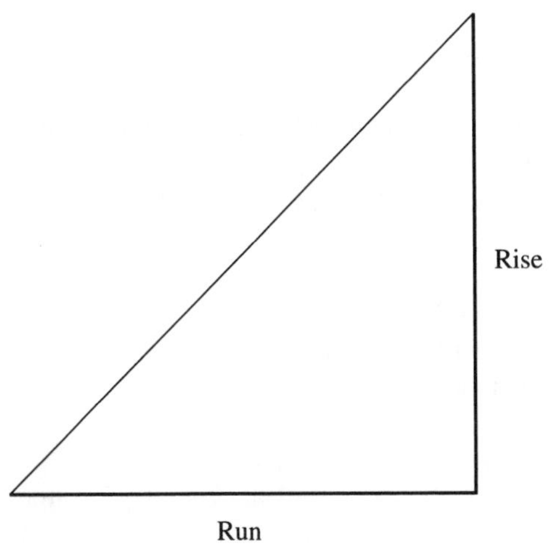

Figure 1.6

Example 1.38 Let us find the slope of the line that passes through the points A and B with coordinates (3,1) and (5,3) respectively. When we refer to coordinates of a point, the first entry corresponds to the value on the horizontal axis and the second entry to the value on the vertical axis.

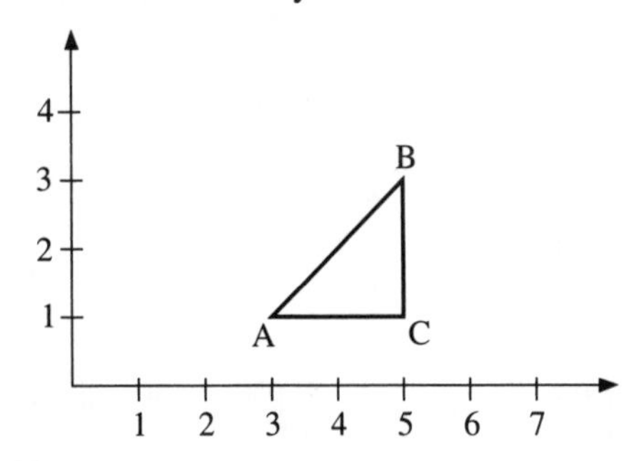

Figure 1.7

The slope of the above line is given by $(3-1)/(5-3) = 2/2 = 1$. This is because $AC = 5-3 = 2$ and $BC = 3-1 = 2$. See Figure 1.7. ■

Example 1.39 Find the slope of the line segment joining points A and B with coordinates $(3, -1)$ and $(5, 3)$ respectively.

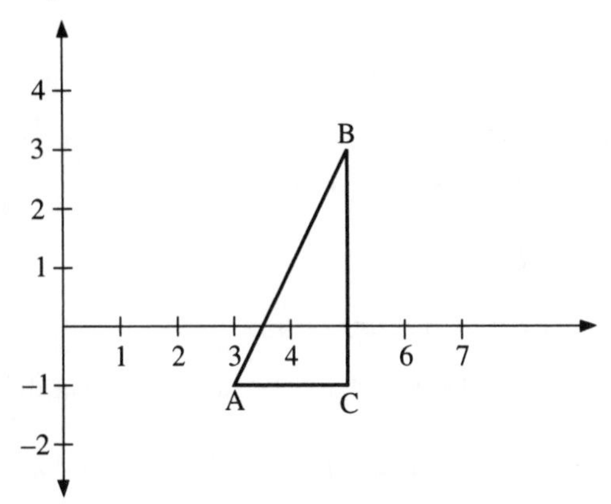

Figure 1.8

The slope is given by the ratio of BC to AC. In this case $BC = 3-(-1) = 3+1 = 4$ and $AC = 5-3 = 2$. Therefore, $BC/AC = 2$. See Figure 1.8. ■

In general we denote the vertical axis by y and the horizontal by x. Then the slope is defined as $\Delta y/\Delta x$ or the change in y divided by the change in x. The symbol Δ denotes change. Therefore if y changes from -1 to 3, the change in y is given by $\Delta y = 3-(-1) = 4$. In other words y jumps from -1 by 4 units to reach 3.

Example 1.40 Find the slope of the line through points A and B with coordinates (3,1) and (-2,1) respectively.

In this case, the change in x, $\Delta x = BA = 3-(-2) = 5$, whereas the change in y, Δy is $1-1=0$. Therefore, $\Delta y/\Delta x = 0/5 = 0$.

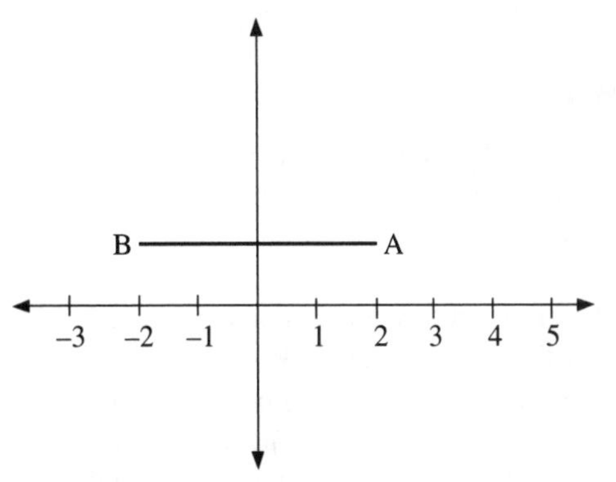

Figure 1.9

Therefore, the slope of a horizontal line is zero. See Figure 1.9. ■

Example 1.41 Find the slope of a line through A and B, with coordinates (1,3) and $(1, -2)$ respectively.

In this case, the slope of the line is computed using $\Delta x = 1-1 = 0$ and $\Delta y = 3-(-2) = 5$. Then $\Delta y/\Delta x = 5/0$, which is not defined. In fact the slope is infinitely large, since a "nearly" vertical line will have $a\,\Delta x$ be an arbitrarily small number, which is nevertheless not zero. Therefore, Δy divided by this small number will be arbitrarily large, depending on how small Δx is. As the "nearly" vertical line becomes more and more vertical, the slope will become larger and larger as well. Therefore in the limit it will tend to infinity (which is another way of saying it becomes indefinitely large). See Figure 1.10.

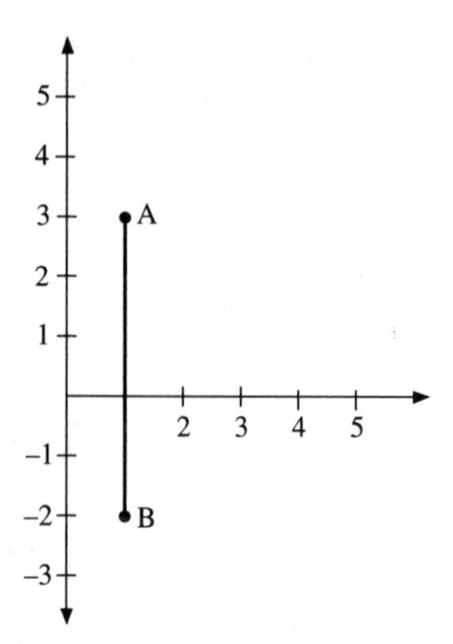

Figure 1.10

■

Linear equations

Once we have found the slope of a line we can use it with an arbitrary point on this line to obtain the equation that describes the line.

Example 1.42 Determine the equation of a line through the point (2,1) with slope $\Delta y/\Delta x = 3$.

Denote $\Delta y/\Delta x$ by m. Let (x, y) be any point on the line other than (2,1). In that case we have that the points (x, y) and (2,1) would have to satisfy the condition that

$$(y - 1)/(x - 2) = 3$$

Multiplying both sides by $(x - 2)$ we obtain

$$\begin{aligned} y - 1 &= 3(x - 2) \\ y &= 3x - 6 + 1 \\ y &= -5 + 3x \end{aligned}$$

■

The coefficient attached to x is the slope, whereas the number -5 is known as the **intercept** of the line. It yields the origin of the line when x takes the value zero. (More accurately we refer to it as the y-intercept to distinguish it from the x-intercept which is obtained by setting the y value to zero).

In general, given a point on the line (x_1, y_1) and the slope of the line m, the equation of the line is expressed as

$$\begin{aligned} (y - y_1)/(x - x_1) &= m \\ y - y_1 &= m(x - x_1) \\ y &= mx + (y_1 - x_1) \end{aligned}$$

In this case m, the slope, is attached to x, whereas $(y_1 - x_1)$ is the number that corresponds to the y-intercept. (From now onwards we will refer to the y-intercept as intercept, unless otherwise stated).

Conventionally, we express the variable that corresponds to the vertical axis, y, in terms of the other variable x, (multiplied by the slope) and the intercept. We can also determine the equation of a line given two points on the line.

Example 1.43 Find the equation of the line that passes through the points with coordinates $(-1, -2)$ and (2,3).

We know that the slope of the line is given by $\Delta y/\Delta x$. In our case $\Delta y = -2 - 3 = -5$ and $\Delta x = -1 - 2 = -3$. Therefore, $\Delta y/\Delta x = -5/-3 = 5/3$.

Now we have the slope $m = 5/3$ and a point on the line, either $(-1, -2)$ or (2,3). Suppose that we try $(-1, -2)$. Then we have that for any other point (x, y) the condition holds that

$$\frac{y - (-2)}{x - (-1)} = \frac{y + 2}{x + 1} = \frac{5}{3}$$

$$\begin{aligned} y + 2 &= (5/3)(x + 1) \\ y &= (5/3)x + 5/3 - 2 \\ y + 2 &= (5/3)(x + 1) \\ y &= (5/3)x + 5/3 - 2 \\ y &= (5/3)x - 1/3 \end{aligned}$$

In this case $-1/3$ is the intercept.

■

Example 1.44 Find the equation that passes through the points with coordinates $(-1, -3)$ and (3,-3).

We first obtain the slope of the line.

$$\frac{\Delta y}{\Delta x} = m = \frac{-3 - (-3)}{-1 - 3} = \frac{-3 + 3}{-4} = \frac{0}{-4} = 0$$

Now we have obtained the slope $m = 0$ and the point on the line $(-1, -3)$ or $(3, -3)$. Let us take $(-1, -3)$. Then choosing any other point on the line (x, y) we get

$$\begin{aligned} \frac{y - (-3)}{x - (-1)} &= 0 \\ y + 3 &= 0 \\ y &= -3 \end{aligned}$$

In other words, the line is parallel to the horizontal axis and passes through the point $x = 0$ and $y = -3$.

■

Parallel and perpendicular lines

Two lines are parallel if they have the same slope but different intercepts.

Two lines are perpendicular if the product of their slopes is -1.

Example 1.45 Determine the equation of the line through $(-4, 1)$ and perpendicular to $y = -3 + 2x$.

We know that since the line is perpendicular to $y = -3 + 2x$ its slope must be $-1/2$, since $2(-1/2) = -1$. Now, we have the slope of this equation and the fact that it passes through $(-4, 1)$. Let us take another point (x, y) on the line. Then

$$\begin{aligned} y - 1 &= (-1/2)(x - (-4)) \\ y - 1 &= (-1/2)(x + 4) \\ y &= -1 - (1/2)x \end{aligned}$$

■

Example 1.46 Determine the equation of the line that passes through $(-2, 4)$ and is parallel to $y = 4 - 2x$.

The slope of the line must be the same as the slope of $y = 4 - 2x$, (-2), since the lines are parallel. Then we have the point (-2,4) and another point (x, y). That results in

$$\begin{aligned} y - 4 &= (-2)[x - (-2)] \\ y &= 4 - 2(x + 4) \\ y &= -4 - 2x \end{aligned}$$

■

Length of a line segment

A horizontal line segment is determined by joining two points with the same y-coordinates but different x-coordinates. Suppose these two points are (x_1, y) and (x_2, y). The length of a horizontal line segment is then determined by calculating the absolute value of the difference between the x-coordinates, i.e., the positive value of the difference of these coordinates. Recall that in general, the absolute value of a real number x is denoted as $|x|$ and is defined as the positive value of the number. For example, if $x = -3$, then $|-3| = 3$. The length of the horizontal line segment is then defined as $|x_2 - x_1|$.

Example 1.47 Find the length of the segment that passes through the points A and B with coordinates $(-5, 2)$ and (4,2).

Then the length of AB is given by

$$AB = |4 - (-5)| = |4 + 5| = |9| = 9$$

See Figure 1.11.

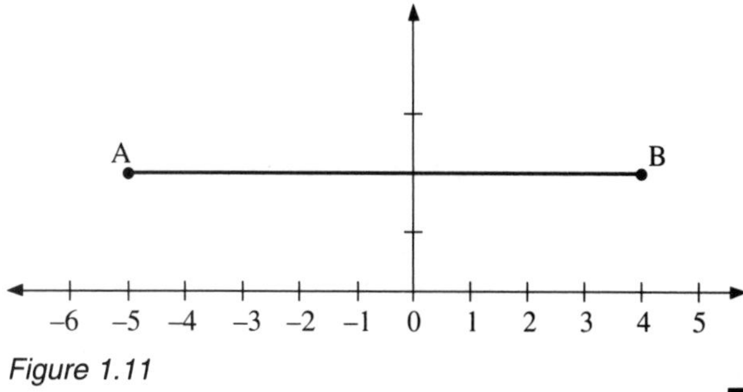

Figure 1.11

■

A vertical line segment is defined as one that passes through two points with the same x-coordinate but with different y-coordinates. Suppose that these points are (x, y_1) and (x, y_2). Then the length of the vertical segment is defined as the absolute value of the difference of the y-coordinates. It is given by $|y_2 - y_1|$.

Example 1.48 Find the length of the line segment that passes through the points A and B with coordinates (3,1) and $(3, -2)$.

The length AB is given by

$$AB = |y_2 - y_1| = |-2 - 1| = |-3| = 3$$

See Figure 1.12.

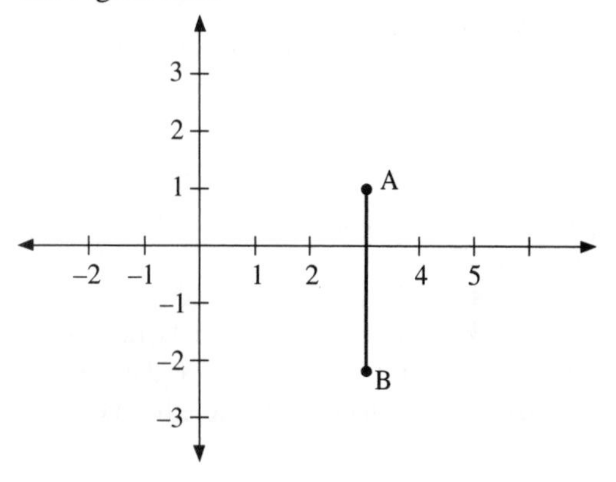

Figure 1.12

The length of a general line segment joining two points A and B can be found by using Pythagoras' theorem. Let us look at the example below.

Example 1.49 Find the length of the line segment joining point A and B with coordinates (3,2) and (5,3).

The Pythagorean theorem states that $(AB)^2 = (AC)^2 + (BC)^2$. Therefore

$$AB = \sqrt{[(AC)^2 + (BC)^2]}$$

The length AC is given by $(5 - 3) = 2$ and the length of BC by $(3 - 2) = 1$. Then

$$AB = \sqrt{(2^2 + 1)} = \sqrt{5}$$

See Figure 1.13.

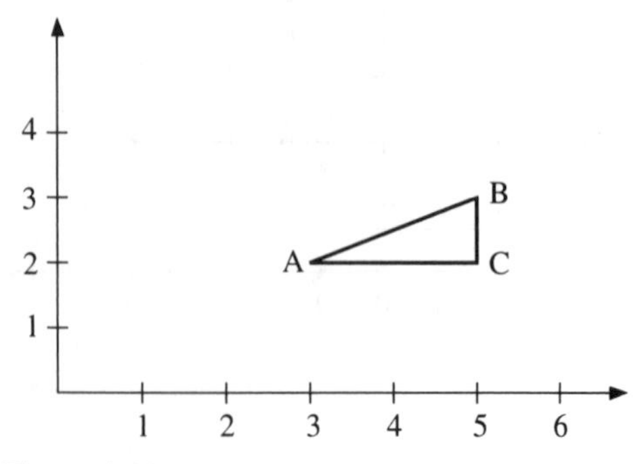

Figure 1.13

■

Functions

We have defined a relation as a set of ordered pairs. We will now look at a special type of relation known as **function.**

A function is a relation such that for *every* element of the first set (the set that contains the first entries of the pairs) there is only *one* element of the second set. For example the pairs of the following relation constitute a function:

$$\{(2, 1), (3, 5), (4, 7), (5, -3)\}$$

For each of the members of the set $\{2, 3, 4, 5\}$ there is a unique member in the set $\{1, 5, 7, -3\}$ corresponding to it. However, the relation

$$\{(2, 1), (3, 5), (3, 7), (7, 3)\}$$

is *not* a function, since to the number 3 there are two corresponding entries 5 and 7.

The set that consists of the first entries of the pairs that make up the relation is known as the **domain** of the function, whereas the set of the second entries is known as the **range**. Then the definition of a function becomes more formally as follows:

Definition 1.2: A **function** is a relation that satisfies the following requirement. For each element in the domain, there is a single corresponding element in the range.

Note that there may be the same element in the range corresponding to two distinct elements in the domain. For example

$$\{(2, 1), (3, 1), (4, 5), (5, 5)\}$$

is a function, since to the two elements 2 and 3 of the domain there corresponds the same element 1 of the range and for elements 4 and 5 of the domain there is the corresponding element 5 of the range.

In a more abstract way a function may be expressed for example, in the following manner:

$$f : x \rightarrow 3x - 1$$

This states that any element x of the domain will be transformed to an element $3x - 1$ in the range. For example if $x = 1$ in the domain, it will correspond to $3 - 1 = 2$ in the range.

In general, a function is also known as a **map** or **transformation**, that transforms the element x into y or

$$f : x \rightarrow y$$

Two specific forms that f can take are, for example

$$\begin{aligned} y &= 2x + 3 \\ y &= x^2 - 1 \end{aligned}$$

In the first case for each x there will be a corresponding y defined as $2x + 3$ and in the second case for each x there will be a y defined as $x^2 - 1$.

A more convenient way of expressing a function is to write

$$f(x) = 3x - 1$$

where $f(x)$ is read as "f of x". The statement means that for a specific value that x takes, it is transformed into $f(x)$. For example

$$\begin{aligned} f(1) &= 3(1) - 1 = 2 \quad \text{for } x = 1 \\ f(2) &= 3(2) - 1 = 5 \quad \text{for } x = 2 \\ f(5) &= 3(5) - 1 = 14 \quad \text{for } x = 5 \end{aligned}$$

Here, $f(2)$ denotes the value of the function when $x = 2$, or using the previous definition of the function, it denotes the value in the the range corresponding to the value 2 of the domain.

Alternative ways of expressing the functional relationship $f(x) = 3x - 1$ are

$$\begin{aligned} y &= 3x - 1 \\ & \{(x, y) | y = 3x - 1\} \\ f : x &\rightarrow 3x - 1 \\ & (x, 3x - 1) \end{aligned}$$

All of these are equivalent.

An important type of functional relationship is one that expresses a function not of a variable but of another function. For example, suppose that

$$g(x) = 2x + 1$$

and

$$f(x) = 3x$$

Then

$$f(g(x)) = f(2x + 1) = 3(2x + 1) = 6x + 3$$

In this case $f(g(x))$ simply relates the range of the function g to the range of the function f. In other words $g(x)$ serves as the domain for f.

Systems of Equations

Graphical solution

An equation in one variable, such as $x + 3 = 1$, has one real value of x for a solution, that is $x = -2$. In that case, the left-hand side (LHS) of the equation is made equal to the right-hand side (RHS). For an equation of two variables, say

$$x + y = 1$$

the solution set consists of an infinite number of ordered pairs (x, y) that satisfy the above equation, such as $(1, 0)$, $(0, 1)$, $(2, -1)$, $(-1, 2)$ etc. All these pairs would make the LHS equal to the RHS of the equation.

When we have two equations in two unknowns, if these two equations can be represented by intersecting lines, then the solution that satisfies both of these two equations simultaneously is simply the pair of the (x, y) values that corresponds to the intersection of the two lines.

Example 1.50 Find the point of intersection of the equations defined by

$$\begin{aligned} y &= x + 3 \\ y &= 4 - 2x \end{aligned}$$

For the first equation $y = x + 3$, we can take any two points that satisfy it, say A with coordinates $(x = 1, y = 4)$ and B with coordinates $(x = 2, y = 5)$ and draw the line through these two points. See Figure 1.14.

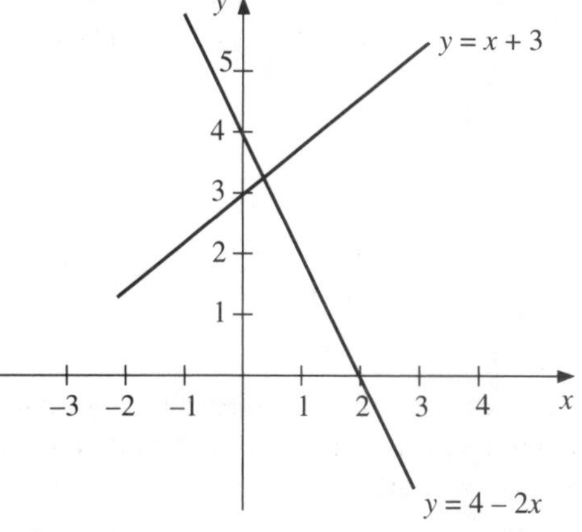

Figure 1.14

For the equation $y = 4 - 2x$, we also draw the line that goes through any two points that satisfy the equation, say C with coordinates $(x = 1, y = 2)$ and D with coordinates $(x = 2, y = 0)$.

The solution that is the pair of x and y that satisfies both equations can be read from the diagram as $x = 1/3$ and $y = 10/3$. These two values satisfy both $y = x + 3$ and $y = 4 - 2x$ simultaneously. In that case the LHS and the RHS are made equal for both equations.

■

Algebraic solution

The solution to a pair of equations may not be read very accurately from a graph. Also if the coordinates of the solution pair are very large, then reading off the solution from the graph may not prove practically possible. An alternative way of obtaining a solution is by using algebra.

Example 1.51 Obtain the solution algebraically of the pair of equations

$$\begin{aligned} y &= 2x + 5 \\ y &= 2 - x \end{aligned}$$

The algebraic approach is based on obtaining the coordinates of the point of intersection of the two lines by algebraic means. If the lines intersect they *must* have the same y-coordinate at the point of intersection. Therefore

$$2x + 5 = 2 - x$$

Solving for x yields $3x = 2 - 5$, or $x = -1$.

Substituting back the value of $x = -1$ into any of the two equations would then yield $y = 2(-1) + 5 = 3$ or $y = 2 - (-1) = 3$.

■

The principle that underlies the method of the algebraic solution of a pair of equations relies on the fact that the y-coordinate of the first equation has to equal the y-coordinate of the second equation and the x-coordinate of the first equation has to equal the x-coordinate of the second equation at the point of intersection.

Example 1.52 Solve $y = 2x + 3$ and $y = x + 4$

We first equalize the y-coordinates of the two equations, which implies that

$$\begin{aligned} 2x + 3 &= x + 4 \\ 2x - x &= 4 - 3 \\ x &= 1 \end{aligned}$$

If we substitute $x = 1$ into either of the above equations $y = 2x + 3$ and $y = x + 4$, we obtain $y = 5$. The pair (x, y) that solves the equations simultaneously is given by $(1, 5)$.

■

Classifying and interpreting linear systems

A system of two linear equations in two variables will satisfy one of the following cases:

1. The system has exactly one solution, when the lines associated with the equations in question intersect at a single point. A system of equations is said to be **consistent** in that case.

2. When the lines are parallel, then there is no point of intersection. Therefore, there is no solution in that case and the system is said to be **inconsistent**.
3. If the two equations in the system define the same straight line, then there is an infinite number of solutions. In fact, in that case, one of the equations is redundant and the system is said to be **dependent**.

Example 1.53 Analyze the following two equation systems and determine whether they are consistent, inconsistent, or dependent.

(i)

$$\begin{aligned} x + y &= 4 \\ x + y &= 3 \end{aligned}$$

(ii)

$$\begin{aligned} x + y &= 4 \\ 2x + y &= 3 \end{aligned}$$

(iii)

$$\begin{aligned} x + y &= 4 \\ 2x + 2y &= 8 \end{aligned}$$

For system (i) we can express both equations in the form $y = b + mx$, where b stands for the intercept and m for the slope as

$$\begin{aligned} y &= 4 - x \\ y &= 3 - x \end{aligned}$$

In that case, since the two equations have the same slope and they only differ with respect to their intercept, they will be parallel to each other. Therefore, the system will be inconsistent, since the two lines will never intersect.

For system (ii) the equations can be rewritten as

$$\begin{aligned} y &= 4 - x \\ y &= 3 - 2x \end{aligned}$$

In this case we can obtain the algebraic solution by solving for the coordinates of (x, y) that define the point of intersection of the two lines: $4 - x = 3 - 2x$ or $x = -1$.

Substituting the value $x = -1$ into either of the two equations yields $y = 4 - (-1) = 5$. Therefore $(-1, 5)$ is the unique solution to the system and corresponds to the point of intersection of the two lines.

For system (iii), we notice that if we were to multiply the first equation of the system by 2, the LHS would equal the RHS and the equation would not change. However, the first equation would become exactly the same as the second equation and therefore it becomes clear that one of them is redundant. The system in that case is dependent, since the two lines are one and the same, with an infinite number of (x, y) pairs that satisfies this single line. ■

Linear inequalities in two variables: A graphical solution

A linear equation given by an expression $y = b + mx$, such as $1 + x$, divides the (x, y) plane into 3 regions as can be seen from Figure 1.15.

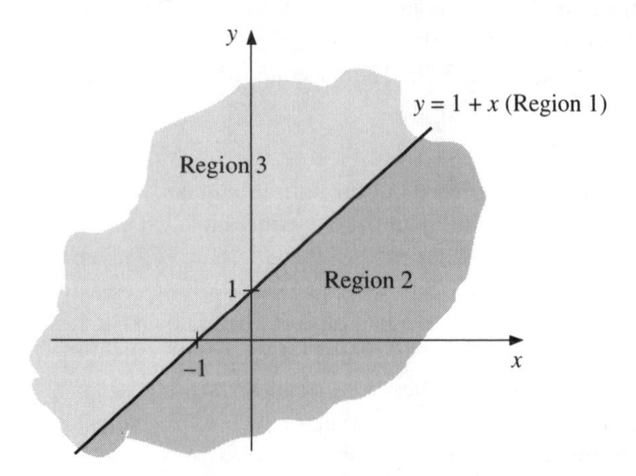

Figure 1.15

Region 1: The set of all points that lie on the line $y = 1 + x$. Therefore, Region 1 refers to all (x, y) pairs that satisfy the equation.

Region 2: The set of all the points that satisfy the condition that $y < x + 1$. In other words, Region 2 consists of all points that are below the line.

Region 3: The set of all points that satisfy the condition that $y > x + 1$. Region 3 consists of the points that lie above the line.

Example 1.54 Sketch the graph of $y = 3 - x$ and indicate on the graph the following regions:

(1) $y = 3 - x$
(2) $y < 3 - x$
(3) $y > 3 - x$
See Figure 1.16.

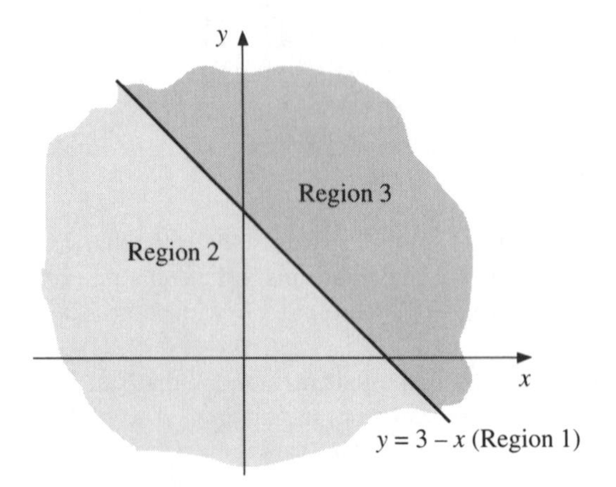

Figure 1.16

All the points on the line correspond to Region 1. For Region 2, choose any point not on the line such as (1, 1). In that case for $x = 1$, the value of y that would be needed for the point to be on the line is $y = 2$. Therefore if $y = 1$, the point (1, 1) must lie below the line. All points below the line correspond to Region 2 and all points above the line such as (1, 4) belong to Region 3.

■

In general for $y > b + mx$ the required region is above the line and for $y < b + mx$ the required region in below the line.

EXIT TEST

1. Simplify $(x^2 + x - 6)/(x - 2)$.
   a) $x - 3$
   b) $x + 2$
   c) $x + 3$
   d) $x - 2$
   e) $2x + 2$

2. Simplify $(3a^2 + ab - 2b^2)/(a + b)$.
   a) $3a + 2b$
   b) $2b - 3a$
   c) $3a - b$
   d) $2b + 3a$
   e) $3a - 2b$

3. Simplify $[1/(x-1)-1/(x-2)]/[1/(x-2)-1/(x-3)]$.
   a) $(x - 3)/(x - 1)$
   b) $(x - 1)/(x - 3)$
   c) $(x - 1)^2/(x - 3)^2$
   d) $(x - 3)^2/(x - 1)^2$
   e) $1/(x - 2)^2$

4. Solve for x in $2(x + 3) = (3x + 5) - (x - 5)$.
   a) $x = 6$
   b) $x = 10$
   c) $x = 0$
   d) $x = 1$
   e) There is no real number x that satisfies this.

5. Solve for x in $4(3x + 2) - 11 = 3(3x - 2)$.
   a) $x = -3$
   b) $x = -1$
   c) $x = 2$
   d) $x = 3$
   e) $x = 8$

6. Solve for x and y in

$$\begin{aligned} x + 2y &= 8 \\ 3x + 4y &= 20 \end{aligned}$$

   a) $x = 3, y = 5$
   b) $x = 4, y = 2$
   c) $x = 1, y = 0$
   d) $x = 0, y = 1$
   e) There are no real numbers x and y that satisfy these equations.

7. Solve algebraically for x and y in

$$\begin{aligned} 4x + 2y &= -1 \\ 5x - 3y &= 7 \end{aligned}$$

   a) $x = 1/2, y = -3/2$
   b) $x = 2, y = 3$
   c) $x = 1, y = 1$
   d) $x = 0, y = 1$
   e) There are no real numbers x and y that satisfy these equations.

8. Solve algebraically for x and y in

$$\begin{aligned} 5x + 6y &= 4 \\ 3x - 2y &= 1 \end{aligned}$$

   a) $x = 3, y = 6$
   b) $x = 1/2, y = 1/4$
   c) $x = -3, y = 6$
   d) $x = 2, y = 4$
   e) $x = 1/3, y = 3/2$

9. Solve algebraically for x and y in

$$\begin{aligned} 4x - 2y &= -14 \\ 8x + y &= 7 \end{aligned}$$

   a) $x = 0, y = 7$
   b) $x = 2, y = -7$
   c) $x = 2, y = 7$
   d) $x = 7, y = 0$
   e) There are no real numbers x and y that satisfy these equations.

10. Solve algebraically for x and y in

$$\begin{aligned} 6x - 3y &= 1 \\ -9x + 5y &= -1 \end{aligned}$$

   a) $x = 1, y = -1$
   b) $x = 2/3, y = 1$
   c) $x = 1, y = 2/3$
   d) $x = -1, y = 2/3$
   e) $x = 1, y = 2/3$

11. Factor the quadratic $x^2 + 8x + 15$.
   a) $(x+3)(x+5)$
   b) $(x-3)(x-5)$
   c) $(x+3)(x-5)$
   d) $(x-3)(x+5)$
   e) $(x+3)^2$

12. Factor the quadratic $z^2 - 2z - 3$.
   a) $(z-3)(z+1)$
   b) $(z+3)(z-1)$
   c) $(z-3)(z-1)$
   d) $(z+3)(z+1)$
   e) $(z-1)^2$

13. Factor the quadratic $2x^2 - 11x - 6$.
   a) $(2x+1)(x-6)$
   b) $(x+2)(x-6)$
   c) $(2x-1)(x-6)$
   d) $(2x-1)(x+6)$
   e) $(x-1)(x-3)$

14. Find the solutions (roots) of $x^2 - 8x + 16 = 0$.
   a) $8, 2$
   b) $1, 16$
   c) $4, 4$
   d) $-2, 4$
   e) $4, -4$

15. Find the solutions (roots) of $6x^2 - x - 2 = 0$.
   a) $2, 3$
   b) $1/2, 1/3$
   c) $-1/2, 2/3$
   d) $2/3, 3$
   e) $2, -1/3$

16. Find the set of x's that satisfies the inequality $2x + 5 > 9$.
   a) $\{x|x > 2\}$ (all x, such that x is greater than 2)
   b) $\{x|x < 2\}$ (all x, such that x is less than 2)
   c) $\{x|x > 1\}$ (all x, such that x is greater than 1)
   d) $\{x|x > 0\}$ (all x, such that x is greater than 0)
   e) $\{\emptyset\}$ (no such x exists)

17. Find the set of x's that satisfies the inequality $4x + 3 < 6x + 8$.
   a) $\{x|x > -5/2\}$
   b) $\{x|x > 5/2\}$
   c) $\{x|x > 2/5\}$
   d) $\{x|x < 5/2\}$
   e) $\{x|x < 5/2\}$

18. Solve the system of equations for the values of x, y and z that satisfy

$$\begin{aligned} 2x + 3y - 4z &= -8 \\ x + y - 2z &= -5 \\ 7x - 2y + 5z &= 4 \end{aligned}$$

   a) $x = -1, y = 2, z = 3$
   b) $x = 1, y = 0, z = 0$
   c) $x = 0, y = 1, z = 0$
   d) $x = 1, y = -2, z = -3$
   e) $x = -1, y = 2, z = -3$

19. Find the solutions (roots) to $12x^2 + 5x = 3$.
   a) $1/3, -1/4$
   b) $4, -3$
   c) $4, 1/6$
   d) $1/3, -4$
   e) $-3/4, 1/3$

20. Find the solutions (roots) to $3x^2 + 3x = 6$.
   a) $3, -6$
   b) $2, 3$
   c) $-3, 2$
   d) $1, -3$
   e) $1, -2$

Answers to Self-Tests

Arithmetic pretest

1. d	6. b	11. e	16. b
2. c	7. d	12. e	17. c
3. b	8. e	13. e	18. b
4. c	9. b	14. e	19. d
5. b	10. d	15. c	20. e

Arithmetic exit test

1. c	6. c	11. d	16. c
2. d	7. b	12. b	17. b
3. d	8. c	13. b	18. a
4. c	9. a	14. c	19. d
5. c	10. a	15. d	20. a

Algebra pretest

1. d	6. d	11. c	16. d
2. b	7. d	12. d	17. b
3. d	8. a	13. e	18. c
4. c	9. d	14. a	19. b
5. e	10. c	15. e	20. a

Algebra exit test

1. c	6. b	11. a	16. a
2. e	7. a	12. a	17. a
3. a	8. b	13. a	18. a
4. e	9. a	14. c	19. e
5. b	10. b	15. c	20. e

CHAPTER 2

Section 2.1 (page 17)

1. In both cases, of course, x is contained in X, but in the first case x is thought of as an element, in the second case as a subset, i.e., a set.

3. There are 32 possible subsets:

$$B_1 = \emptyset,\ B_2 = \{1\},\ B_3 = \{2\},$$
$$B_4 = \{3\},\ B_5 = \{4\},\ B_6 = \{5\}$$
$$B_7 = \{1, 2\},\ B_8 = \{1, 3\},\ B_9 = \{1, 4\}$$
$$B_{10} = \{1, 5\},\ B_{11} = \{2, 3\}$$
$$B_{12} = \{2, 4\},\ B_{13} = \{2, 5\}$$
$$B_{14} = \{3, 4\},\ B_{15} = \{3, 5\}$$
$$B_{16} = \{4, 5\},\ B_{17} = \{1, 2, 3\}$$
$$B_{18} = \{1, 2, 4\},\ B_{19} = \{1, 2, 5\}$$
$$B_{20} = \{1, 3, 4\},\ B_{21} = \{1, 3, 5\}$$
$$B_{22} = \{1, 4, 5\},\ B_{23} = \{2, 3, 4\}$$
$$B_{24} = \{2, 3, 5\},\ B_{25} = \{2, 4, 5\}$$
$$B_{26} = \{3, 4, 5\},\ B_{27} = \{1, 2, 3, 4\}$$
$$B_{28} = \{1, 3, 4, 5\},\ B_{29} = \{1, 2, 4, 5\}$$
$$B_{30} = \{1, 2, 3, 5\},\ B_{31} = \{2, 3, 4, 5\}$$
$$B_{32} = \{1, 2, 3, 4, 5\}$$

5. Yes, the order of the elements in a set is not important (compare with Definition 2.3).

7. Figure 2.1.7 illustrates. Set B is given by the horizontally shaded area. Set C is shaded diagonally. Set $B \cup C$ is all the shaded area. Set $B \cap C$ is shaded vertically. The interpretation is as follows:

(a) B is the set of combinations of goods 1 and 2 which the consumer can afford to buy. C is the set of quantities of goods 1 and 2 which the consumer is physically capable of consuming.

(b) $B \cup C$ is the set of quantities of goods 1 and 2 which the consumer can afford to buy or is physically capable of consuming.

(c) $B \cap C$ is the set of quantities of goods 1 and 2 which the consumer is physically capable of consuming and can afford to buy. It is the set most relevant to economics since it expresses the problem of choice under scarcity. Consumer theory tries to resolve which element of this set is chosen by the consumer.

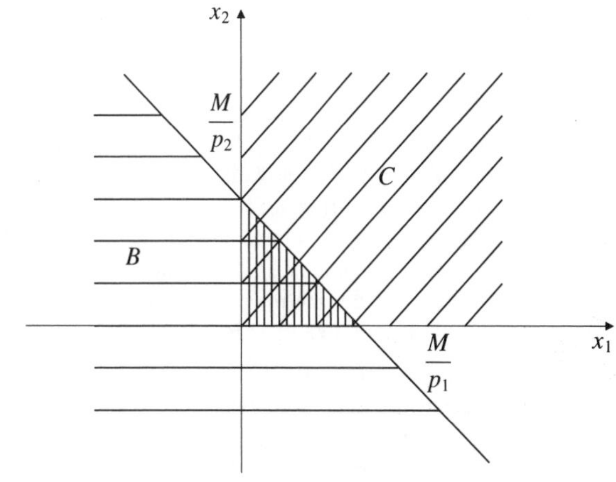

Figure 2.1.7

9. Figure 2.1.9 illustrates. P are all technologically feasible input-output combinations. $\bar{x}$ is the maximum labor supply.

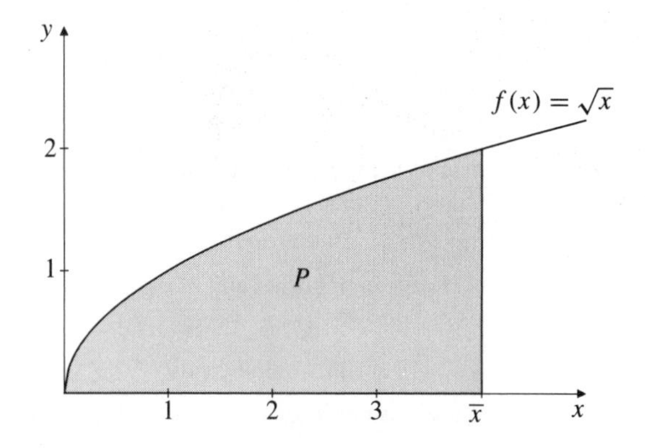

Figure 2.1.9

Section 2.2 (page 26)

1. (a) Z_+ is bounded below by 1 which also is its infimum. There is no upper bound since for all $x \in Z_+$ we have $(x + 1) \in Z_+$.

(b) Z is unbounded. For all $x \in Z$ also $(x - 1)$ and $(x + 1)$ are elements of Z.

(c) $\mathbb{R}_+$ is bounded below by 0 which by definition is its infimum. There is no upper bound since for all $x \in \mathbb{R}_+$ also $(x + 1) \in \mathbb{R}_+$.

(d) $\mathbb{R}_+$ is bounded above by its supremum 0. There is no lower bound.

(e) S is bounded below by its infinum 0 and bounded above by its supremum $\sqrt{2}$.

3. (a) The wage rate is usually defined as wages per hour of work. Therefore, it has the dimension $\frac{\text{dollars}}{\text{quantity of input}}$. The marginal product refers to the extra output produced by one additional unit of input, here an hour of work. Its dimension is $\frac{\text{quantity of output}}{\text{quantity of input}}$. Consequently, λ has the dimension $\frac{\text{dollars}}{\text{quantity of output}}$.

(b) National income and investment can be measured in nominal terms, i.e., in dollars, or in real terms, or in quantities of goods. Changes of these variables have the same dimension as the variables themselves since they are simply the difference between two values of the variables. Thus, they can also be recorded in nominal or real terms. When relating these changes they must be measured in the same manner if the result is to be meaningful. Therefore λ must be a pure number.

(c) Profits are measured in dollars. Amount of labor used is the quantity of input used. Hence, the dimension of λ is $\frac{\text{dollars}}{\text{quantity of input}}$.

(d) Solving the equation for λ yields

$$\lambda = \frac{\text{elasticity of demand for the good}}{\text{tax per unit of a good}}$$

An elasticity is always a pure number because it is the ratio of the relative change of two variables, i.e., the ratio of two pure numbers. For example, the elasticity of demand for the good is defined as

$$\frac{\dfrac{\text{change in quantity demanded}}{\text{quantity demanded (at the initial price)}}}{\dfrac{\text{change in price}}{\text{initial price}}}$$

The dimension of tax per unit of a good is simply

$$\frac{\text{dollars}}{\text{quantity of the good}}.$$

Hence, λ must have the dimension

$$\frac{\text{quantity of the good}}{\text{dollars}}.$$

(e) Profits are recorded in dollars. An import quota defines the maximum quantity of goods that are allowed to be imported. Changes of these variables have the same dimension as the variables themselves. Thus, the dimension of λ is $\frac{\text{dollars}}{\text{quantity of input good}}$.

5. Sets with a maximum are $\mathbb{R}_- = \{x \in \mathbb{R} : x \leq 0\}$, $T = \{x \in \mathbb{R} : x \leq 5\}$. They are both bounded because only bounded sets can have a maximum. Sets without a maximum are $\mathbb{R}_+ = \{x \in \mathbb{R} : x \geq 0\}$, $S = \{x \in \mathbb{R} : x < 5\}$. $\mathbb{R}_+$ is unbounded above and therefore cannot have a maximum. S is bounded above but does not contain its supremum.

Section 2.3 (page 35)

1. (a) $\{1, 2, 3, 4, 5, 6\} \otimes \{7, 8, 9\}$ amounts to

$$\{\{1,7\},\{1,8\},\{1,9\},\{2,7\},\{2,8\},\{2,9\}$$
$$\{3,7\},\{3,8\},\{3,9\},\{4,7\},\{4,8\},\{4,9\}$$
$$\{5,7\},\{5,8\},\{5,9\},\{6,7\},\{6,8\},\{6,9\}\}$$

as shown in Figure 2.3.1 (a)

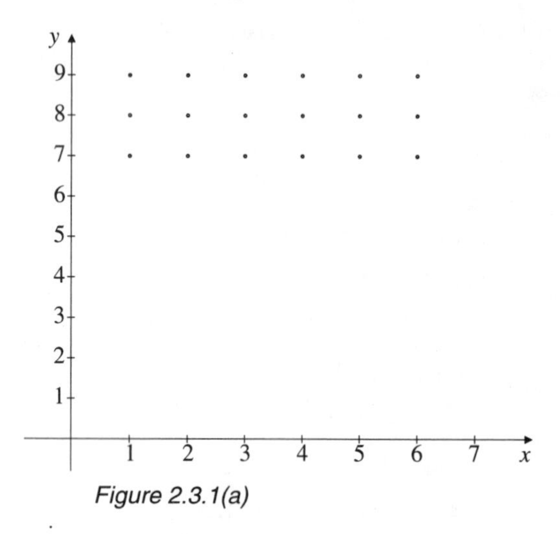

Figure 2.3.1(a)

(b) $Z_+ \otimes Z_+ = \{(x, y) : x \in Z_+, y \in Z_+\}$ which is shown in Figure 2.3.1 (b)

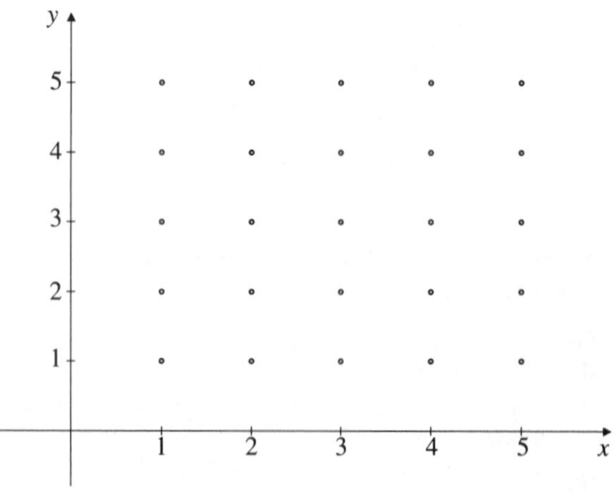

Figure 2.3.1(b)

(c)

$$\left\{(x, y) : x \in Z_+ \text{ and } \frac{x}{2} \in Z_+, \; y \in Z_+ \text{ and } \frac{y+1}{2} \in Z_+\right\}$$

which is shown is Figure 2.3.1 (c)

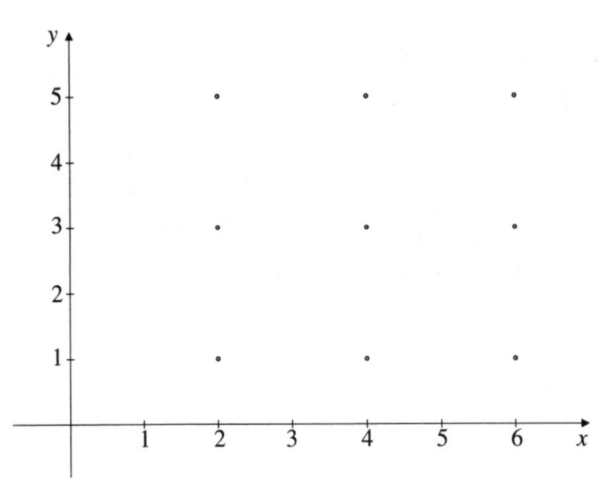

Figure 2.3.1(c)

3. B is closed (the boundary points of B, i.e., the sides of the triangle CDE form part of B)

B is bounded since for any $x_0 \in B$ and $\epsilon > m/p_1$ and $\epsilon > m/p_2$ it is true that $B \subset N_\epsilon(x_0)$.

B is convex as Figure 2.3.3 (a) reveals.

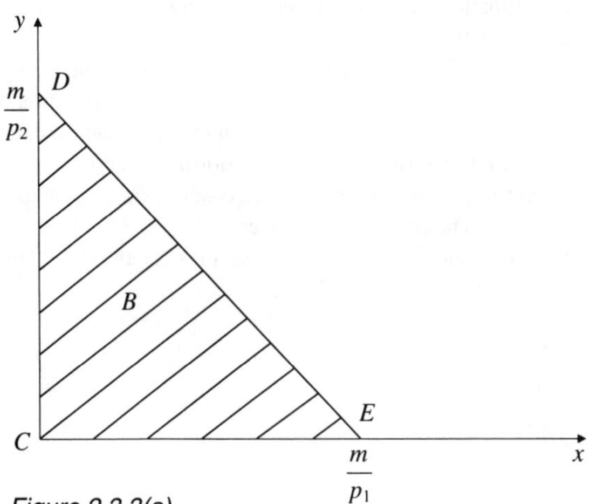

Figure 2.3.3(a)

If we interpret x' and y' as subsistence consumption, the case $x = \emptyset$ signifies that the consumer cannot afford the consumption bundle necessary for his or her survival.

x is a closed set (the sides of the triangle FGH form part of x).

Since x is a subset of the bounded set B, x must be bounded.

x is convex as shown in Figure 2.3.3 (b).

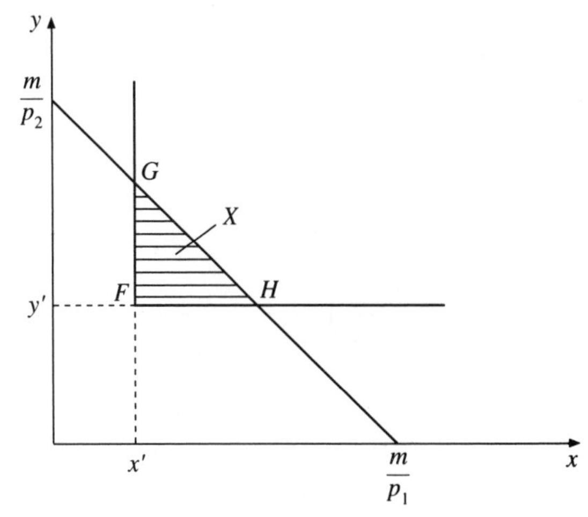

Figure 2.3.3(b)

5. (a) 9

(b) 16.64

(c) 13.71

7. (a) $N_\epsilon(-1) = \left\{x \in \mathbb{R} : \sqrt{(x+1)^2} < \epsilon\right\}$

For $\epsilon = 0.1$, $N_\epsilon(-1)$ is the open interval $(-1.1, -0.9)$.
For $\epsilon = 10$, $N_\epsilon(-1)$ is the open interval $(-11, 9)$

(b) $N_\epsilon(-1, 1) = \left\{(x, y) \in \mathbb{R}^2 : \sqrt{(x+1)^2 + (y-1)^2} < \epsilon\right\}$

$N_\epsilon(-1, 1)$ is the set of points of $\mathbb{R}^2$ lying inside a circle centered on $(-1, 1)$ with radius ϵ.

(c)

$$N_\epsilon(-1, 1, -1) = \Big\{(x, y, z) \in \mathbb{R}^3 : \sqrt{(x+1)^2 + (y-1)^2 + (z+1)^2} < \epsilon\Big\}$$

$N_\epsilon(-1, 1, -1)$ is the set of points of $\mathbb{R}^3$ lying within a sphere centered at (-1,1,-1) and with radius ϵ.

Section 2.4 (page 49)

1. (a) $y = 1 - 2x$

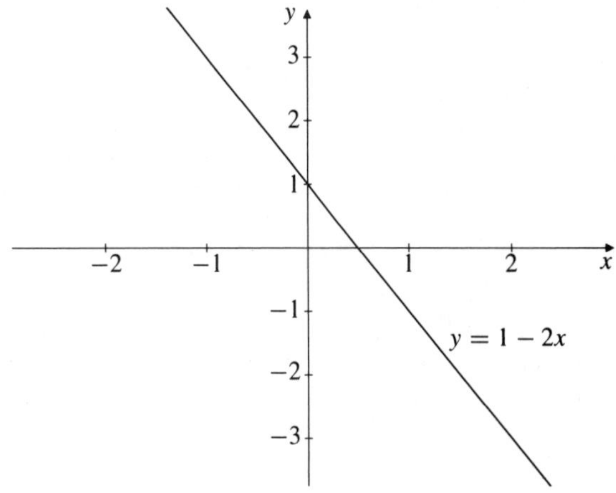

Figure 2.4.1(a)

(b) $y = -8 - 5x$

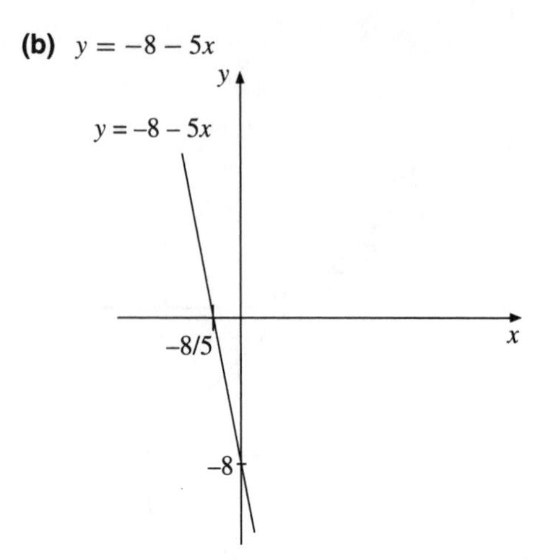

Figure 2.4.1(b)

(c) $y = -3 - \frac{3}{2}x$

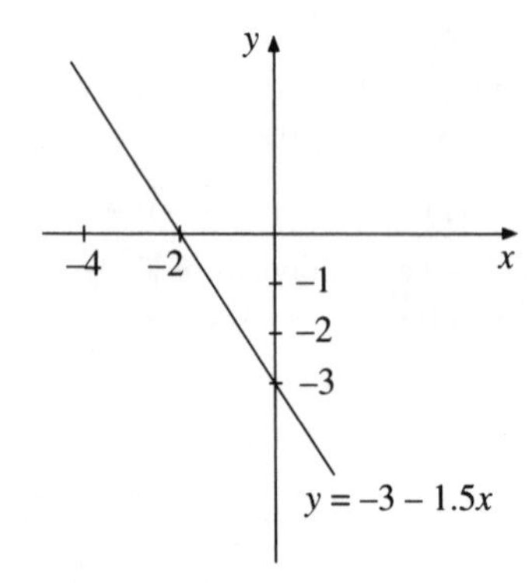

Figure 2.4.1(c)

3. (a) $\bar{x} = (1 - 3\lambda, -2 + 2\lambda, 2 - \lambda), \lambda \in [0, 1]$

−3 −2 −1 0 1 2 3 4 x

Convex combination of (−2) and 4

Figure 2.4.3(a)

(b) $\bar{x} = (3 - 4\lambda, 4 - 3\lambda), \lambda \in [0, 1]$

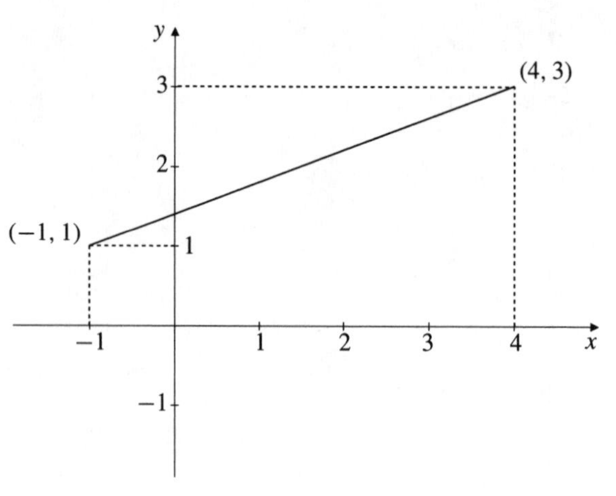

Figure 2.4.3(b)

(c) $\bar{x} = (1 - 3\lambda, -1 + 2\lambda, 2 - \lambda), \lambda \in [0, 1]$

5. There are no output levels at which the firm can break even at a price of \$10.

To find the range of prices that imply a loss for the firm at all output levels we must find the minimum of the average cost function. All prices below the corresponding average cost form the solution.

In the book you learned that a function of the type $y = ax^2 + bx + c$ where $a > 0$ has its unique minimum at the point $x^* = -b/2a$ (see the section on quadratic functions). In the case of the average cost function $a = 1$ and $b = (-20)$. Therefore, the minimum average cost occurs at an output of $x^* = 10$. The corresponding average cost is \$20. Thus, if the price is below \$20 the firm makes a loss at all output levels. This is shown in Figure 2.4.5.

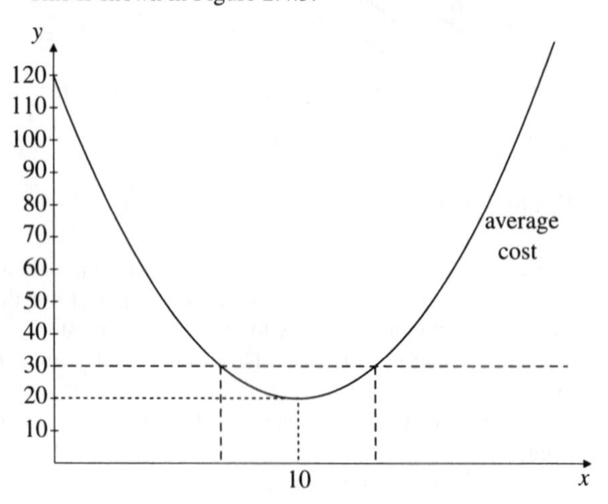

Figure 2.4.5

7. (a) See Figure 2.4.7 (a).

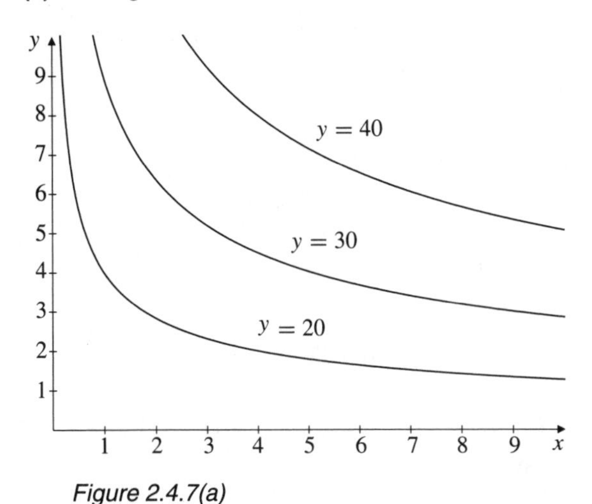

Figure 2.4.7(a)

(b) See Figure 2.4.7 (b).

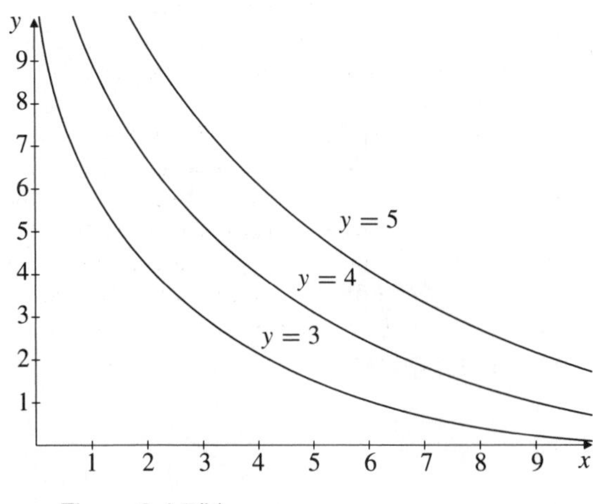

Figure 2.4.7(b)

(c) See Figure 2.4.7 (c).

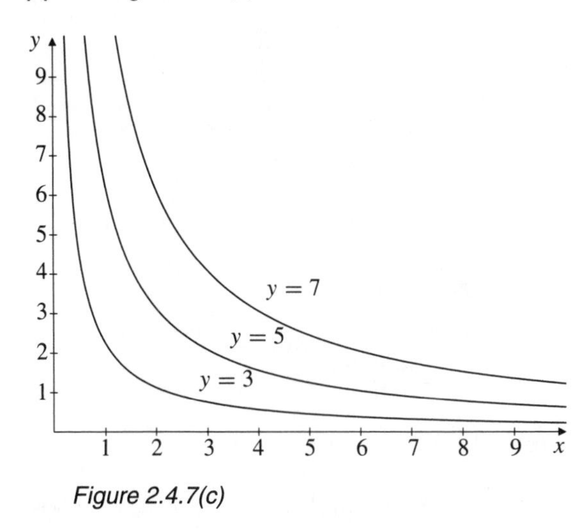

Figure 2.4.7(c)

9. The function $y = x_1^2 x_2^2$ is strictly quasiconcave but not concave. The strict quasiconcavity can immediately be seen by noting that the level sets are described by $c = x_1^2 x_2^2$ or $x_2 = c^{0.5}/x_1$ and the fact that the function is increasing in both variables.

To prove that the function is not concave we select a counterexample. Take the two points (0, 0) and (1, 1). Their convex combination is

$$\bar{x} = \lambda(0,0) + (1-\lambda)(1,1) = (1-\lambda, 1-\lambda)$$

The value of the function of this convex combination is $f(\bar{x}) = (1-\lambda)^2(1-\lambda)^2 = (1-\lambda)^4$ For the convex combination of the function value of each point we obtain

$$\lambda f[(0,0)] + (1-\lambda) f[(1,1)] = 0 + 1 - \lambda = 1 - \lambda$$

For $\lambda = 0.5$ we have that

$$\begin{aligned} f(\bar{x}) &= 0.0625 < 0.5 f[(0,0)] + \\ &\quad (1-0.5) f[(1,1)] = 0.5 \end{aligned}$$

Thus, the function cannot be concave. The graph in Figure 2.4.9 demonstrates the properties of the function.

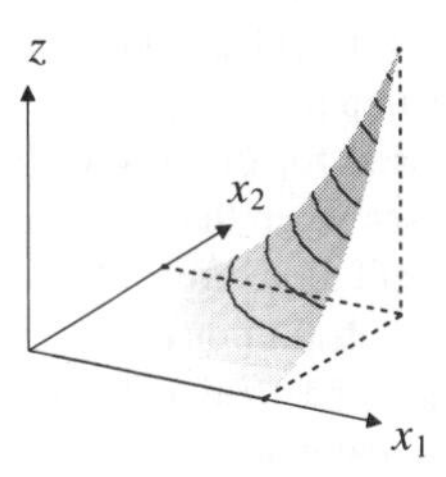

Figure 2.4.9

11. From Definition 2.27, the function $y = x^{1/2}$ is strictly concave if

$$[\lambda x_1 + (1-\lambda)x_2]^{1/2} > \lambda x_1^{1/2} + (1-\lambda)x_2^{1/2}$$

for $\lambda \in [0, 1]$. Squaring both sides, this is equivalent to

$$\lambda x_1 + (1-\lambda)x_2 > \lambda^2 x_1 + 2\lambda(1-\lambda)x_1^{1/2}x_2^{1/2} + (1-\lambda)^2 x_2$$

or

$$\lambda(1-\lambda)x_1 + \lambda(1-\lambda)x_2 > 2\lambda(1-\lambda)x_1^{1/2}x_2^{1/2}$$

Simplifying this and squaring both sides gives

$$x_1^2 + 2x_1x_2 + x_2^2 > 4x_1x_2$$

or

$$(x_1 - x_2)^2 > 0$$

which is certainly true. Therefore, the function is strictly concave.

Section 2.5 (page 53)

1. Setting demand equal to supply we obtain the equilibrium price $p^* = \frac{a-\alpha}{b+\beta}$. Since p^* cannot be negative, an equilibrium price will only exist when $a \geq \alpha$. Hence, proposition 1 is false and proposition 2 is true.

3. *Proof of the first statement:*

A sufficient condition for the demand for a good to increase when its price falls is that it is a normal good. Denoting

A: 'The good is normal' and B: 'The demand for the good increases when its price falls' the statement can formally be written as A⇒B. Now we are asked to prove it in the following ways: (i) direct proof: A⇒ B, (ii) proof by contrapositive proposition: not B ⇒ not A, and (iii) proof by contradiction: A and not B leads to a contradiction.

(i) *Direct proof:* A⇒ B When the price of the good falls, the consumer's real income increases. Since the good is normal by assumption this signifies that the income effect is an increase in the demand for the good. The substition effect has the same consequence: The fall in price leads to increased demand. Thus, the total effect must be that the demand of the good rises.

(ii) *Proof of the contrapositive proposition:* not B ⇒ not A. If the demand for the good decreases when its price falls, t must be the income effect that causes decreased demand because the substitution effect always leads to increased demand. Since the fall in prices increases real income, the good must be inferior, i.e., not normal.

(iii) *Proof by contradiction:* A and not B leads to a contradiction. A fall in the price of the good leads to an increase in real income. Because the good is normal by assumption the income effect leads to an increase in demand for that good. If the demand for the good decreases when the price falls, this can only be due to the substitution effect. This contradicts the assumption that the substitution effect always leads to an increase in demand for a good whose price has fallen. Thus it must be true that when the good is normal, the demand rises when its price falls.

Proof of the second statement:

A necessary but not sufficient condition for the demand for the good to decrease when its price falls is that it is an inferior good.

Review Exercises (page 55)

1. (a) $\lambda(-2) + (1 - \lambda)2 = 2 - 4\lambda$

(b) $[\lambda(-2) + (1 - \lambda)(-3), \lambda 2 + (1 - \lambda)3]$

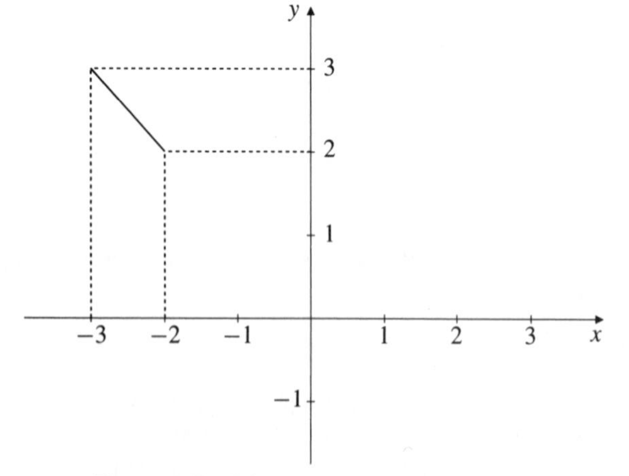

Figure 2.R.1(b)

(c) $[\lambda(0) + (1 - \lambda)x_1, \lambda(0) + (1 - \lambda)x_2] = [(1 - \lambda)x_1, (1 - \lambda)x_2]$

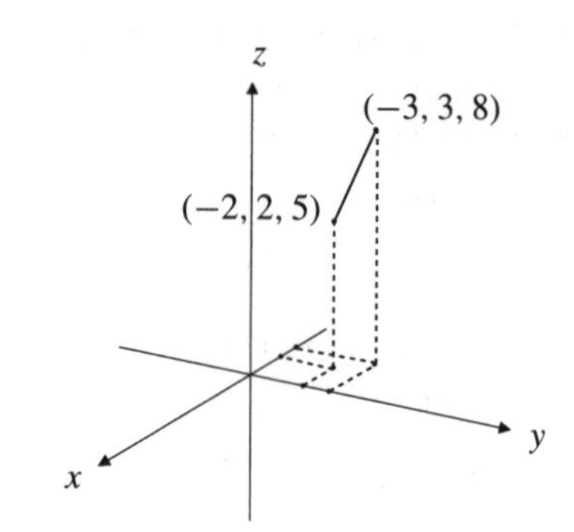

Figure 2.R.1(c)

(d) $[\lambda(-2) + (1 - \lambda)(-3), \lambda 2 + (1 - \lambda)3, \lambda 5 + (1 - \lambda)8]$

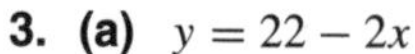

3. (a) $y = 22 - 2x$

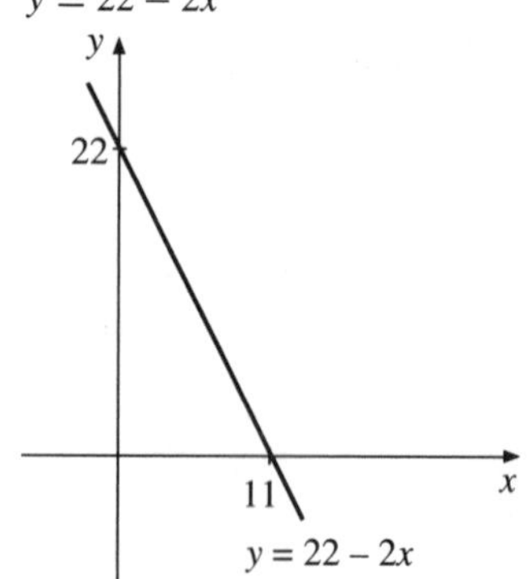

Figure 2.R.3(a)

(b) $y = 2.5 + 0.75x$

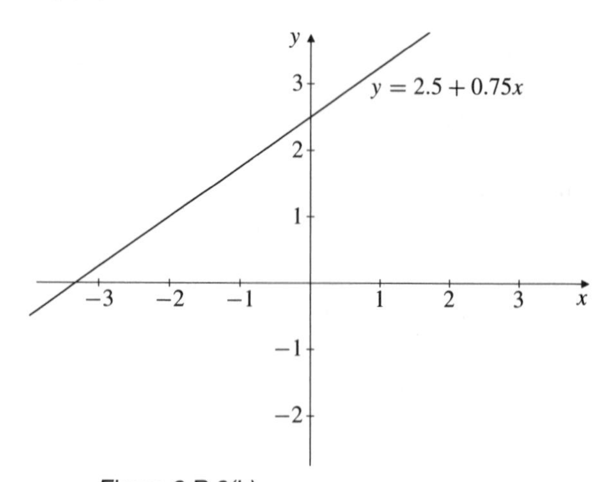

Figure 2.R.3(b)

5. (a) $\dfrac{(ab)^3}{a^2b} = \dfrac{a^3b^3}{a^2b} = ab^2$

(b) $a\left(\dfrac{b}{a}\right)^q = a.a^{-q}b^q = a^{1-q}b^q$

(c) $y^{1/8} = 5x^{0.25} \Rightarrow \left(y^{1/8}\right)^8 = (5x^{0.25})^8 \Rightarrow y = 5^8x^2$

(d) $\log_b(b^x)^3 = \log_b(b^{3x}) = 3x\log_b b = 3x$

7. Example 2.16 on page 45 of the text shows that x^2 is strictly convex. Therefore, $-x^2$ is strictly concave. Therefore, $10 - x^2$ is strictly concave, since adding a constant to the function makes no difference to its shape.

9. We must show that

$$[\lambda x_1 + (1-\lambda)x_1' + \lambda x_2 + (1-\lambda)x_2']^{1/2}$$

$$> \lambda(x_1 + x_2)^{1/2} + (1-\lambda)(x_1' + x_2')^{1/2}$$

Squaring both sides gives

$$\lambda(x_1 + x_2) + (1-\lambda)(x_1' + x_2')$$

$$> \lambda^2(x_1 + x_2) + 2\lambda(1-\lambda)(x_1 + x_2)^{1/2}(x_1' + x_2')^{1/2} +$$

$$(1-\lambda)^2(x_1' + x_2')$$

This is equivalent to

$$(x_1 + x_2) + (x_1' + x_2') > 2(x_1 + x_2)^{1/2}(x_1' + x_2')^{1/2}$$

Squaring both sides and simplifying gives

$$(x_1 + x_2)^2 - 2(x_1 + x_2)(x_1' + x_2') + (x_1' + x_2')^2 > 0$$

or

$$[(x_1 + x_2) - (x_1' + x_2')]^2 > 0$$

which is certainly true. Thus, the function is strictly concave.

CHAPTER 3

Section 3.1 (page 59)

1. (a) The first 10 terms of the sequence $f(n) = 5 + 1/n$ are: 6, 5.5, 5.33, 5.25, 5.20, 5.17, 5.14, 5.125, 5.11, 5.1

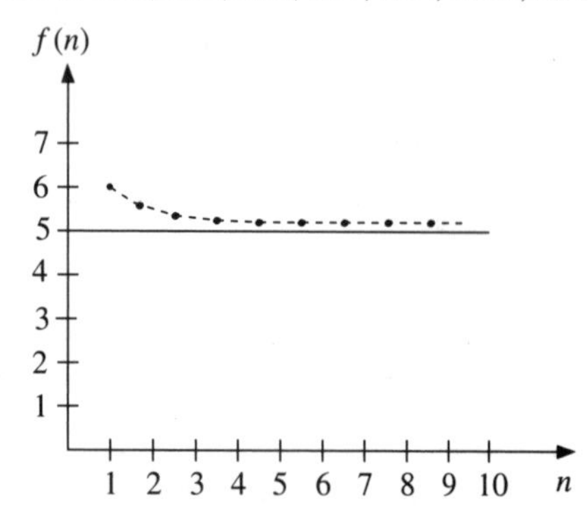

Figure 3.1.1 (a)

(b) The first 10 terms of the sequence $f(n) = 5n/(2^n)$ are: 2.5, 2.5, 1.875, 1.25, 0.78, 0.47, 0.27, 0.16, 0.088, 0.049

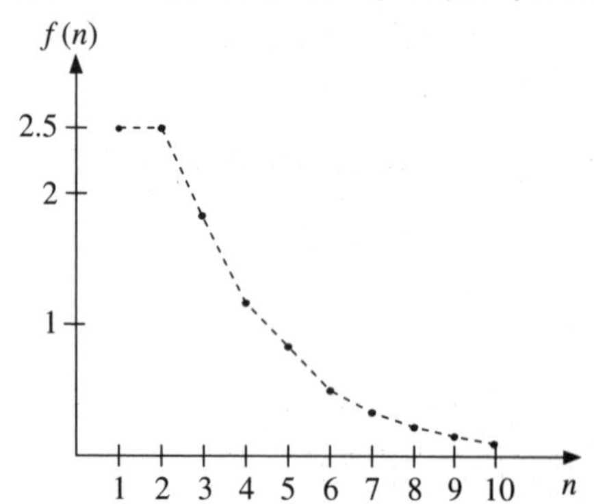

Figure 3.1.1 (b)

(c) The first 10 terms of the sequence $f(n) = (n^2 + 2n)/n$ are: 3, 4, 5, 6, 7, 8, 9, 10, 11, 12

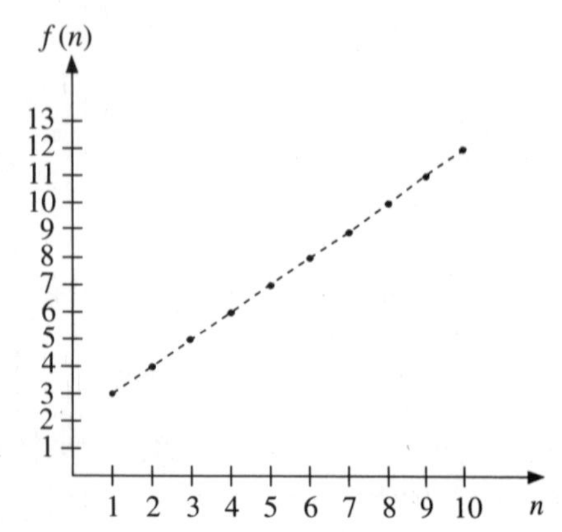

Figure 3.1.1 (c)

3. The sequence $\hat{f}(n) = 2(n-1)$, $n = 1, 2, 3, \ldots$ is identical to the sequence $f(n) = 2n$, $n = 0, 1, 2, \ldots$. (Check the first five or so terms of each.)

5. The sequence $\hat{f}(n) = (1+r)^{n+25}$, $n = 1, 2, 3, \ldots$ is identical to the sequence $f(n) = (1+r)^n$, $n = 1, 2, 3, \ldots$, starting with the 26th term; i.e., $f(n) = (1+r)^n$, $n = 26, 27, 28, \ldots$.

Section 3.2 (page 63)

1. (a) We need to show that for any $\epsilon > 0$ there must be some value N such that

$$\left|\frac{n}{n+1} - 1\right| < \epsilon$$

for every $n > N$. That is

$$\left|\frac{n-(n+1)}{n+1}\right| < \epsilon$$

or

$$\left|-\frac{1}{n+1}\right| < \epsilon$$

That is

$$\frac{1}{n+1} < \epsilon$$

or

$$1 < \epsilon(n+1)$$

or

$$\frac{1}{\epsilon} < n+1$$

which holds for any $n > \frac{1}{\epsilon} - 1$. Thus, we can satisfy the condition $\left|\frac{n}{n+1} - 1\right| < \epsilon$ for any $n > N$ by choosing N to be the next integer greater than the number $\frac{1}{\epsilon} - 1$.

(b) We need to show that for any $\epsilon > 0$ there must be some value N such that

$$\left|5 + \frac{1}{n} - 5\right| < \epsilon$$

for every $n > N$. That is

$$\left|\frac{5n+1-5n}{n}\right| < \epsilon$$

or

$$\left|\frac{1}{n}\right| < \epsilon$$

That is

$$\frac{1}{n} < \epsilon$$

or

$$1 < \epsilon n$$

which holds for any $n > \frac{1}{\epsilon}$. Thus, we can satisfy the condition $\left|5 + \frac{1}{n} - 5\right| < \epsilon$ for any $n > N$ by choosing N to be the next integer greater than the number $\frac{1}{\epsilon}$.

(c) We need to show that for any $\epsilon > 0$ there must be some value N such that

$$\left|\left(-\frac{1}{2}\right)^n - 0\right| < \epsilon$$

for every $n > N$. That is

$$\left|\left(-\frac{1}{2}\right)^n\right| < \epsilon$$

Since $\left|\left(-\frac{1}{2}\right)^n\right| = \left(\frac{1}{2}\right)^n = \frac{1}{2^n}$, we can write this as

$$\frac{1}{2^n} < \epsilon$$

or

$$\frac{1}{\epsilon} < 2^n$$

If we take the $\log_2$ of each side, where we choose base of 2 for convenience, we get

$$\log_2\left(\frac{1}{\epsilon}\right) < n, \quad \text{since} \quad \log_2 2^n = n \log_2 2 = n$$

Thus, we can satisfy the condition $\left|\left(-\frac{1}{2}\right)^n - 0\right| < \epsilon$ for any $n > N$ by choosing N to be the next integer greater than the number $\log_2\left(\frac{1}{\epsilon}\right)$.

3. (a)

$$\lim_{n\to\infty} n^2 = \infty$$

To see that this sequence is definitely divergent, notice that for any value K, no matter how large, it is always possible to find an N large enough that $n^2 > K$ for every $n > N$. Choosing N to be the next integer greater than the number $\sqrt{K}$ will satisfy this condition.

(b)

$$\lim_{n\to\infty} (-n)^3 = -\infty$$

To see that the sequence is definitely divergent, notice that for any value K, no matter how large, it is always possible to find an N large enough that $(-n)^3 < -K$ for every $n > N$. Since $(-n)^3 = -n^3$ we can see that, upon multiplying by -1, this inequality becomes

$$n^3 > K$$

and so choosing N to be the next integer greater than the number $K^{1/3}$ will satisfy this condition.

(c) If $|c| > 1$ the sequence $(-c)^n$ is divergent and is not definitely divergent. If $|c| < 1$ then

$$\lim_{n\to\infty} (-c)^n = 0$$

If $c > 0$ then $(-c)^n > 0$ for even and < 0 for n odd. If $c < 0$ then $(-c)^n > 0$ for n even and < 0 for n odd. Therefore, in the case that $|c| > 1$ then for any N, the sequence $(-c)^n$ will contain both arbitrarily large positive and negative values for $n > N$. (This follows from the fact $|(-c)^n| = |c|^n$ and $\lim_{n\to\infty} |c|^n = +\infty$ for $|c| > 1$.) Therefore, the sequence would be divergent and not definitely divergent. If $|c| < 1$ then we can show the sequence converges to the limit 0 in the same way as we did for question 1 (c). That is, we need to show that for any $\epsilon > 0$ there must be some value N such that

$$|(-c)^n - 0| < \epsilon$$

for every $n > N$. That is

$$|(-c)^n| < \epsilon$$

or

$$|c|^n < \epsilon$$

If we take logs to the base e, i.e., ln, of each side of the inequality we get

$$n \ln|c| < \ln \epsilon$$

or, noting that $\ln|c| < 0$ (since $|c| < 1$)

$$n > \ln\epsilon / \ln|c|$$

Thus, we can satisfy the condition $|(-c)^n - 0| < \epsilon$ for any $n > N$ by choosing N to be the next integer greater than the number $\ln\epsilon/\ln|c|$.

Section 3.3 (page 71)

1. According to equation (3.3) in the text we have

$$PV_1 = \frac{V}{(1+r)^1} = \frac{V}{1+r} = \frac{500}{1+0.08} = \frac{500}{1.08} = \$462.96$$

3. (a) If interest is compounded annually and the present value is the same in each case then

$$\frac{V_2}{(1+r)^{t_2}} = \frac{V_1}{(1+r)^{t_1}}$$

and so

$$\frac{V_2}{V_1} = \frac{(1+r)^{t_2}}{(1+r)^{t_1}}$$

Now, since $t_2 > t_1$ we have $(1+r)^{t_2} > (1+r)^{t_1}$ and so

$$\frac{(1+r)^{t_2}}{(1+r)^{t_1}} > 1$$

which implies that

$$\frac{V_2}{V_1} > 1$$

or

$$V_2 > V_1$$

(b) We need to show that

$$\frac{V_2}{(1+r)^{t_2+k}} = \frac{V_1}{(1+r)^{t_1+k}}$$

or

$$\frac{V_2}{(1+r)^{t_2}(1+r)^k} = \frac{V_1}{(1+r)^{t_1}(1+r)^k}$$

The terms $(1+r)^k$ divide out and we are left with the original equality from part (a). That is

$$\frac{V_2}{(1+r)^{t_2}} = \frac{V_1}{(1+r)^{t_1}}$$

5. Let Z_t be the population t years from now. Under continuous compounding we have that

$$Z_t = 100e^{0.02t} \quad \text{(million)}$$

(a) $Z_5 = 100e^{0.02(5)} = 100e^{0.10} = 110.52$ million

(b) $Z_{10} = 100e^{0.02(10)} = 100e^{0.20} = 122.14$ million

(c) $Z_{20} = 100e^{0.02(20)} = 100e^{0.40} = 149.18$ million

Section 3.4 (page 76)

1. From Theorem 3.3 we know that a monotonic sequence is convergent if and only if it is bounded. Letting $a_t = \frac{V}{(1+r)^t}$ represent a general term in the sequence we first show it is a monotonic sequence.

$$\frac{a_{t+1}}{a_t} = \frac{V/(1+r)^{t+1}}{V/(1+r)^t} = \frac{(1+r)^t}{(1+r)^{t+1}} = \frac{1}{1+r}$$

and so

$$a_{t+1} = \frac{a_t}{1+r}$$

Therefore, the sequence is monotonically decreasing if $r > 0$ while it is monotonically increasing if $-1 < r < 0$. In the case with $r > 0$, $a_t = V/(1+r)^t$ is bounded below by 0 and above by V and so the sequence is convergent. In the case with $-1 < r < 0$ we have $0 < 1+r < 1$ and $\lim_{t\to\infty} V/(1+r)^t = +\infty$, that is, the sequence is not bounded and so is not convergent.

3. Result (iii) of Theorem 3.2 states that if $\lim_{n\to\infty} a_n = L^a$ and $\lim_{n\to\infty} b_n = +\infty$ then $\lim_{n\to\infty}(a_n - b_n) = -\infty$.

Now, $\lim_{n\to\infty} b_n = +\infty$ means that for (arbitrarily large) $\hat{K} > 0$, there is a value N_1 large enough that

$$b_n > \hat{K} \quad \text{for every} \quad n > N_1$$

$\lim_{n\to\infty} a_n = L^a$ means that for any $\epsilon > 0$ (no matter how small) there is a value N_2 large enough that

$$|a_n - L^a| < \epsilon \quad \text{for every} \quad n > N_2$$

and so, in particular, if $a_n > L^a$

$$a_n - L^a < \epsilon \quad \text{for every} \quad n > N_2$$

We need to show that for any (arbitrarily large) value $K > 0$, there is a value N large enough that

$$a_n - b_n < -K \quad \text{for every} \quad n > N$$

Well, since for any $\hat{K} > 0$ we can find a value N_1 such that

$$b_n > \hat{K}$$

or

$$-b_n < -\hat{K}$$

and for any $\epsilon > 0$ we can find a value N_2 such that

$$a_n < L^a + \epsilon$$

then, taking N to be the larger of N_1 and N_2 we can see that, upon adding the left and right sides of the above pairs of inequalities we have

$$a_n - b_n < -\hat{K} + L^a + \epsilon \quad \text{for every} \quad n > N$$

Since $\hat{K}$ is any arbitrarily large value we can see that $K = \hat{K} - L^a - \epsilon$ also represents an arbitrarily large value and so we can write

$$a_n - b_n < -K \quad \text{for every} \quad n > N$$

Section 3.5 (page 92)

1. Since $a_n = c$, and $a_{n+1} = c$ it is trivial to show that

$$\lim_{n\to\infty}\left|\frac{a_{n+1}}{a_n}\right| = \lim_{n\to\infty}\left|\frac{c}{c}\right| = 1$$

The series formed by this sequence is

$$\begin{aligned} S_n = \sum_{t=1}^{n} a_t &= a_1 + a_2 + \ldots + a_n \\ &= c + c + \ldots + c = nc \end{aligned}$$

Since $c > 0$ it is clear that $\lim_{n\to\infty} S_n = \lim_{n\to\infty} nc = +\infty$. That is, the series diverges.

3.

$$s_n = a\rho + a\rho^3 + a\rho^5 + \ldots + a\rho^{2n-3} + a\rho^{2n-1}$$

Multiplying this expression by ρ^2 gives

$$\rho^2 s_n = a\rho^3 + a\rho^5 + a\rho^7 + \ldots + a\rho^{2n-1} + a\rho^{2n+1}$$

Upon subtraction of these two expressions we get

$$s_n - \rho^2 s_n = a\rho - a\rho^{2n+1}$$

which implies that

$$(1-\rho^2)s_n = a\rho - a\rho^{2n+1}$$

and so

$$s_n = \frac{a\rho - a\rho^{2n+1}}{1-\rho^2}$$

Taking the limit as $n \to \infty$ and noting that $\lim_{n\to\infty} \rho^{2n+1} = 0$ since $|\rho| < 1$, we obtain the result that

$$\lim_{n\to\infty} s_n = \frac{a\rho}{1-\rho^2}$$

5. (a) Beginning at the end of a year the project generates net revenue of \$500, 000 per year (i.e., gross revenue of \$1.5 million less running costs of \$1 million). If this stream of net revenue had begun immediately (i.e. in the first year) its present value would be

$$\frac{500,000}{r} = \frac{500,000}{0.08} = \$6,250,000$$

However, this sum must be discounted by one year since it does not begin until the end of the year following the one-year construction period. Therefore, the present value of net benefits, excluding building costs, is

$$\begin{aligned} PV \text{ of net operating revenue} &= \$6,250,00/(1+r) \\ &= \$6,250,00/(1.08) \\ &= \$5,787,037 \end{aligned}$$

Since the building costs are incurred immediately we have that

$$PV \text{ of building costs} = \$10 \text{ million}$$

Therefore, the present value of building costs exceeds the present value of operating revenues and so this project is not profitable.

(b) For a general interest rate r the PV of net operating revenues is

$$\frac{500,000}{r}/(1+r) = \frac{500,000}{r(1+r)}$$

The value of r that equates the present value of net operating revenues to building costs satisfies the equation

$$\frac{500,000}{r(1+r)} = 10,000,000$$

which implies

$$20r^2 + 20r - 1 = 0$$

Using the formula for solving quadratic equations we find the solution to be

$$r = \frac{-20 \pm \sqrt{480}}{40} = 0.048 \quad \text{or} \quad -1.048$$

Only the positive value is relevant and so the project is profitable if the interest rate is less than 0.048 (i.e., 4.8%) but is unprofitable if the interest rate is greater than 0.048.

7. The first five round of effects are

1st round effect	20.00 billion	
2nd round effect	14.00 billion	(20×0.7)
3rd round effect	9.80 billion	(14×0.7)
4th round effect	6.86 billion	(9.8×0.7)
5th round effect	4.802 billion	(6.86×0.7)

The overall impact is

$$20 + 20(0.7) + 20(0.7)^2 + 20(0.7)^3 + \ldots = \frac{20}{1-0.7} = 66.67 \text{ billion}$$

Review Exercises (page 95)

1. (a) 1, 0.25, 0.111, 0.063, 0.04

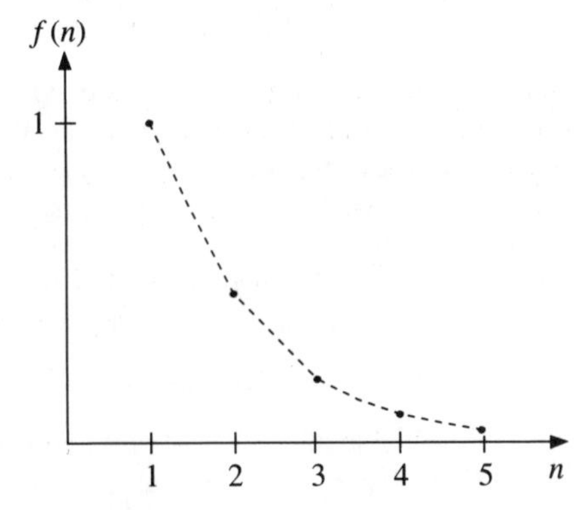

Figure 3.R.1 (a)

(b) 1, 2.5, 1.67, 2.25, 1.8

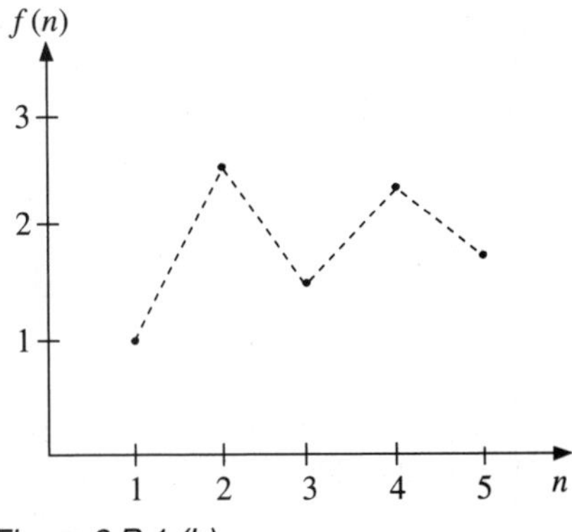

Figure 3.R.1 (b)

(c) 0.2, 0.25, 0.273, 0.286, 0.294

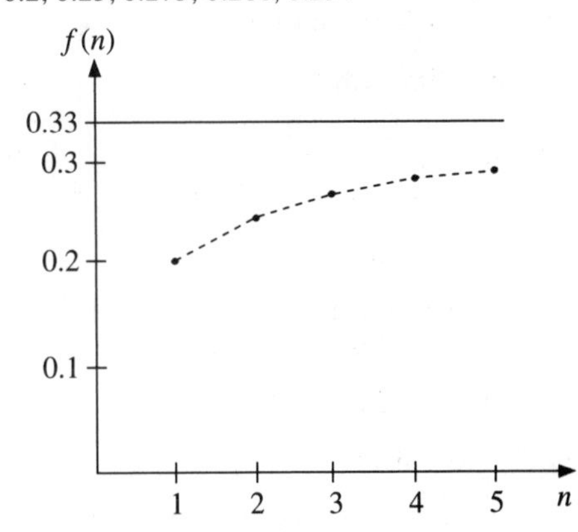

Figure 3.R.1 (c)

(d) −1, −4, −9, −16, −25

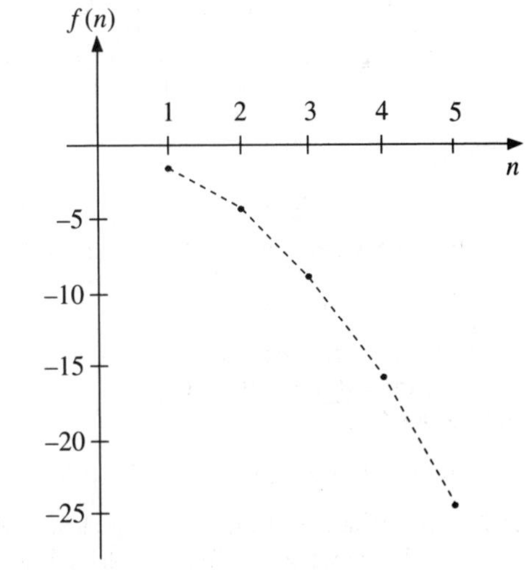

Figure 3.R.1 (d)

(e) 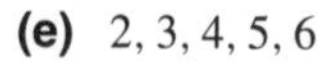 2, 3, 4, 5, 6

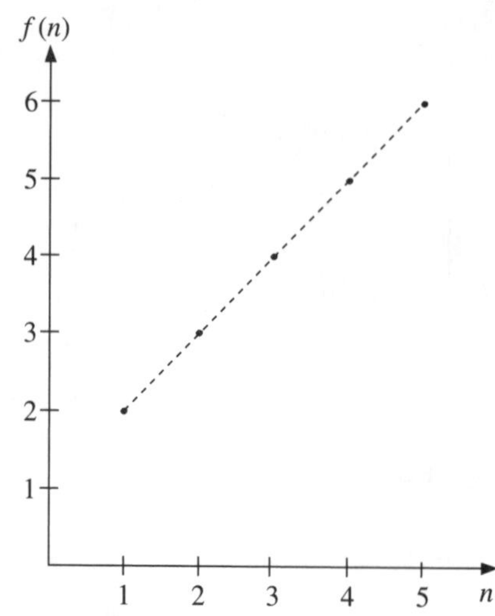

Figure 3.R.1 (e)

(f) 6, 5.5, 5.33, 5.25, 5.2

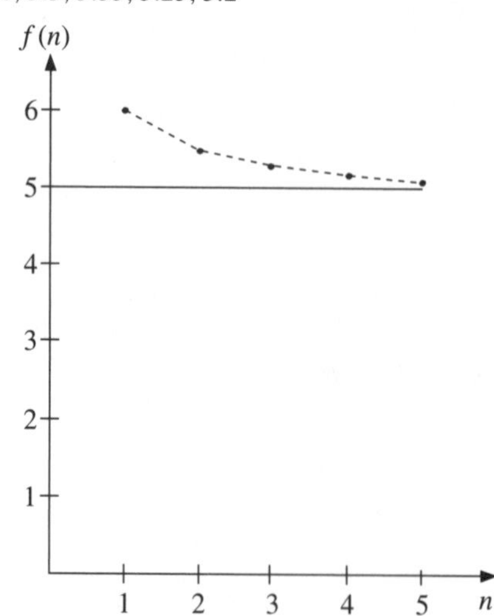

Figure 3.R.1 (f)

(g) 4, 4.5, 4.67, 4.75, 4.8

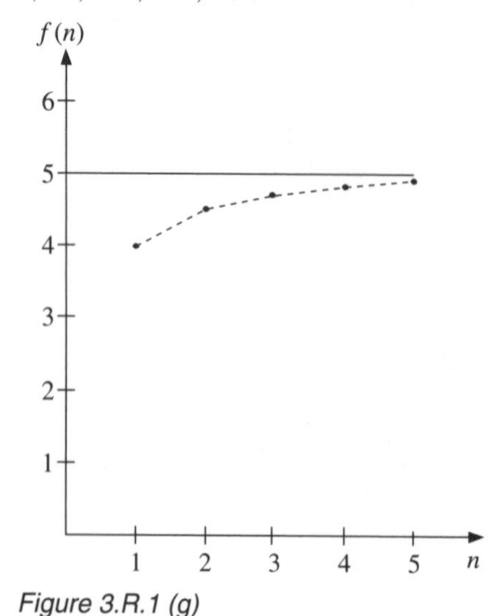

Figure 3.R.1 (g)

3. (a) Using the formula $PV_t = \frac{V}{(1+r)^t}$ to determine the present value of \$$V$ received in t years time if the interest rate is r we get, for $r = 0.08$ (8%)

(i) $PV_1 = \frac{100}{1.08} = \$92.59$

(ii) $PV_5 = \frac{150}{(1.08)^5} = \$102.09$

(b) The individual should rank alternative (ii) as better. Even if the individual wants the money well in advance of 5 years she should rank alternative (ii) as the better one because at $r = 0.08$ (8%), she could borrow more money now and pay it back after 5 years with the \$150 received at that point in time than she could from receiving the \$100 in one year's time.

5. (a) We use the formula $PV = V/r$ to determine that

$$PV = \frac{100}{0.10} = \$1,000$$

is the net present value of \$100 per year received in perpetuity at the interest rate $r = 0.10$ (10%).

(b) The present value of \$$V$ received each year for T years at the interest rate r is

$$P_T = \sum_{t=1}^{T} \frac{V}{(1+r)^t}$$

This is a (finite) geometric series with first term $a = \frac{V}{1+r}$ and $\rho = \frac{1}{1+r}$ and so, from equation (3.4) in the text, we have

$$P_T = \frac{a(1-\rho^T)}{1-\rho} = \frac{\frac{V}{1+r}\left[1 - (\frac{1}{1+r})^T\right]}{1 - (\frac{1}{1+r})}$$

which simplifies to

$$P_T = \frac{V\left[1 - \frac{1}{(1+r)^T}\right]}{r}$$

For $V = \$100,\ T = 25,\ r = 0.10$ we get

$$P_T = \frac{100\left[1 - \frac{1}{(1.10)^{25}}\right]}{0.10} = 907.70$$

(c) Since in part (a) we found the value of \$100 per year received in perpetuity at interest rate $r = 0.10$ equals \$1000 and in part (b) we found the value of \$100 per year received for 25 years at interest rate $r = 0.10$ equals \$907.70 it follows that the value of \$100 per year received after 25 years (in perpetuity), at interest rate $r = 0.10$ is equal to the difference

$$\$1,000 - \$907.70 = \$92.30$$

Another way of finding this value is to note that, since receiving \$100 per year every year (forever) at interest rate $r = 0.10$ is like receiving immediately a single lump sum of \$1000, then receiving such a stream of payments but commencing only after 25 years have passed has the same present value as does receiving a single lump sum amount of \$1000 after 25 years have passed. This has a present value of

$$PV = \frac{1000}{(1+0.10)^{25}} = 92.30$$

7. Overall impact is

$$50 + 50(0.75) + 50(0.75)^2 + 50(0.75)^3 + ...$$
$$= \frac{50}{1 - 0.75} = \frac{50}{0.25} = \$200 \text{ billion}$$

9. The expected utility of this gamble is

$$EU = \lim_{N \to \infty} \sum_{n=1}^{N} p_n u_n$$

where $p_n = \frac{1}{2^n}$ and $u_n = y_n^2 = (2^n)^2 = 2^{2n}$. Therefore, we get

$$\begin{aligned} EU &= \lim_{N \to \infty} \sum_{n=1}^{N} \frac{1}{2^n} 2^{2n} \\ &= \lim_{N \to \infty} \sum_{n=1}^{N} 2^n = +\infty \end{aligned}$$

CHAPTER 4

Section 4.1 (page 105)

1. (a) $f(x) = 5x$ and so f_n or $f(x_n) = 5x_n$. Therefore, for the sequence of x-values, $x_n = 2 - \frac{1}{n}$, we have

$$\begin{aligned} f(x_n) &= 5\left(2 - \frac{1}{n}\right) \\ &= 10 - \frac{5}{n} \quad n = 1, 2, 3, \ldots \end{aligned}$$

Since $\lim_{n\to\infty} \frac{5}{n} = 0$, it follows that $\lim_{n\to\infty} f(x_n) = 10$. To see this more formally, note that $|f(x_n) - 10| < \epsilon$ for any value of n such that $\frac{5}{n} < \epsilon$, or $n > \frac{5}{\epsilon}$. This suggests, therefore, that

$$\lim_{x\to 2^-} f(x) = 10$$

Notice that $n > \frac{5}{\epsilon} \Leftrightarrow \frac{1}{n} < \frac{\epsilon}{5}$ and so the x_n values lie in the interval $2 - \frac{\epsilon}{5} < x < 2$ whenever $n > \frac{5}{\epsilon}$. In other words, using the notation of Definition 4.1, these x_n values lie in the interval $2 - \delta < x < 2$ for $\delta = \frac{\epsilon}{5}$ and so a $\delta > 0$ exists such that $|f(x) - 10| < \epsilon$, $\forall x$, satisfying $2 - \delta < x < 2$.

(b) $f(x) = -3x + 4$ and so f_n or $f(x_n) = -3x_n + 4$. Therefore, for the sequence of x-values, $x_n = 2 - \frac{1}{n}$, we have

$$\begin{aligned} f(x_n) &= -3\left(2 - \frac{1}{n}\right) + 4 \\ &= -2 + \frac{3}{n} \quad n = 1, 2, 3, \ldots \end{aligned}$$

Since $\lim_{n\to\infty} \frac{3}{n} = 0$, it follows that $\lim_{n\to\infty} f(x_n) = -2$. To see this more formally, note that $|f(x_n) - (-2)| = |f(x_n) + 2| < \epsilon$ for any value of n such that $\frac{3}{n} < \epsilon$, or $n > \frac{3}{\epsilon}$. This suggests, therefore, that

$$\lim_{x\to 2^-} f(x) = -2$$

Notice that $n > \frac{3}{\epsilon} \Leftrightarrow \frac{1}{n} < \frac{\epsilon}{3}$ and so the x_n values lie in the interval $2 - \frac{\epsilon}{3} < x < 2$ whenever $n > \frac{3}{\epsilon}$. In other words, using the notation of Definition 4.1, these x_n values lie in the interval $2 - \delta < x < 2$ for $\delta = \frac{\epsilon}{3}$ and so a $\delta > 0$ exists such that $|f(x) - (-2)| < \epsilon$, $\forall x$, satisfying $2 - \delta < x < 2$.

(c) $f(x) = mx + b$ and so f_n or $f(x_n) = mx_n + b$. Therefore, for the sequence of x-values, $x_n = 2 - \frac{1}{n}$, we have

$$f(x_n) = m\left(2 - \frac{1}{n}\right) + b = 2m + b - \frac{m}{n}, \quad n = 1, 2, 3, \ldots$$

Since $\lim_{n\to\infty} \frac{m}{n} = 0$, it follows that $\lim_{n\to\infty} f(x_n) = 2m + b$. To see this more formally, note that $|f(x_n) - (2m + b)| < \epsilon$ for any value of n such that $\frac{m}{n} < \epsilon$, or $n > \frac{m}{\epsilon}$. This suggests, therefore, that

$$\lim_{x\to 2^-} f(x) = 2m + b$$

Notice that $n > \frac{m}{\epsilon} \Longleftrightarrow \frac{1}{n} < \frac{\epsilon}{m}$ and so the x_n values lie in the interval $2 - \frac{\epsilon}{m} < x < 2$ whenever $n > \frac{m}{\epsilon}$. In other words, using the notation of Definition 4.1, these x_n values lie in the interval $2 - \delta < x < 2$ for $\delta = \frac{\epsilon}{m}$ and so a $\delta > 0$ exists such that $|f(x) - (2m + b)| < \epsilon$, $\forall x$, satisfying $2 - \delta < x < 2$.

(d) $f(x) = x^2$ and so f_n or $f(x_n) = x_n^2$. Therefore, for the sequence of x-values, $x_n = 2 - \frac{1}{n}$, we have

$$\begin{aligned} f(x_n) &= \left(2 - \frac{1}{n}\right)^2 \\ &= 4 - \frac{4}{n} + \frac{1}{n^2} \quad n = 1, 2, 3, \ldots \end{aligned}$$

Since $\lim_{n\to\infty} \frac{4}{n} = 0$ and $\lim_{n\to\infty} \frac{1}{n^2} = 0$, it follows that $\lim_{n\to\infty} f(x_n) = 4$. To see this more formally, note that $|f(x_n) - 4| < \epsilon$ for any value of n such that $\frac{4}{n} < \epsilon$ and $\frac{1}{n^2} < \epsilon$, or $n > \frac{4}{\epsilon}$ and $n^2 > \frac{1}{\epsilon}$, that is, $n > \frac{4}{\epsilon} (> \frac{1}{\sqrt{\epsilon}})$ if $\epsilon < 1$. This suggests, therefore, that

$$\lim_{x\to 2^-} f(x) = 4$$

Notice that $n > \frac{4}{\epsilon} \Longleftrightarrow \frac{1}{n} < \frac{\epsilon}{4}$ and so the x_n values lie in the interval $2 - \frac{\epsilon}{4} < x < 2$ whenever $n > \frac{4}{\epsilon}$. In other words, using the notation of Definition 4.1, these x_n values lie in the interval $2 - \delta < x < 2$ for $\delta = \frac{\epsilon}{4}$ and so a $\delta > 0$ exists such that $|f(x) - 4| < \epsilon$, $\forall x$, satisfying $2 - \delta < x < 2$.

3. (a) This function

$$f(x) = \begin{cases} 2x + 3, & \text{if } x < 1 \\ x + 5, & \text{if } x \geq 1, \end{cases}$$

is not continuous at the point $x = 1$ because the left-hand limit

$$\lim_{x\to 1^-} f(x) = \lim_{x\to 1^-} 2x + 3 = 5$$

is not equal the right-hand limit

$$\lim_{x\to 1^+} f(x) = \lim_{x\to 1^+} x + 5 = 6$$

Therefore, the second part of condition (i) in Definition 4.3 is not satisfied at the point $x = 1$.

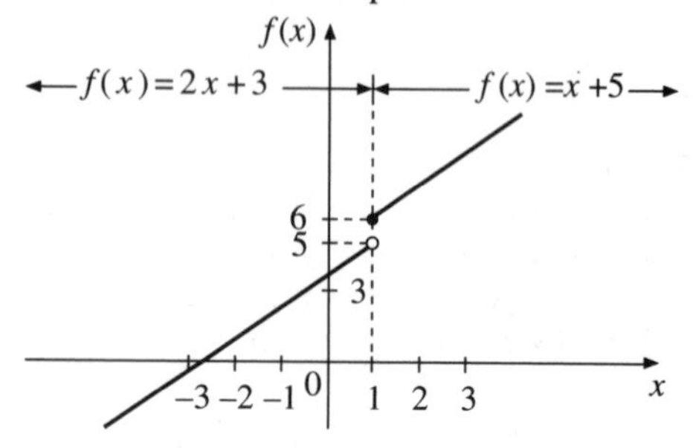

Figure 4.1.3 (a)

(b) The function $f(x) = 1/x$ is not continuous at the point $x = 0$ because it is not defined there (i.e., $1/0$ is not defined). Moreover, the left-hand limit at $x = 0$

$$\lim_{x\to 0^-} \frac{1}{x} = -\infty$$

is not equal to the right-hand limit

$$\lim_{x\to 0^+} \frac{1}{x} = +\infty$$

Therefore, neither part of condition (i) in Definition 4.3 is satisfied at the point $x = 0$.

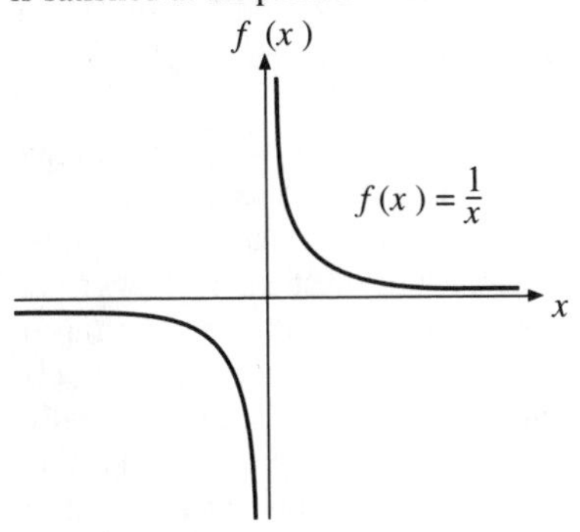

Figure 4.1.3 (b)

(c) The function $f(x) = 1/(x-3)^2$ is not continuous at the point $x = 3$ because it is not defined there (i.e., $1/0^2$ is not defined). However, note that the left-hand limit at $x = 3$

$$\lim_{x\to 3^-} \frac{1}{(x-3)^2} = +\infty$$

is equal to the right-hand limit

$$\lim_{x\to 3^+} \frac{1}{(x-3)^2} = +\infty$$

because the operation of squaring makes the denominator positive whether $x - 3 > 0$ or $x - 3 < 0$. Therefore, it is the first part of condition (i) in Definition 4.3 that is not satisfied at the point $x = 3$.

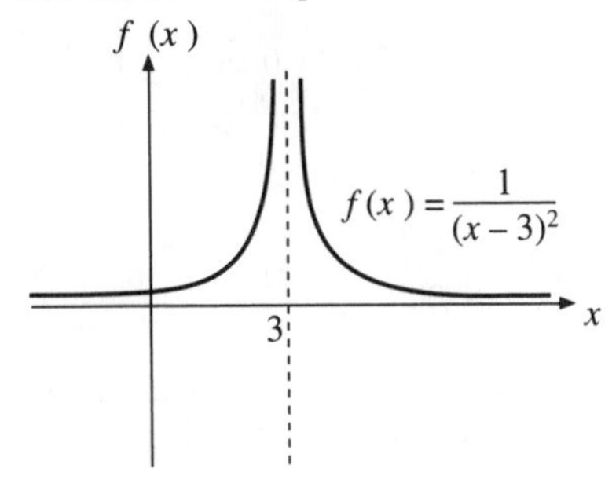

Figure 4.1.3 (c)

(d) The function $f(x) = (x-2)(x^2-x-2)$ is not continuous at the point $x = 2$ because it is not defined there (i.e., since $0/0$ is not defined) and it is also not continuous at the point $x = -1$ (i.e., since $-3/0$ is not defined). Note that by factoring we get

$$f(x) = \frac{x-2}{(x-2)(x+1)} = \frac{1}{x+1}$$

for $x \neq 2$. We can use this result to determine that the left-hand limit at $x = 2$

$$\lim_{x\to 2^-} \frac{x-2}{x^2-x-2} = \lim_{x\to 2^-} \frac{1}{x+1} = \frac{1}{3}$$

is equal to the right-hand limit

$$\lim_{x\to 2^+} \frac{x-2}{x^2-x-2} = \lim_{x\to 2^+} \frac{1}{x+1} = \frac{1}{3}$$

Therefore, it is the first part of condition (i) in Definition 4.3 which is not satisfied at the point $x = 2$.
At the point $x = -1$ we see that the left-hand limit is

$$\lim_{x\to -1^-} \frac{x-2}{x^2-x-2} = \lim_{x\to -1^-} \frac{1}{x+1} = -\infty$$

which is not equal to the right-hand limit, which is

$$\lim_{x\to -1^+} \frac{x-2}{x^2-x-2} = \lim_{x\to -1^+} \frac{1}{x+1} = +\infty$$

and so both parts of condition (i) in Definition 4.3 which are not satisfied at the point $x = -1$.

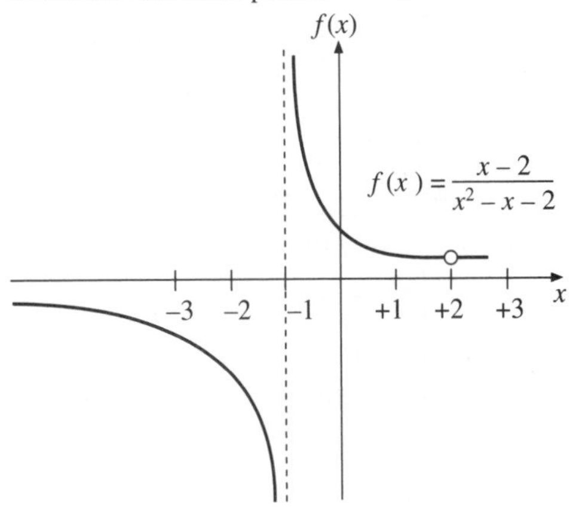

Figure 4.1.3 (d)

5. (a) The function $f(x) = 4x+3$ is continuous at every point $x \in \mathbb{R}$. To see this is so, according to Definition 4.4, we need to show that at any point $x = a$ there is always some $\delta > 0$ such that

$$|f(x) - f(a)| < \epsilon$$

whenever $|x - a| < \delta$ for any $\epsilon > 0$. For this function, that means finding some δ such that

$$|(4x+3) - (4a+3)| < \epsilon$$

whenever $|x - a| < \delta$ for any $\epsilon > 0$. Notice that

$$\begin{aligned} |(4x+3) - (4a+3)| < \epsilon \quad &\Leftrightarrow \quad |4x - 4a| < \epsilon \\ &\Leftrightarrow \quad 4|x-a| < \epsilon \\ &\Leftrightarrow \quad |x-a| < \frac{\epsilon}{4} \end{aligned}$$

Therefore, by choosing $\delta = \frac{\epsilon}{4}$ we see that we can satisfy the condition of Definition 4.4 for continuity.

(b) The function $f(x) = mx + b$ is continuous at every point $x \in \mathbb{R}$. To see this is so, according to Definition 4.4, we need to show that at any point $x = a$, there is always some $\delta > 0$ such that

$$|f(x) - f(a)| < \epsilon$$

whenever $|x-a|<\delta$ for any $\epsilon>0$. For this function, that means finding some δ such that

$$|(mx+b)-(ma+b)|<\epsilon$$

whenever $|x-a|<\delta$ for any $\epsilon>0$. Notice that

$$\begin{aligned}|(mx+b)-(ma+b)|<\epsilon \quad &\Leftrightarrow \quad |mx-ma|<\epsilon\\ &\Leftrightarrow \quad m|x-a|<\epsilon\\ &\Leftrightarrow \quad |x-a|<\frac{\epsilon}{m}\end{aligned}$$

Therefore, by choosing $\delta=\frac{\epsilon}{m}$ we see that we can satisfy the condition of Definition 4.4 for continuity.

7.

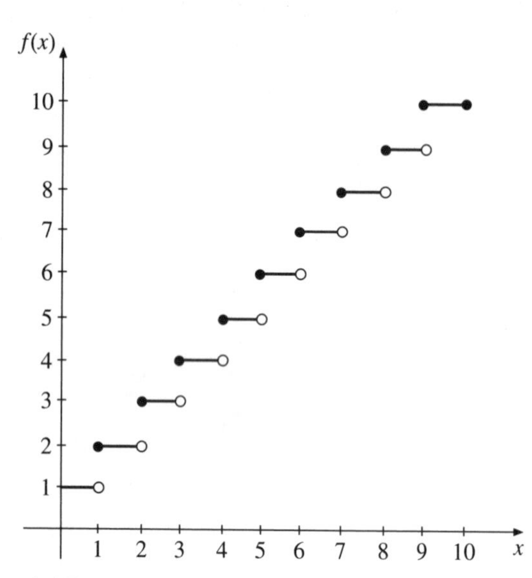

Figure 4.1.7

Although this function is defined on every point $x\in[0,10]$, it is not continuous at the points $x=1,2,...,9$ since at each one of these points the left- and right-hand limits are not equal (See Definition 4.3). However, within each subinterval $[0,1),[1,2),\ldots,[8,9)$ and $[9,10]$ the function is continuous since the function is continuous at any interior point of each interval and the left-hand limit of the function is equal to the function value at the left-most point included in each half-open, half-closed subinterval, $[\cdot,\cdot)$, and the function is continuous also at every point in the closed interval $[9,10]$.

Section 4.2 (page 119)

1. The cost of L units of labor is wL and the inverse of the production function, $L(Q)$, is $L(Q)=Q/b$. Therefore, the labor costs of producing Q units of output is $wL(Q)$ or wQ/b. Given fixed cost of c_0 this means the cost function is

$$C(Q)=c_0+(w/b)Q$$

Letting $\bar{p}$ be the price of the firm's output, we have that the revenue function is $R(Q)=\bar{p}Q$ and the profit function

$$\pi(Q)=R(Q)-C(Q)$$

That is

$$\pi(Q)=\bar{p}Q-c_0-(w/b)Q$$

Now we prove that the production function, $Q(L)=bL$, is continuous. According to Definition 4.4, we need to show that at any point $L=a$, there is always some $\delta>0$ such that

$$|Q(L)-Q(a)|<\epsilon$$

whenever $|L-a|<\delta$ for any $\epsilon>0$. For this function, that means finding some δ such that

$$|bL-ba|<\epsilon$$

whenever $|L-a|<\delta$ for any $\epsilon>0$. Notice that

$$\begin{aligned}|bL-ba|<\epsilon \quad &\Leftrightarrow \quad |b||L-a|<\epsilon\\ &\Leftrightarrow \quad |L-a|<\frac{\epsilon}{|b|}\end{aligned}$$

Therefore, by choosing $\delta=\frac{\epsilon}{|b|}$ we see that we can satisfy the condition of Definition 4.4 for continuity.

Now, since the cost function is a linear function of the inverse of the production function, it follows from results (i), (ii), and (vi) of Theorem 4.1 that the cost function, $C(Q)$, is also continuous. The revenue function, $R(Q)=\bar{p}Q$, is also a continuous function. (To see this formally one needs to complete the same steps as was done above to show that the production function is continuous.) Since the profit function is just the difference between the revenue function and the cost function, both of which are continuous functions, it follows from part (iii) of Theorem 4.1 that the profit function is also continuous.

3. Let S represent sales per month and P the salesperson's pay for the month. It follows that the function describing her salary-sales relationship is

$$P(S)=\begin{cases}\$800+0.15S, & \text{if } S\le \$10,000\\ \$1800+0.15S, & \text{if } \$10,000<S\le \$15,000\\ \$4300+0.15S, & \text{if } S>\$15,000,\end{cases}$$

which is illustrated in the Figure 4.2.3.

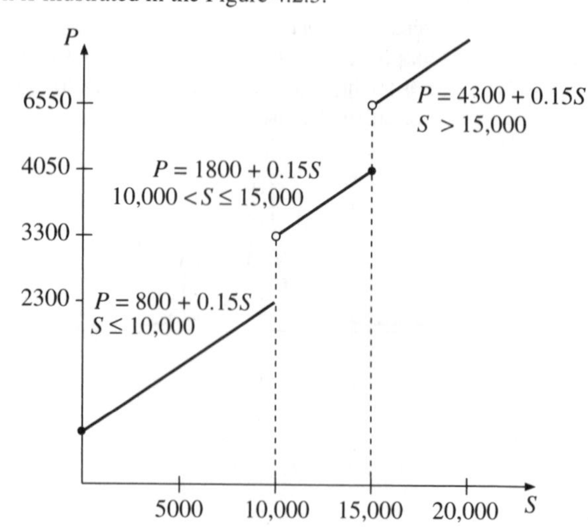

Figure 4.2.3

At $S=10,000$ the left-hand limit of $P(S)$ is

$$\lim_{S\to 10,000^-} P(S)=2300$$

while the right-hand limit is

$$\lim_{S\to 10,000^+} P(S)=3300$$

and so the function $P(S)$ is not continuous at the point $S = \$10,000$. Also, the function $P(S)$ is not continuous at the point $S = \$15,000$ since the left-hand limit

$$\lim_{S \to 15,000^-} P(S) = 4050$$

is not equal to the right-hand limit

$$\lim_{S \to 15,000^+} P(S) = 6550$$

As a salesperson's monthly sales approach either of these two points of discontinuity, one can see that the incentive to increase sales rises dramatically in order to attempt to earn the relevant bonus payment.

5. Let x be income before tax and y be income after tax. If an individual earns less than or equal to \$20,000 he pays no tax and so

$$y = x, \quad x \leq 20,000.$$

However, for every dollar in excess of \$20,000 (i.e., $x - 20,000$) he pays \$0.25 in tax and so if $x > 20,000$, the first \$20,000 are not taxed but he pays $0.25(x - \$20,000)$ in tax (i.e., for that income in excess of \$20,000) and so

$$y = 20,000 + (1 - 0.25)(x - 20,000), \quad x > 20,000$$

or, more simply

$$y = 5,000 + 0.75x, \quad x > 20,000$$

If x equals or exceeds \$60,000, the individual pays an additional lump sum tax of \$1000 and so

$$y = 5,000 + 0.75x - 1000, \quad x \geq 60,000$$

or, more simply

$$y = 4,000 + 0.75x, \quad x \geq 60,000$$

Therefore, the relationship between before tax income, x, and after tax income, $y = f(x)$, is

$$f(x) = \begin{cases} x, & \text{if } x \leq \$20,000 \\ 5000 + 0.75x, & \text{if } \$20,000 < x < \$60,000 \\ 4000 + 0.75x, & \text{if } x \geq \$60,000, \end{cases}$$

and is illustrated in Figure 4.2.5.

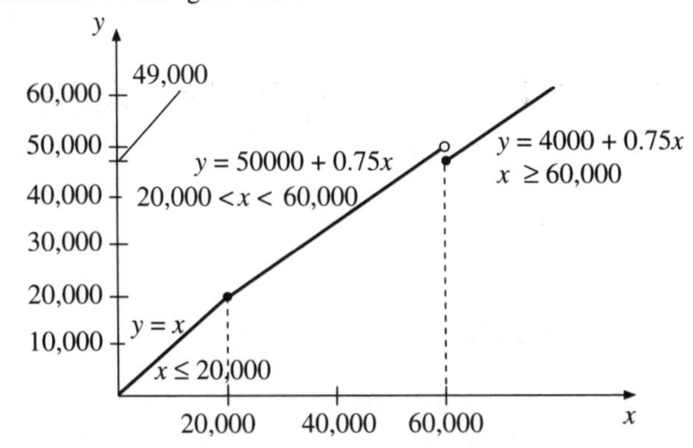

Figure 4.2.5

One can see that at the point $x = 60,000$, the left-hand limit

$$\lim_{x \to 60,000^-} f(x) = 50,000$$

is not equal to the right-hand limit

$$\lim_{x \to 20,000^+} f(x) = 49,000$$

and so the function is not continuous at this point.

If an individual were working, or considering working, a number of hours consistent with earning just under \$60,000, he would not have an incentive to work a bit more to earn a bit more than \$60,000 since the discontinuity in the tax schedule would mean he effectively would incur a discrete drop of \$1,000 in his income. In fact, this discontinuity in the tax schedule induces a disincentive for any increment in hours worked that push an individual from just below \$60,000 income to just above \$60,000.

7. (a) There are $7 \times 24 = 168$ hours in each week. Capacity of the plant is 100 cars per hour or 16,800 cars per week. For every 2000 worker-hours, 100 cars are produced. This implies a marginal product of $1/20$ of a car for every worker hour up to the point that capacity is reached. Capacity is reached after 16,800 cars are produced or, alternatively, after $168 \times 2000 = 336,000$ worker-hours are used.

Thus, the marginal product of labor is

$$\text{MP}(h) = \begin{cases} 1/20, & 0 \leq h \leq 336,000 \\ 0, & h > 336,000 \end{cases}$$

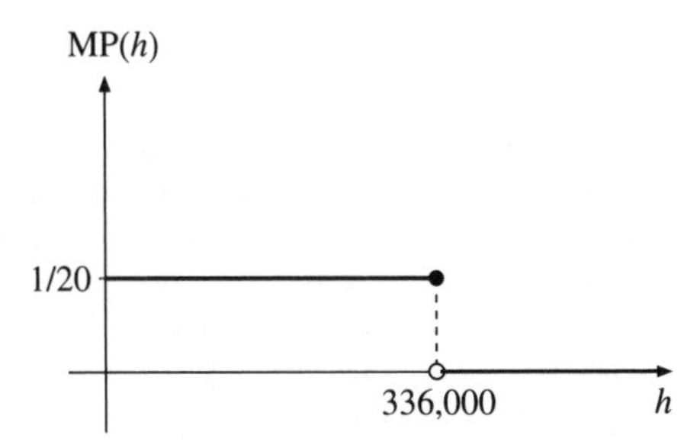

Figure 4.2.7 (a)

According to Definition 4.3, this function is discontinuous at the point $h = 336,000$ as the left-hand limit

$$\lim_{h \to 336,000^-} \text{MP}(h) = \frac{1}{20}$$

is not equal to the right-hand limit

$$\lim_{h \to 336,000^+} \text{MP}(h) = 0$$

Alternatively, we can see that the function is not continuous at the point $h = 336,000$ according to Definition 4.4 since for any value $\delta > 0$, no matter how small, the function $MP(h)$ takes on both the value $1/20$ and 0 for some points in the interval $(336,000 - \delta, 336,000 + \delta)$, and so, for example, one cannot find a $\delta > 0$ small enough such that

$$|\text{MP}(h) - \text{MP}(336,000)| < \epsilon \quad \text{for} \quad \epsilon < \frac{1}{20}$$

at every point h satisfying $|h - 336,000| < \delta$, (i.e., within the interval $(336,000 - \delta, 336,000 + \delta)$).

(b) There are $5 \times 24 = 120$ hours in the weekdays of every week and so $120 \times 100 = 12,000$ can be produced during the weekdays. Labor cost per car on a weekday is $20 \times 30 = \$600$. (i.e., since 2000 worker-hours are needed for each 100 cars we have that each car requires 20 worker-hours of labor and each hour of labor costs \$30 on a weekday.) Once 12,000 cars have been produced, however, the next 4,800 cars that can be produced are produced on the weekend when labor cost per car is $20 \times 60 = \$1200$. Therefore, profit earned per car produced in a week, $\pi(y)$ is

$$\pi(y) = \begin{cases} 1000 - 600 = 400, & 0 \le y \le 12,000 \\ 1000 - 1200 = -200, & 1200 < y \le 16,800 \end{cases}$$

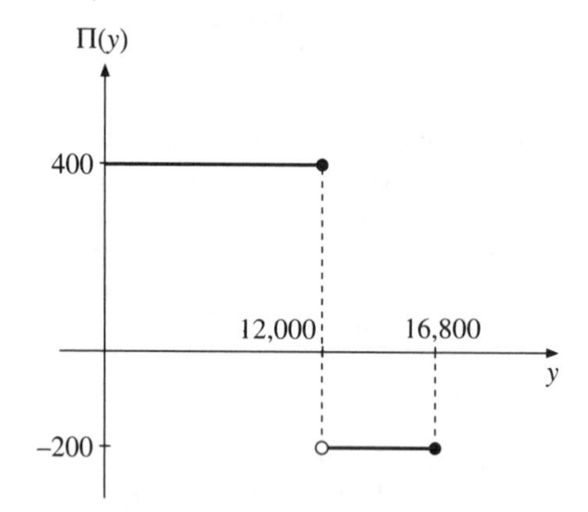

Figure 4.2.7 (b)

This function has a point of discontinuity. At $y = 12,000$ no more cars can be produced on weekdays and so the extra cost of hiring labor for the weekend shift leads to a drop or discontinuity in the profit per car, from \$400 per car to \$ − 200 (a loss). At the point of 16,800 cars per week no more cars can be produced since plant capacity has been reached at this level. Thus, the domain for this function is $y \in [0, 16, 800]$. Notice that in part (a) we do not restrict the domain at $h = 336,000$ worker-hours even though this represents capacity in term of number of cars produced. The reason for this distinction is that one can employ more workers beyond $h = 336,000$ if one wishes. However, no additional cars can be produced using this extra but redundant labor.

9. If $y = kx$ is the production function relating input x to output y then the inverse of the production function, $x = y/k$, indicates the amount of input x needed in order to produce y units of output. Therefore, the variable cost of producing y units of output is

$$VC(y) = \frac{\bar{c}}{k} y$$

Adding in fixed cost, c_0, we get cost function

$$C(y) = c_0 + \frac{\bar{c}}{k} y$$

The revenue function is

$$R(y) = \bar{p} y$$

and the profit function is

$$\pi(y) = R(y) - C(y) = \bar{p} y - c_0 - \frac{\bar{c}}{k} y$$

All of these functions are continuous. One way to see this is to show that the function $y = kx$ is continuous and then note that obtaining the cost function, $C(y)$, involves operations relating to points (vi), (i), and (ii) of Theorem 4.1. If we then were to show that $R(y) = \bar{p}y$ is continuous then obtaining the profit function, $\pi(y)$, then involves subtracting one continuous function from another, $(R(y) - C(y))$, which is operation (iii) of Theorem 4.1.

11. * Firm B locates at $L_B = 0.3$. If firm A locates at a point to the left of 0.3 (i.e., $L_A < 0.3$) it will attract all consumers to the left of its choice, a fraction equal to L_A, plus half of the consumers between L_A and L_B — namely those consumers closer to L_A than L_B — which represents a further fraction of $0.5(L_B - L_A)$. Market share, then, is

$$M^A(L_A) = L_A + 0.5(0.3 - L_A), \quad L_A < 0.3$$

If firm A chooses to locate at the same point as firm B ($L_A = L_B = 0.3$) then consumers are indifferent between purchasing from A or B and so they share the market equally. Thus

$$M^A(L_A) = 0.5, \quad L_A = 0.3$$

If firm A locates just to the right of firm B then virtually all consumers to the right of the point 0.3 will go to firm A. This represents the fraction $(1 - 0.3)$ of the consumers. Further to the right of $L_B = 0.3$, firm A will attract all consumers to the right of its location, representing $1 - L_A$ market share,and also 50% of consumers between firm B's location (0.3) and its own location (L_A), thus representing a further fraction $0.5(L_A - 0.3)$ of the market. Thus

$$M^A(L_A) = (1 - L_A) + 0.5(L_A - 0.3), \quad L_A > 0.3$$

Putting these expressions together gives us

$$M^A(L_A) = \begin{cases} L_A + 0.5(0.3 - L_A), & L_A < 0.3 \\ 0.5, & L_A = 0.3 \\ (1 - L_A) + 0.5(L_A - 0.3), & L_A > 0.3 \end{cases}$$

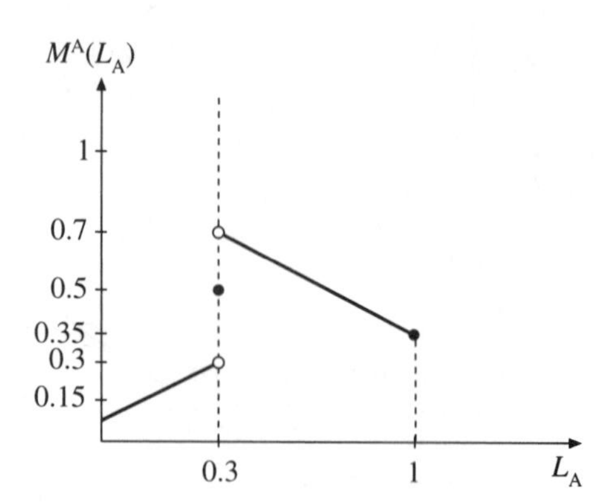

Figure 4.2.11 (a)i

There are 100 consumers in the market and the profit made per consumer is \$4 (i.e., \$10 − \$6) and so the profit function relating to market share is

$$\pi^A(L_A) = 400 M^A(L_A)$$

and so by substituting the function for market share into this equation we get

$$\pi^{A}(L_A) = \begin{cases} 400[L_A + 0.5(0.3 - L_A)], & L_A < 0.3 \\ 200, & L_A = 0.3 \\ 400[(1 - L_A) + 0.5(L_A - 0.3)], & L_A > 0.3 \end{cases}$$

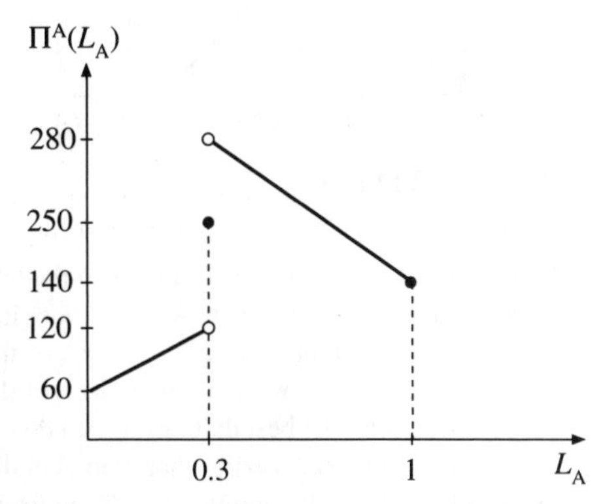

Figure 4.2.11 (a)ii

Firm A's market share and profit change in a discontinuous fashion at the point $L_A = L_B = 0.3$. Locating just to the left of 0.3 means firm A obtains virtually all the consumers to the left of 0.3 and so

$$\lim_{L_A \to 0.3^-} M^A(L_A) = 0.3$$

is the left-hand limit of $M^A(L_A)$ at the point $L_A = 0.3$. However, if firm A locates precisely at the point $L_A = 0.3$ it shares the market equally with firm B and so

$$M^A(0.3) = 0.5$$

If firm A locates just to the right of firm B (i.e., to the right of 0.3) it gets all the consumers to the right of 0.3 and so

$$\lim_{L_A \to 0.3^+} M^A(L_A) = 0.7$$

Thus, at $L_A = 0.3$, the right-hand limit for $M^A(L_A)$ is not equal to the left-hand limit, and neither is equal to the value of $M^A(L_A)$ at $L_A = 0.3$, and so by Definition 4.3 we can see that this function is not continuous at this point.

(b) The same arguments made in part (a) apply to any choice $\bar{L}_B < 0.5$ made by firm B. Thus, firm A's market share function is

$$M^A(L_A) = \begin{cases} L_A + 0.5(\bar{L}_B - L_A), & L_A < \bar{L}_B \\ 0.5, & L_A = \bar{L}_B \\ (1 - L_A) + 0.5(L_A - \bar{L}_B), & L_A > \bar{L}_B \end{cases}$$

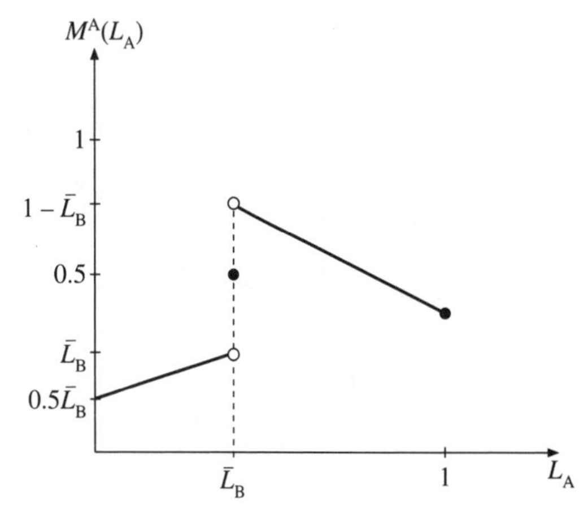

Figure 4.2.11 (b)i

and the profit share function is

$$\pi^{A}(L_A) = \begin{cases} 400[L_A + 0.5(\bar{L}_B - L_A)], & L_A < \bar{L}_B \\ 200, & L_A = \bar{L}_B \\ 400[(1 - L_A) + 0.5(L_A - \bar{L}_B)], & L_A > \bar{L}_B \end{cases}$$

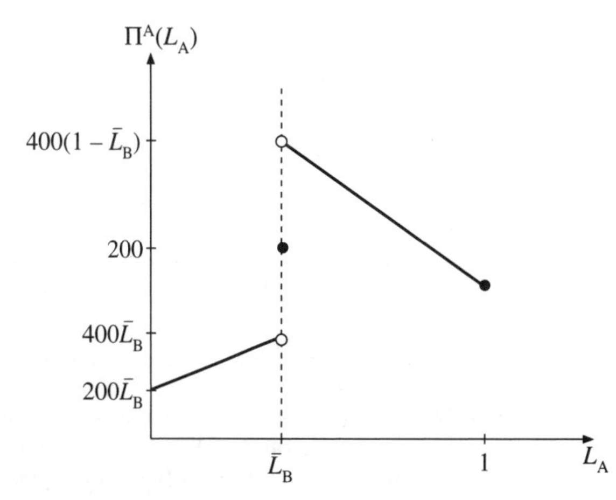

Figure 4.2.11 (b)ii

(c) If firm B locates to the right of midpoint ($\bar{L}_B > 0.5$), then a similar method is used to determine A's market share and profit as was used in part (b). The one important difference is that firm A's market share and profit are higher when it locates just to the left of firm B's location than if it locates just to the right. Thus, the same functions as in part (b) describe the market share and profit for firm A but the appropriate figures are in Figures 4.2.11 (c).

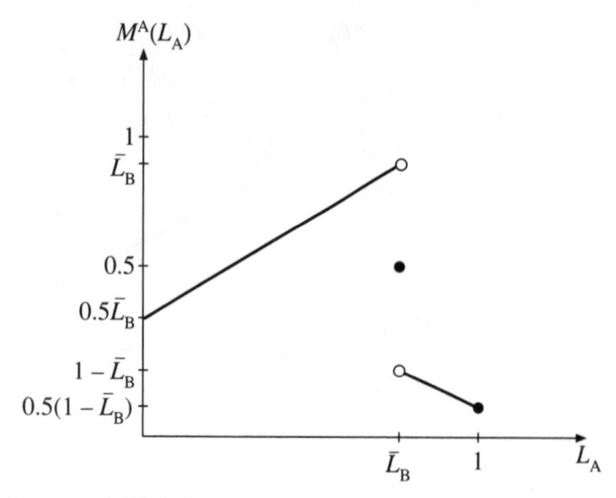

Figure 4.2.11 (c)i

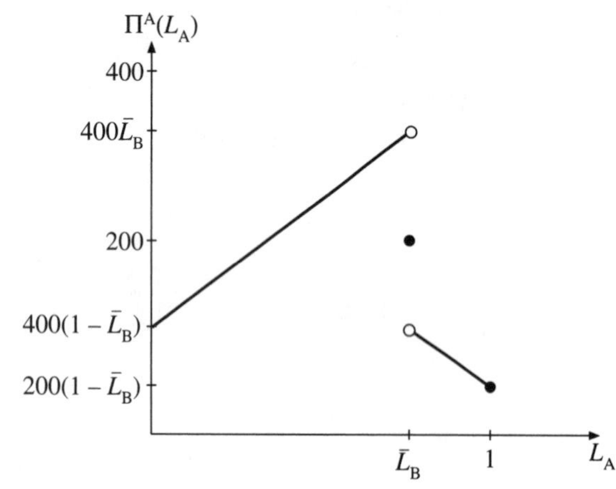

Figure 4.2.11 (c)ii

(d) Once again, the functions in part (b) describe the market share and profit for firm A as a function of its location, L_A. However, since $\bar{L}_B = 0.5$, firm A's market share approaches the value 0.5 as L_A approaches the value 0.5 either from the left ($L_A < 0.5$) or from the right ($L_A > 0.5$), moreover, market share for A is 0.5 when it locates at the point $L_A = \bar{L}_B = 0.5$. Therefore, the market share function is continuous at the point $L_A = \bar{L}_B$, as is the profit function. The appropriate graphs are given in Figures 4.2.11 (d).

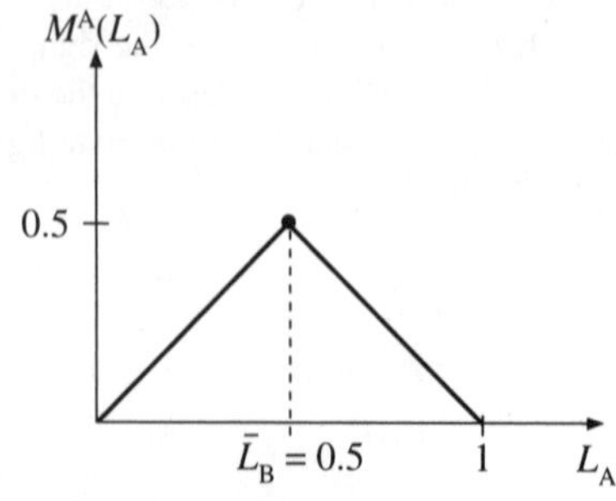

Figure 4.2.11 (d)i

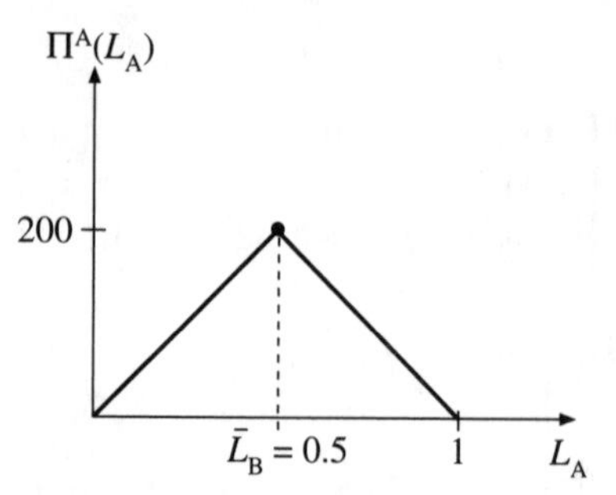

Figure 4.2.11 (d)ii

(e) Firm B knows that if it locates either to the left or right of the midpoint, 0.5, then firm A will choose its location so that firm A obtains greater than 50% of the market share, leaving firm B with less than 50% of the market share. Therefore, the best that firm B can do is to locate at the midpoint, recognizing that firm A will respond by also locating at the midpoint, with the firms sharing the market equally. The reverse argument also applies if we consider where firm A would locate considering the reaction of firm B and so the equilibrium outcome is that both firms locate at the midpoint.

Section 4.3 (page 127)

1. To find the equilibrium price set $D(p) = S(p)$:

$$\begin{aligned} 50 - 2p^e &= -10 + p^e \\ -3p^e &= -60 \\ p^e &= 20 \end{aligned}$$

To find the equilibrium quantity, substitute $p^e = 20$ into either the demand or supply function (since they are equal at the equilibrium price).

$$D(p^e) = 50 - 2p^e \Longrightarrow y^e = 10$$

Also

$$S(p^e) = -10 + p^e \Longrightarrow y^e = 10$$

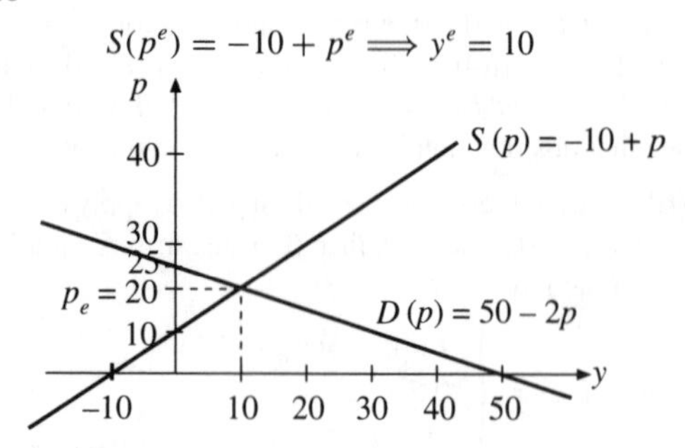

Figure 4.3.1 (a)

The excess demand function is

$$\begin{aligned} z(p) &= D(p) - S(p) = (50 - 2p) - (-10 + p) \\ &= 60 - 3p \end{aligned}$$

Note that $z(p^e) = 0 \Longrightarrow p^e = 20$ is another method of determining the equilibrium price.

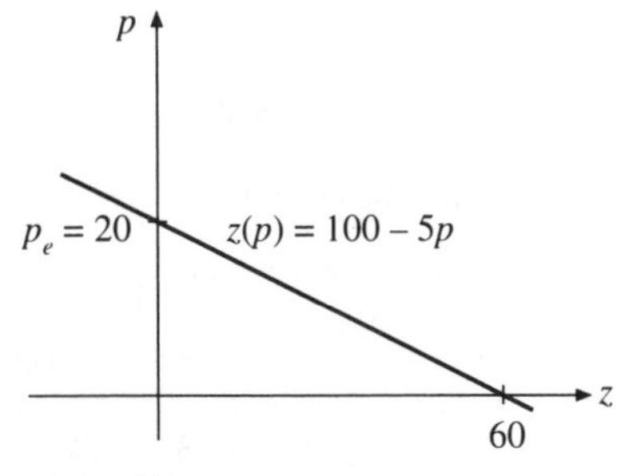

Figure 4.3.1 (b)

The demand and supply functions satisfy Theorem 4.3 since:

(i) $D(0) = 50$, $S(0) = -10$, and so $D(0) > S(0)$.

(ii) Picking price $\bar{p} = 25$ (or any price exceeding 20 in this case) we see that there is indeed a price $p = \bar{p}$ such that $S(\bar{p}) > D(\bar{p})$ (i.e., note that $S(25) = 15$, $D(25) = 0$).

3. $D(p) = 50 - 8p$ is the demand function. $S(p) = -100 + 2p$ is the supply function.

$$z(p) = D(p) - S(p) = (50 - 8p) - (-100 + 2p)$$

and so $z(p) = 150 - 10p$ is the excess demand function.

The equilibrium price, p^e, satisfies $z(p^e) = 0$ and so $p^e = 15$. To find the equilibrium quantity, substitute p^e into either $D(p)$ or $S(p)$ and so we get $y^e = -70$.

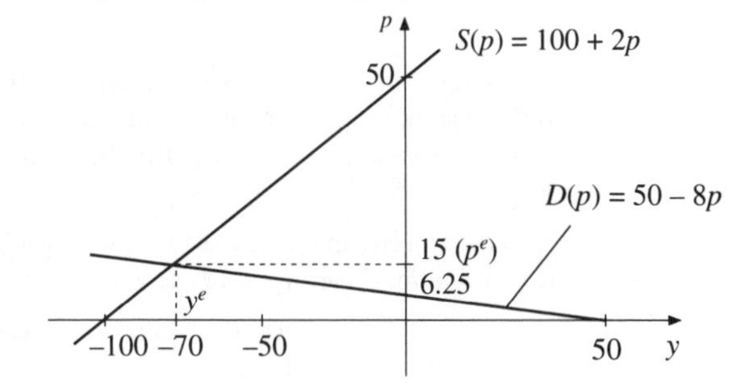

Figure 4.3.3 (a)

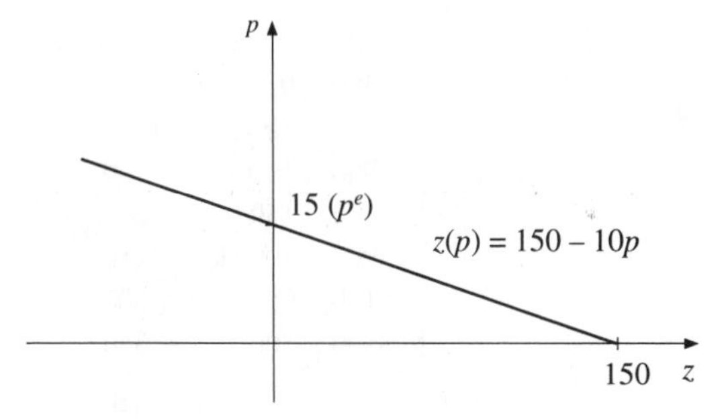

Figure 4.3.3 (b)

Since $D(0) = 50$ and $S(0) = -100$ where $D(0) > S(0)$, i.e., condition (a) of Theorem 4.3 is satisfied, and for any $\bar{p} > 15$ we have $D(\bar{p}) < S(\bar{p})$ or $z(\bar{p}) < 0$, and so condition (b) of Theorem 4.3 is also satisfied. What is peculiar about this example is that using the standard computational approach delivers a negative value for the equilibrium quantity, which makes no economic sense. Upon considering the vertical (p) intercepts for both the supply and demand functions we see that firms require a minimum price of \$50 in order to be induced to produce any of this product while consumers will not purchase any output if price exceeds \$6.25. Therefore, there is no price which would induce any (positive) market transactions and so this market would be inactive.

5. $D(p) = 20 + 2p$, $S(p) = -10 + p$

$$z(p) = D(p) - S(p) = (20 + 2p) - (-10 + p)$$

and so $z(p) = 30 + p$ is the excess demand function.

The equilibrium price, p^e, satisfies $z(p^e) = 0$ and so $p^e = -30$. To find the equilibrium quantity, substitute p^e into either $D(p)$ or $S(p)$ and so we get $y^e = -40$.

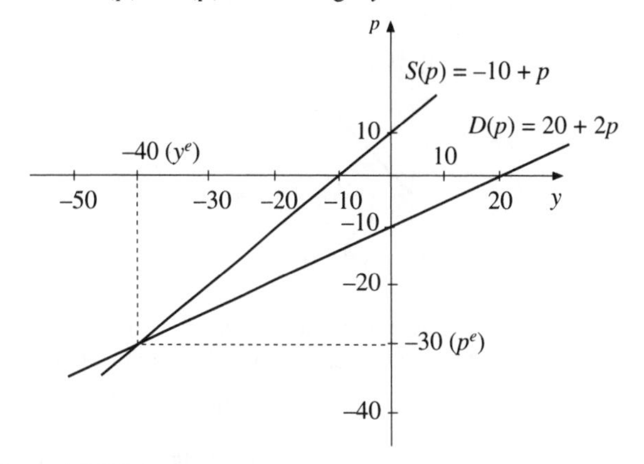

Figure 4.3.5 (a)

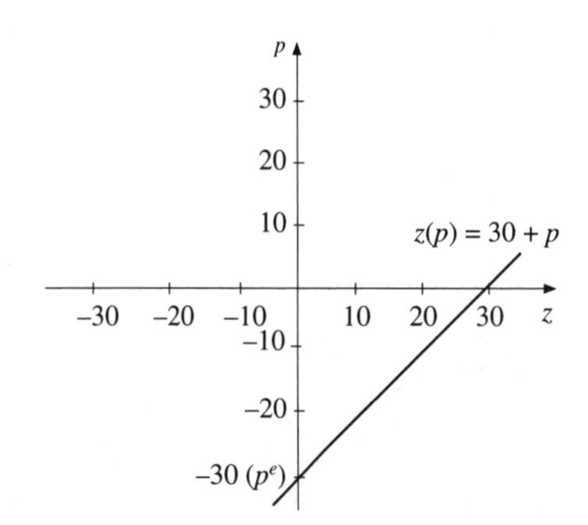

Figure 4.3.5 (b)

Since $D(0) = 20$ and $S(0) = -10$ the first condition of Theorem 4.3, that $D(0) > S(0)$, is satisfied. However, since $S(p) > D(p)$ implies $-10 + p > 20 + 2p$ or $p < 30$, there is no positive price $\bar{p}$ such that $S(\bar{p}) > D(\bar{p})$, and so the second condition of Theorem 4.3 is not satisfied.

In this example, for any price greater than \$10, a positive amount of output will be supplied. However, at every price greater than \$10 demand exceeds supply and so firms could increase price and still sell their entire output. One would expect to see a price increasing without bound. Of course, a commodity with demand increasing in price without bound is not a realistic possibility since consumers would exhaust their incomes if such were the case.

Review Exercises (page 129)

1. (a)

$$f(x) = \begin{cases} 2x, & \text{if } x < 5 \\ x+6 & \text{if } x \geq 5 \end{cases}$$

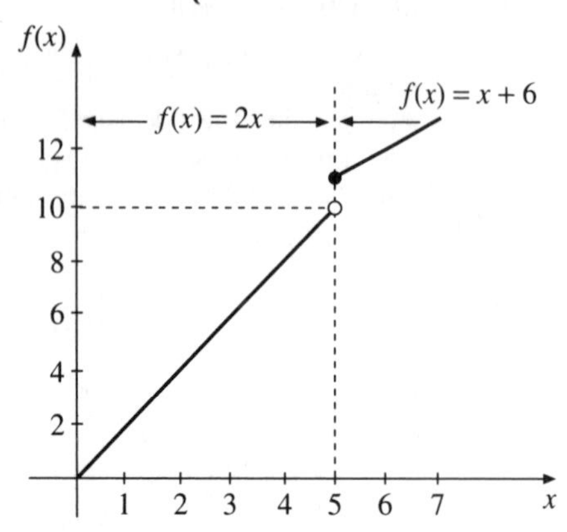

Figure 4.R.1 (a)

At $x = 5$, the left-hand limit of the function is not equal to the right-hand limit of the function. That is

$$\lim_{x\to 5^-} f(x) = 10, \lim_{x\to 5^+} f(x) = 11$$

and so this function is not continuous at the point $x = 5$.

(b)

$$f(x) = \frac{x+1}{x^2-1} = \frac{x+1}{(x+1)(x-1)} = \frac{1}{x-1}, \quad x \neq 1$$

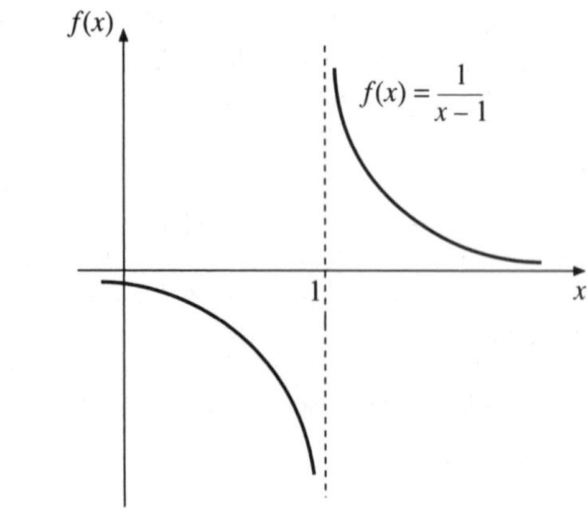

Figure 4.R.1 (b)

For this function we also have that the left-hand and right-hand limits are not equal at the point $x = 1$ as

$$\lim_{x\to 1^-} f(x) = -\infty, \lim_{x\to 1^+} f(x) = +\infty$$

Moreover, the function is not defined at the value $x = 1$. Thus, this function is not continuous at the point $x = 1$.

3.

$$y = \begin{cases} x & x \leq 25{,}000 \\ 0.6x + 10{,}000 & 25{,}000x < 100{,}000 \\ 0.6x + 8{,}000 & x \geq 100{,}000 \end{cases}$$

There is a discontinuity at $x = 100,000$ due to the surtax. (See Figure 4.R.3). The function is discontinuous at the point $x = 100,000$ as the left-hand limit

$$\lim_{x\to 100{,}000^-} = 70,000$$

is not equal to the right-hand limit

$$\lim_{x\to 100{,}000^+} = 68,000$$

This discontinuity in the schedule creates a disincentive to increase hours worked if the result is to push an individual's pre-tax income from just below \$100,000 to just above \$100,000.

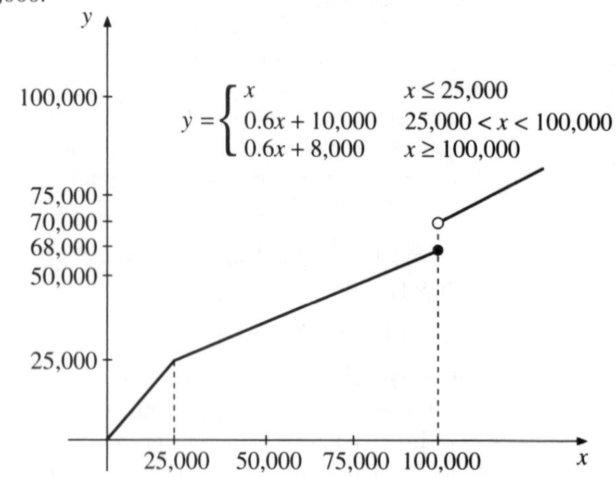

Figure 4.R.3

5. (a) To carry one passenger requires one carriage (cost = \$1000) plus the wage costs of the engineer and guard (\$500) and so the cost of transporting one passenger from A to B is \$1500.

(b) The increase in cost of taking a second passenger is zero since no additional carriages are needed.

(c) To take the 51st passenger requires adding a carriage and so the additional cost is \$800.

(d) The total cost and average cost functions are:

(i) Total cost of taking y passengers is

$$C(y) = \begin{cases} 0 & y = 0 \\ 1500 & 0 < y \leq 50 \\ 2300 & 50 < y \leq 100 \\ 2900 & 100 < y \leq 150 \\ 3100 & 150 < y \leq 200 \\ 3900 & 200 < y \leq 250 \\ 5100 & 250 < y \leq 300 \\ 8500 & 300 < y \leq 350 \end{cases}$$

(ii) Since average cost is just total cost per passenger, $AC(y) = C(y)/y$, we have

$$\text{AC}(y) = \begin{cases} \frac{0}{0}\ (\text{undefined}) & y = 0 \\ 1500/y & 0 < y \leq 50 \\ 2300/y & 50 < y \leq 100 \\ 2900/y & 100 < y \leq 150 \\ 3100/y & 150 < y \leq 200 \\ 3900/y & 200 < y \leq 250 \\ 5100/y & 250 < y \leq 300 \\ 8500/y & 300 < y \leq 350 \end{cases}$$

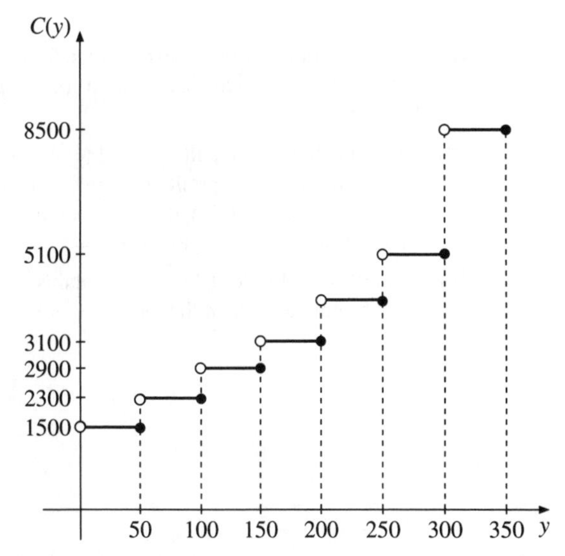

Figure 4.R.5 (a)

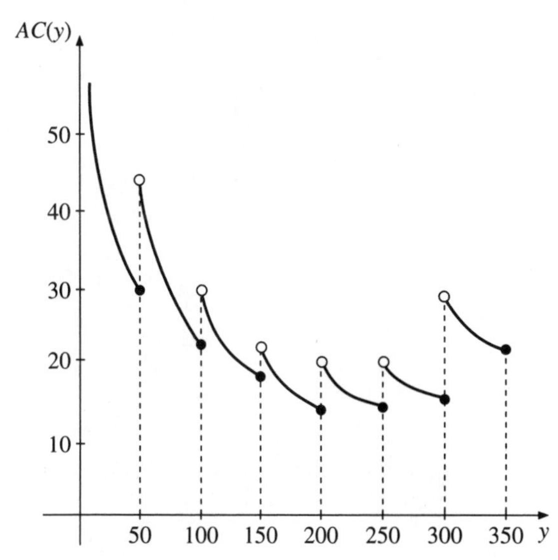

Figure 4.R.5 (b)

Because of the discrete jump in cost associated with the 1st passenger, the 51st passenger, the 101st passenger, etc., the cost function has points of discontinuity at every integer multiple of 50 passengers. Since average cost is just total cost divided by the number of passengers, it follows that the average cost function is also discontinuous at every integer multiple of 50 passengers.

7. Letting y represent quantity for either the demand or supply function and z the excess demand we have

$$D(p) = \begin{cases} 100 - 2p, & p > 45 \\ 120 - 2p, & p \leq 45 \end{cases}$$

$$S(p) = -20 + p$$

$$z(p) = \begin{cases} 120 - 3p, & p > 45 \\ 140 - 3p, & p \leq 45 \end{cases}$$

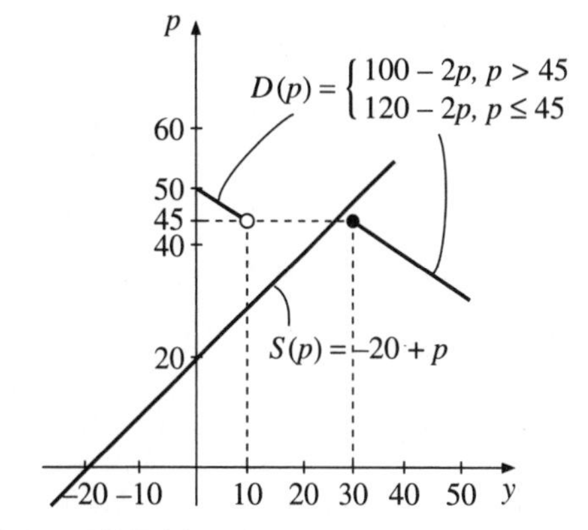

Figure 4.R.7 (a)

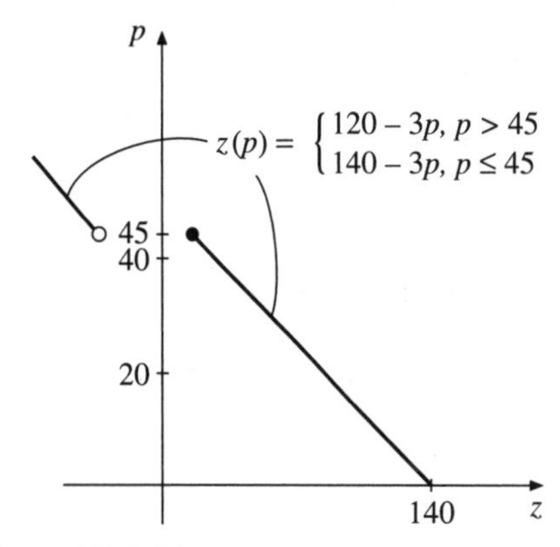

Figure 4.R.7 (b)

Since the demand function is not continuous, and hence neither is the excess demand function, Theorem 4.3 does not apply and so there may not be an equilibrium. This is seen to be the case for this example, as is illustrated by the above graphs.

9. This exercise is similar to Exercise 11 of Section 4.2. The numbers are different, and notice that when firm B chooses a location to the right of the midpoint ($L_B = 0.8$), firm A's largest market share is obtained when it locates just to the *left* of 0.8, as indicated in Figure 4.R.9. Due to the similarity of these questions, we indicate below only how they differ.

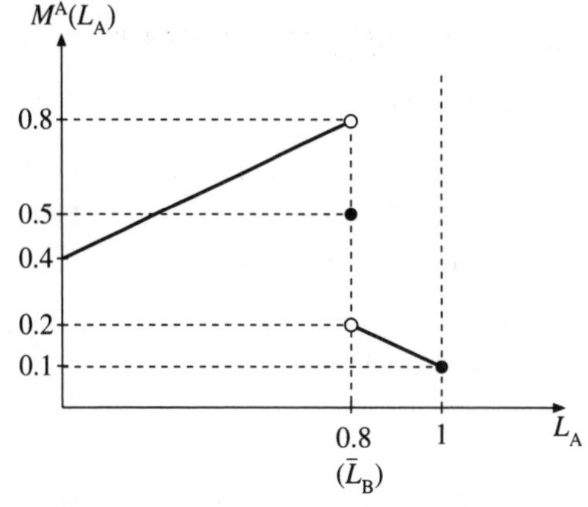

Figure 4.R.9

(a)

$$M^{A}(L_A) = \begin{cases} L_A + 0.5(0.8 - L_A) & L_A < 0.8 \\ 0.5 & L_A = 0.8 \\ (1 - L_A) + 0.5(L_A - 0.8) & L_A > 0.8 \end{cases}$$

Since $\pi^{A}(L_A) = 10,000M^{A}(L_A)$,

$$\pi^{A}(L_A) = \begin{cases} 10,000[L_A + 0.5(0.8 - L_A)] & L_A < 0.8 \\ 5000 & L_A = 0.8 \\ 10,000[(1 - L_A) + 0.5(L_A - 0.8)] & L_A > 0.8 \end{cases}$$

(b) Same functions as in (a), but replace 0.8 by $\bar{L}_B$ everywhere. Notice that when $\bar{L}_B < 0.5$ firm A's market share and profit are highest when it locates just to the right of $\bar{L}_B$

(c) Same functions as in (b) except that firm A's market share and profit are highest when it locates just to the left of $\bar{L}_B$.

(d) Same functions as in parts (b) and (c) except that firm A's market share and profit are highest when it locates just at $L_A = \bar{L}_B = 0.5$ and the market share and profit functions are continuous at this point.

(e) Firm B would locate at $L_B = 0.5$ and this would be the equilibrium outcome of the model.

CHAPTER 5

Section 5.1 (page 137)

1. Yes, the sequence values look like they will converge.

Q_i	(25, 625)	(24, 576)	(23, 529)
Δx	5	4	3
Δy	225	176	129
$\Delta y/\Delta x$	45	44	43

Q_i	(22, 484)	(21, 441)
Δx	2	1
Δy	84	41
$\Delta y/\Delta x$	42	41

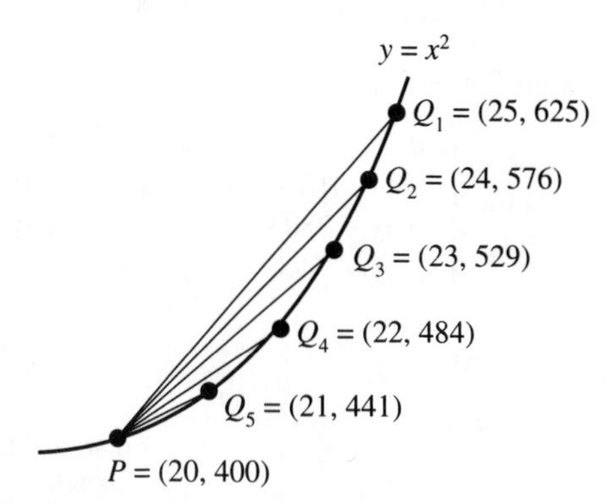

Figure 5.1.1

3. Since the slope of the tangent line for the function $y = x^2$ is $2x$, its value at the point $x = 3$ is 6. Therefore, the equation for the tangent line, written in the form $y = mx + b$, is

$$y = 6x + b$$

Since this line passes through the point $x = 3$, $y = 9$ the value b must satisfy the equation

$$9 = 6(3) + b \Leftrightarrow b = -9$$

and so the equation for the tangent line is

$$y = 6x - 9$$

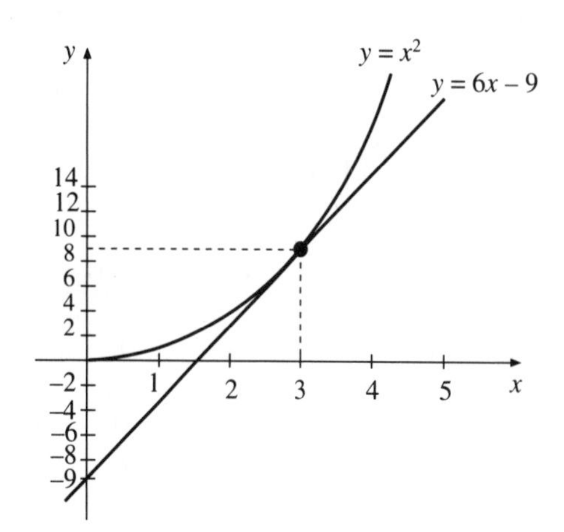

Figure 5.1.3

Section 5.2 (page 143)

1. **(a)**

$$\begin{aligned} f'(x) &= \lim_{\Delta x\to 0} \frac{f(x+\Delta x) - f(x)}{\Delta x} \\ &= \lim_{\Delta x\to 0} \frac{[3(x+\Delta x) - 5] - [3x - 5]}{\Delta x} \\ &= \lim_{\Delta x\to 0} \frac{3\Delta x}{\Delta x} = 3 \end{aligned}$$

(b)

$$\begin{aligned} f'(x) &= \lim_{\Delta x\to 0} \frac{f(x+\Delta x) - f(x)}{\Delta x} \\ &= \lim_{\Delta x\to 0} \frac{8(x+\Delta x) - 8x}{\Delta x} \\ &= \lim_{\Delta x\to 0} \frac{8\Delta x}{\Delta x} = 8 \end{aligned}$$

(c)

$$\begin{aligned} f'(x) &= \lim_{\Delta x\to 0} \frac{f(x+\Delta x) - f(x)}{\Delta x} \\ &= \lim_{\Delta x\to 0} \frac{3(x+\Delta x)^2 - 3x^2}{\Delta x} \\ &= \lim_{\Delta x\to 0} \frac{3(x^2 + 2x\Delta x + \Delta x^2)^2 - 3x}{\Delta x} \\ &= \lim_{\Delta x\to 0} \frac{3x^2 + 6x\Delta x + 3\Delta x^2 - 3x^2}{\Delta x} \\ &= \lim_{\Delta x\to 0} \frac{6x\Delta x + 3\Delta x^2}{\Delta x} = 6x \end{aligned}$$

3.

Q_i	(25, 625)	(24,576)	(23, 529)	(22, 484)	(21, 441)
Δx	5	4	3	2	1
Δy	225	176	129	84	41
$dy = f'(x)dx$	200	160	120	80	40
ϵ	11.1%	9.1%	7.0%	4.8%	2.4%

The size of the error, in absolute value, from using the differential as an estimate of the actual change in y falls as one takes smaller changes in x. Of course, for some functions this will not be a monotonic relationship but the reduction in the error will occur at least eventually as $\Delta x \to 0$.

Section 5.3 (page 148)

1. (a) To the left of the point $x = 5$ the function is defined by the equation $f(x) = 3x + 2$ and so

$$\lim_{\Delta x \to 0^-} \frac{f(x + \Delta x) - f(x)}{\Delta x} =$$

$$\lim_{\Delta x \to 0^-} \frac{[3(x + \Delta x) + 2] - [3x + 2]}{\Delta x} = 3$$

while to the right of the point $x = 5$ the function is defined by the equation $f(x) = x + 12$ and so

$$\lim_{\Delta x \to 0^+} \frac{f(x + \Delta x) - f(x)}{\Delta x} =$$

$$\lim_{\Delta x \to 0^+} \frac{[(x + \Delta x) + 12] - [x + 12]}{\Delta x} = 1$$

Since the left-hand and right-hand derivatives are not equal at $x = 5$, the function is not differentiable at this point.

(b) To the left of the point $x = 0$ the function is defined by the equation $f(x) = -x$ and so

$$\lim_{\Delta x \to 0^-} \frac{f(x + \Delta x) - f(x)}{\Delta x} =$$

$$\lim_{\Delta x \to 0^-} \frac{-(x + \Delta x) - x}{\Delta x} = -1$$

while to the right of the point $x = 0$ the function is defined by the equation $f(x) = x$ and so

$$\lim_{\Delta x \to 0^+} \frac{f(x + \Delta x) - f(x)}{\Delta x} =$$

$$\lim_{\Delta x \to 0^+} \frac{(x + \Delta x) - x}{\Delta x} = 1$$

Since the left-hand and right-hand derivatives are not equal at $x = 0$, the function is not differentiable at this point.

(c) To the left of the point $x = 2$ the function is defined by the equation $f(x) = 4x + 1$ and so

$$\lim_{\Delta x \to 0^-} \frac{f(x + \Delta x) - f(x)}{\Delta x} =$$

$$\lim_{\Delta x \to 0^-} \frac{[4(x + \Delta x) + 1] - [4x + 1]}{\Delta x} = 4$$

while to the right of the point $x = 2$ the function is defined by the equation $f(x) = 11 - x$ and so

$$\lim_{\Delta x \to 0^+} \frac{f(x + \Delta x) - f(x)}{\Delta x} =$$

$$\lim_{\Delta x \to 0^+} \frac{[11 - (x + \Delta x)] - [11 - x]}{\Delta x} = -1$$

Since the left-hand and right-hand derivatives are not equal at $x = 2$, the function is not differentiable at this point.

3. Let $T(y)$ represent the income tax schedule for this example. The marginal tax rate depends on income according to the following schedule:

$$\text{marginal tax rate} = \begin{cases} 0 & \text{for } 0 < y \le 6000 \\ 0.2 & \text{for } 6000 < y \le 16,000 \\ 0.3 & \text{for } 16,000 < y \le 46,000 \\ 0.4 & \text{for } y > 46,000 \end{cases}$$

(a) The tax function is

$$T(y) = \begin{cases} 0 & \text{for } 0 < y \le 6000 \\ 0.2y - 1200 & \text{for } 6000 < y \le 16,000 \\ 0.3y - 2800 & \text{for } 16,000 < y \le 46,000 \\ 0.4y - 7400 & \text{for } y > 46,000 \end{cases}$$

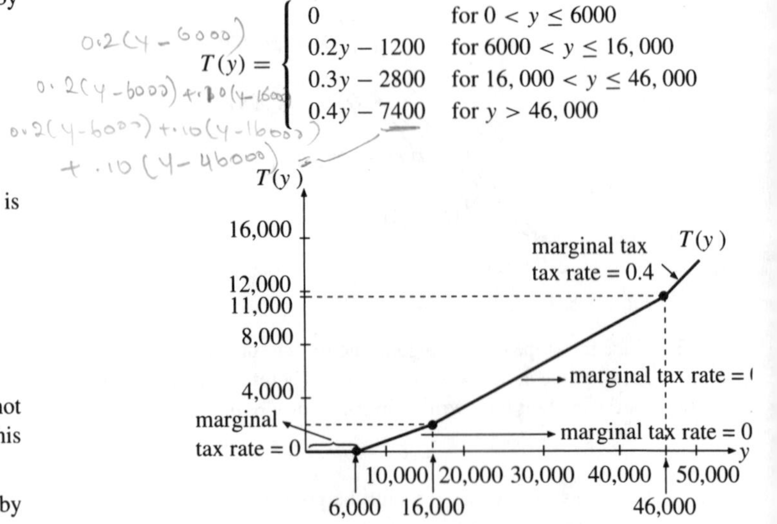

Figure 5.3.3 (a)

(b) The points of nondifferentiability occur each time there is an increase in the marginal tax rate. These occur at income levels $y = 6000$, $y = 16,000$, and $y = 46,000$. The slope of the tax schedule to the left and the right of each of these points differs.

(c) The marginal tax function is given in part (a) while the average tax function is $\text{AT}(y) = T(y)/y$ and so

$$\text{AT}(y) = \begin{cases} 0 & \text{for } 0 < y \leq 6000 \\ 0.2 - 1200/y & \text{for } 6000 < y \leq 16,000 \\ 0.3 - 2800/y & \text{for } 16,000 < y \leq 46,000 \\ 0.4 - 7400/y & \text{for } y > 46,000 \end{cases}$$

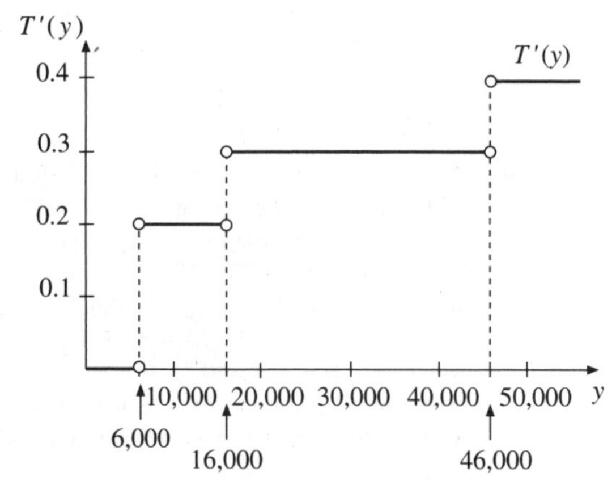

Figure 5.3.3 (c)i

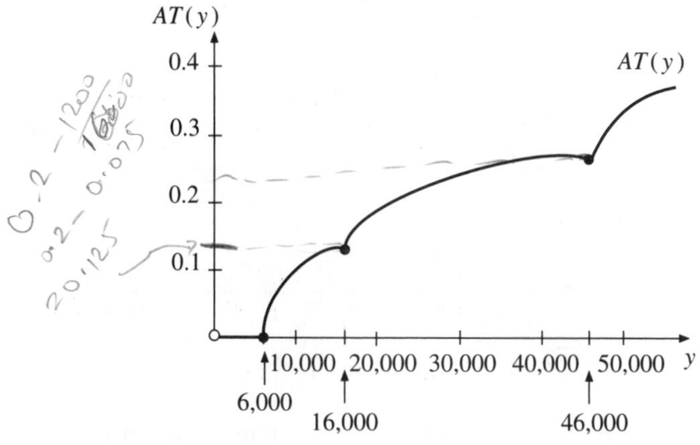

Figure 5.3.3 (c)ii

5. (a) The function relating sales, S, to pay, P, is

$$P(S) = \begin{cases} 600 & S = 0 \\ 600 + 0.1S & 0 < S \leq 10,000 \\ 600 + 0.1S + 0.1(S - 10,000) & S \geq 10,000 \end{cases}$$

which can be simplified to

$$P(S) = \begin{cases} 600 & S = 0 \\ 600 + 0.1S & 0 < S \leq 10,000 \\ -400 + 0.2S & S \geq 10,000 \end{cases}$$

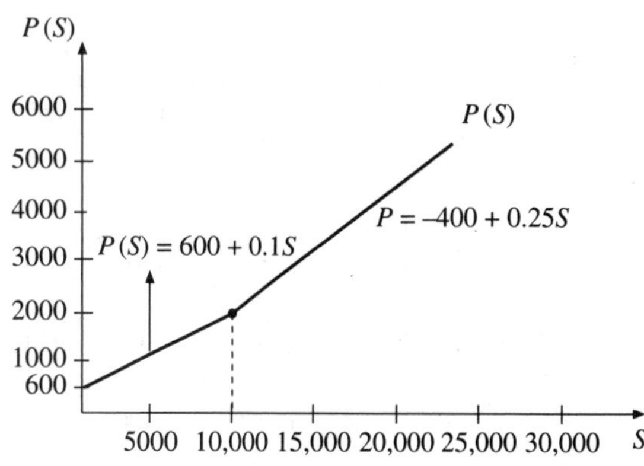

Figure 5.3.5

(b) There is a single point of nondifferentiability at $S = 10,000$. This occurs because at $S = \$10,000$ the commission rate jumps from 10% on any sales less than \$10,000 to 20% on all sales in excess of \$10,000. Thus, the left- and right-hand derivatives of the function $P(S)$ differ at that point.

Section 5.4 (page 175)

1. (a) $f(L) = 10L \Rightarrow f'(L) = 10 \Rightarrow f''(L) = 0$. The second derivative, which gives the slope of the derivative function, is zero. This means that the rate at which output rises with respect to more input being used does not change. This is illustrated by Figures 5.4.1 (a).

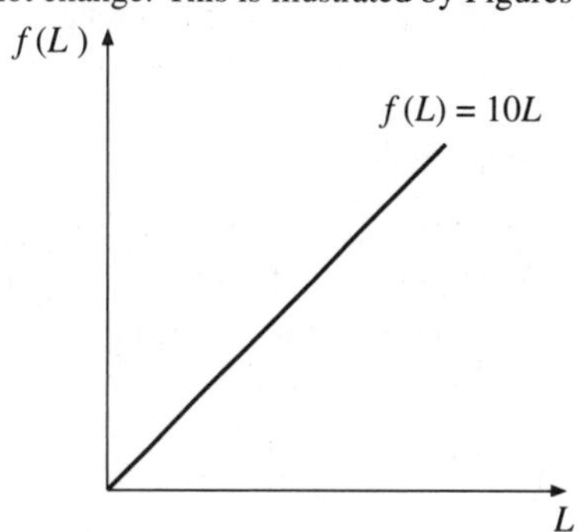

Figure 5.4.1 (a)i

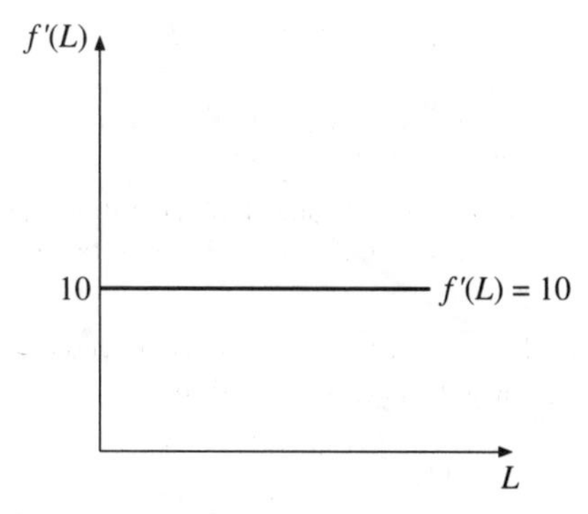

Figure 5.4.1 (a)ii

(b) $f(L) = 8L^{1/3} \Rightarrow f'(L) = (8/3)L^{-2/3} \Rightarrow$ $f''(L) = (-16/9)L^{-5/3} < 0$. The second derivative, which gives the slope of the derivative function, is negative. This means that the rate at which output rises with respect to more input being used is decreasing in L. This is consistent with the so-called law of diminishing marginal product of labor and is illustrated by Figures 5.4.1 (b).

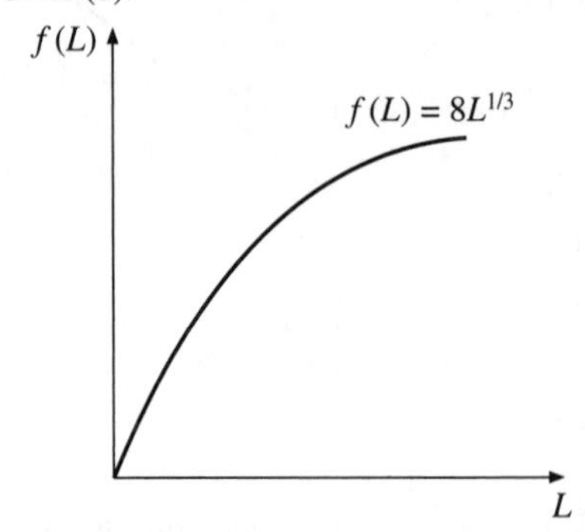

Figure 5.4.1 (b)i

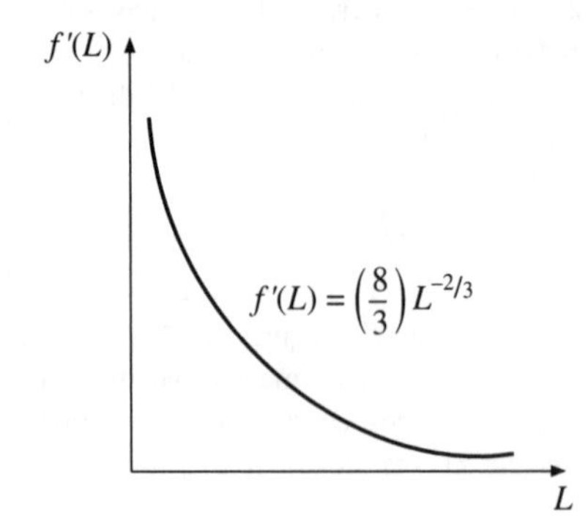

Figure 5.4.1 (b)ii

(c) $f(L) = 3L^4 \Rightarrow f'(L) = 12L^3 \Rightarrow f''(L) = 36L^2 > 0$. The second derivative, which gives the slope of the derivative function, is positive. This means that the rate at which output rises with respect to more input being used is increasing in L. This is inconsistent with the so-called law of diminishing marginal product of labor and is illustrated by Figures 5.4.1 (c).

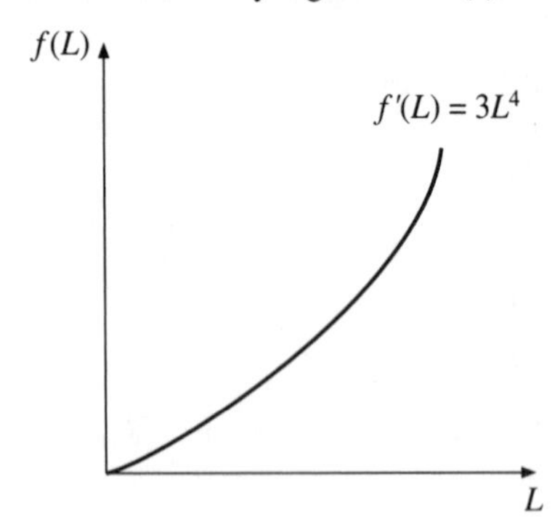

Figure 5.4.1 (c)i

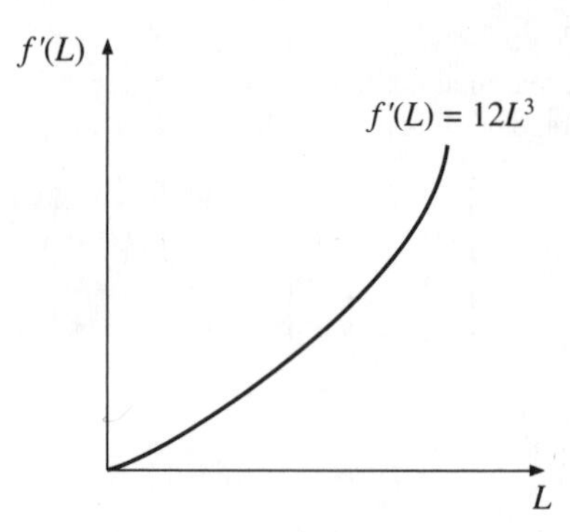

Figure 5.4.1 (c)ii

3. (a) The supply curve for each firm is simply the inverse of its marginal cost curve, and so $\text{MC}^A = dC^A/dq = 10 + 4q \Rightarrow p = 10 + 4q$ is A's supply curve in its inverse form (i.e., $q_s^A = 0.25p - 2.5$ is firm A's supply curve with $q_s^A \geq 0 \Rightarrow p \geq 10$). $\text{MC}^B = dC^B/dq = 15 + 2q \Rightarrow p = 15 + 2q$ is B's supply curve in its inverse form (i.e., $q_s^B = 0.5p - 7.5$ is firm B's supply curve with $q_s^B \geq 0 \Rightarrow p \geq 15$). These functions are differentiable on every point of their domains ($q \geq 0$), as is seen in Figure 5.4.3 (a).

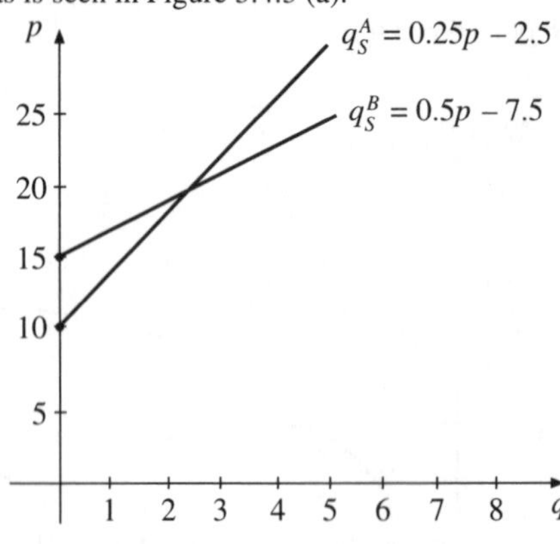

Figure 5.4.3 (a)

(b) To get total supply of the two firms we have to *add* the supplies from each firm at each price. Firm A will supply positive output as soon as price exceeds 10 while firm B will supply positive output as soon as price exceeds 15. Thus, the joint supply for these two firms is

$$q = \begin{cases} 0 & p < 10 \\ q_s^A & 10 \leq p < 15 \\ q_s^A + q_s^B & p \geq 15 \end{cases}$$

and so

$$q = \begin{cases} 0 & p < 10 \\ 0.25p - 2.5 & 10 \leq p < 15 \\ 0.75p - 10 & p \geq 15 \end{cases}$$

Thus, the supply function is differentiable on the range of prices $0 \leq p < 15$ (i.e., $0 \leq q < 1.25$). However, once price reaches the level 15, firm B begins to supply some output and this generates a nondifferentiability in the market supply function, as is seen in Figure 5.4.3 (b).

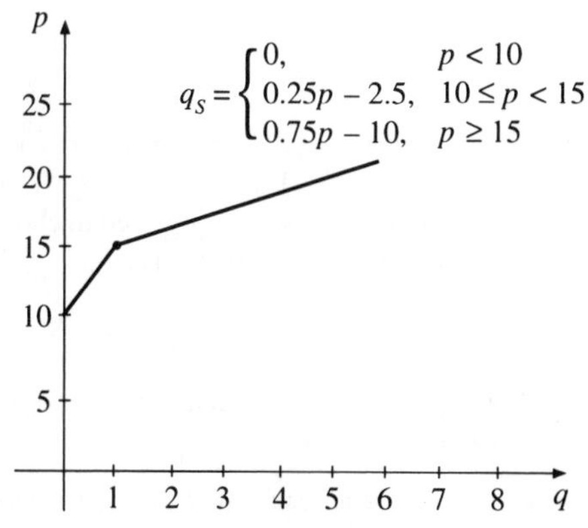

Figure 5.4.3 (b)

5. Since $q = 16L^2$ is the production function, its inverse is $L(q) = (1/4)q^{1/2}$ and so the cost function is $C(q) = c_0 + wL(q) = c_0 + (1/4)wq^{1/2}$. The derivative of the production function is

$$\frac{dq}{dL} = 32L$$

which is increasing in L. The derivative of the cost function is

$$\frac{dC}{dq} = (1/8)q^{-1/2} = \frac{1/8}{q^{1/2}}$$

which is falling in q. Since in general $dC/dq = w\,dL/dq$, which, according to the inverse function rule for differentiation, implies $dC/dq = \frac{w}{dq/dL}$, it follows that if dq/dL is increasing in L (and hence q) then dC/dq is falling in q.

7. The total revenue function is

$$\text{TR}(q) = pq = [a - bq]q = aq - bq^2$$

and so the marginal revenue function is

$$\text{MR}(q) = \text{TR}'(q) = a - 2bq$$

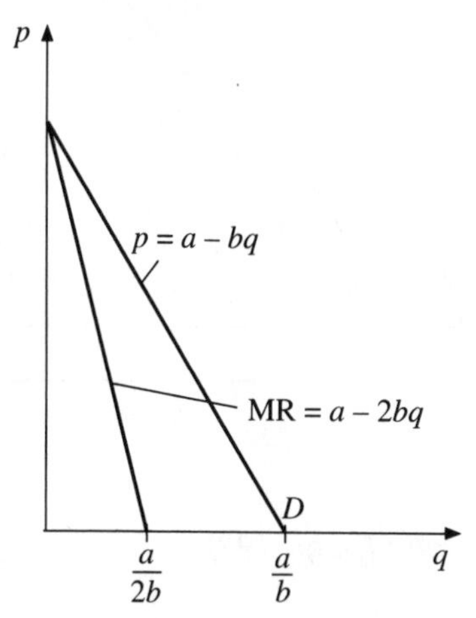

Figure 5.4.7

9. (a) Taking natural logs for the demand function $y = 100/p^2$ we get

$$\ln[y] = \ln[100] - 2\ln[p]$$

and so the (own) price elasticity of demand is $\epsilon = -d\ln[y]/d\ln[p] = 2$.

(b) To write this demand function in units of 'number of bagels,' rather than 'dozens of bagels' simply note that $\hat{y} = 12y$, where $\hat{y}$ is measured in number of bagels and y is measured in number of dozens of bagels and, similary, $\hat{p} = p/12$. This implies that $y = \hat{y}/12$ and $p = 12\hat{p}$ and so we can write the demand function $y = 100/p^2$ as

$$\hat{y}/12 = 100/(12\hat{p})^2 \quad \text{or} \quad \hat{y} = 100/12\hat{p}^2$$

Taking natural logs of both sides we get

$$\ln[\hat{y}] - \ln[12] = \ln[100] - (\ln[12] + \ln[p^2])$$

which gives us

$$\ln[\hat{y}] = \ln[100] - 2\ln[\hat{p}]$$

and so the (own) price elasticity of demand is $\epsilon = -d\ln[y]/d\ln[p] = 2$ which is the same answer as in part (a).

Section 5.5 (page 186)

1. $f'(x) = 4x^3$ and $f''(x) = 12x^2 \geq 0$ for every $x \in R$ and $f''(x) = 12x^2 = 0$ only at the single point $x = 0$. Therefore, $f(x) = x^4$ is a strictly convex function.

3. $y = x^{1/3} \Rightarrow x = y^3$ and so $C(y) = c_0 + ry^3$. The production function $y = x^{1/3}$ is strictly concave since $\frac{d^2y}{dx^2} = -\frac{2}{9}x^{-5/3} < 0$. The cost function is strictly convex since $C''(y) = 6ry > 0$.

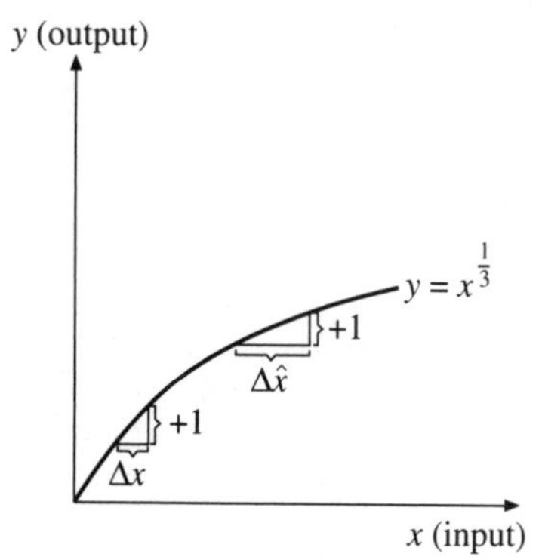

Figure 5.5.3i

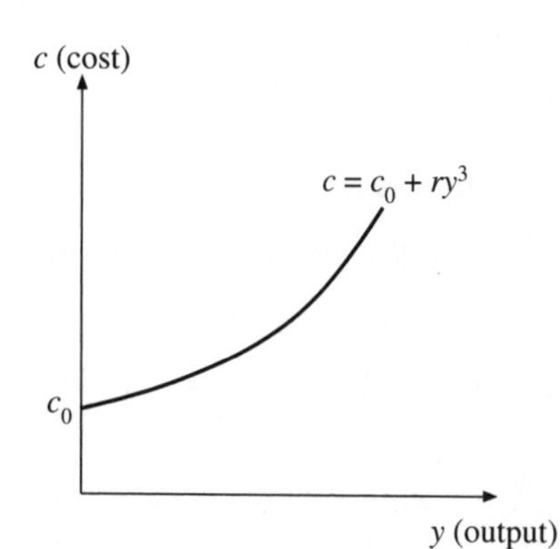

Figure 5.5.3ii

The strict concavity of the production function means that an extra unit of output requires more input the greater is the scale of output, as indicated by the accompanying figures. Thus, to produce one more unit of output requires a larger increment of input be used, and hence higher cost, the greater is the scale of output.

5. $\pi(y) = 45y - (y^3 - 9y^2 + 60y + 10) \Rightarrow \pi(y) = -15y - y^3 + 9y^2 - 10$. Upon differentiating twice we get: $\pi'(y) = -15 - 3y^2 + 18$ and $\pi''(y) = -6y + 18$ Thus, we have

$$\text{(strictly convex)}: \pi''(y) > 0 \Leftrightarrow -6y + 18 > 0 \Leftrightarrow y < 3$$

$$\text{(strictly concave)}: \pi''(y) < 0 \Leftrightarrow -6y + 18 < 0 \Leftrightarrow y > 3$$

Thus, on the interval (0,3) the profit function is stricly convex and on the interval $(3, +\infty)$ the profit function is strictly concave

y	$\pi(y)$	$\pi'(y)$	$\pi''(y)$
0	−10	−15	18
1	−17	0	12
2	−12	9	6
3	−1	12	0
4	10	9	−6
5	15	0	−12
6	8	−15	−18
7	−17	−36	−24
8	−66	−63	−30
9	−145	−96	−36
10	−260	−135	−42

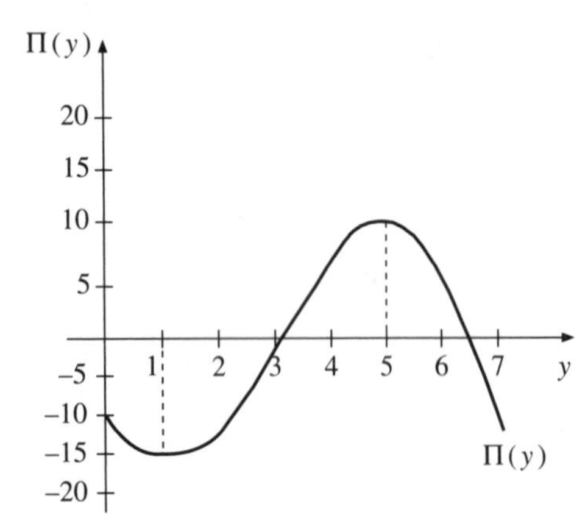

Figure 5.5.5

Section 5.6 (page 191)

1. Taking successive derivatives of $f(x) = e^{-x}$ gives us

$$f'(x) = -e^{-x}, \quad f''(x) = e^{-x},$$

$$f'''(x) = -e^{-x}, \quad f^{(4)}(x) = e^{-x}, \quad \ldots$$

Thus, we have

$$f^{(n)}(x) = (-1)^n e^{-x}$$

and so, since $e^{-0} = e^0 = 1$, we have

$$f^{(n)}(0) = (-1)^n$$

Using this result, and the result that $f(0) = -1$, the Taylor series expansion for e^{-x} evaluated about the point $x = 0$ gives the following result

$$e^{-x} = -1 - x + \frac{x^2}{2!} - \frac{x^3}{3!} + \frac{x^4}{4!} - \frac{x^5}{5!} +$$

$$\cdots + (-1)^{n-1}\frac{x^{n-1}}{(n-1)!} + R_n$$

where $|R_n| = \xi^n/n!$ and ξ is between 0 and x. Thus, if $x \in [0, 1]$ then $\xi \in [0, 1]$ and so $\xi^n \leq 1$. Therefore, to be correct to within 0.001 we simply need to choose n large enough to ensure that $|R_n| \leq 0.001$. That is

$$\frac{1}{n!} \leq 0.001 \quad \text{or} \quad n! \geq 1000$$

The smallest value of n for which this holds is $n = 7$. (Since $6! = 720$ while $7! = 5040$.)

3. For $n = 2$ the Taylor series formula can be written

$$f(x) - f(x_0) = f'(x_0)(x - x_0) + \frac{f''(\xi)(x - x_0)^2}{2}$$

which is in the form

$$\Delta y = dy + \epsilon$$

where $\epsilon = \frac{f''(\xi)(x-x_0)^2}{2}$ is the error term when using dy as an estimate of Δy. So, for the function $f(x) = x^{1/2}$ we have $f'(x) = 1/2x^{1/2}$ and $f''(x) = -1/4x^{3/2}$ and so

$$\epsilon = -\frac{(x - x_0)^2}{8\xi^{3/2}} \leq 0$$

and so $\Delta y \leq dy$ or $dy \geq \Delta y$. Therefore, using the total differential leads to an overestimate of the impact of a change in x on y as illustrated in Figure 5.6.3.

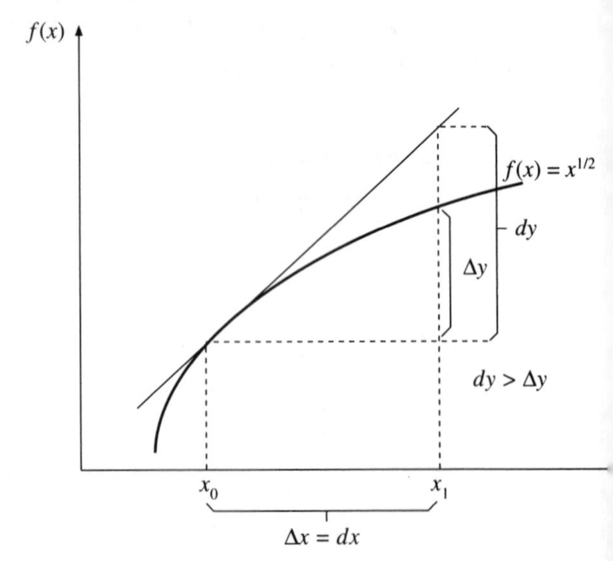

Figure 5.6.3

Review Exercises (page 192)

1.

$$\begin{aligned}
f'(x) &= \lim_{\Delta x \to 0} \frac{f(x + \Delta x) - f(x)}{\Delta x} \\
&= \lim_{\Delta x \to 0} \frac{[(x + \Delta x)^2 + 3(x + \Delta x) - 4] - [x^2 + 3x - 4]}{\Delta x} \\
&= \lim_{\Delta x \to 0} \frac{[x^2 + 2x\Delta x + \Delta x^2 + 3x + 3\Delta x - 4] - [x^2 + 3x - 4]}{\Delta x} \\
&= \lim_{\Delta x \to 0} \frac{2x\Delta x + \Delta x^2 + 3\Delta x}{\Delta x} \\
&= \lim_{\Delta x \to 0} 2x + \Delta x + 3 = 2x + 3
\end{aligned}$$

3. (a) $f(L) = 64L^{1/4} \Rightarrow f'(L) = 16L^{-3/4} \Rightarrow f''(L) = -12L^{-7/4} < 0$. The second derivative, which gives the slope of the derivative function, is negative. This means that the rate at which output rises with respect to more input being used is decreasing in L. This is consistent with the so-called law of diminishing marginal product of labor and is illustrated by the graphs below. This is illustrated by Figures 5.R.3 (a)i-ii.

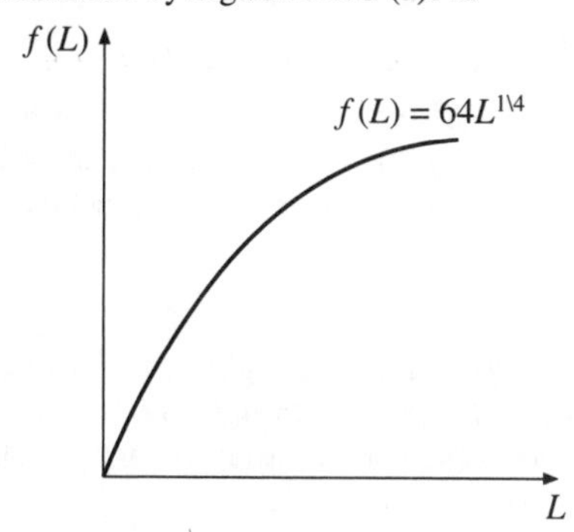

Figure 5.R.3 (a)i

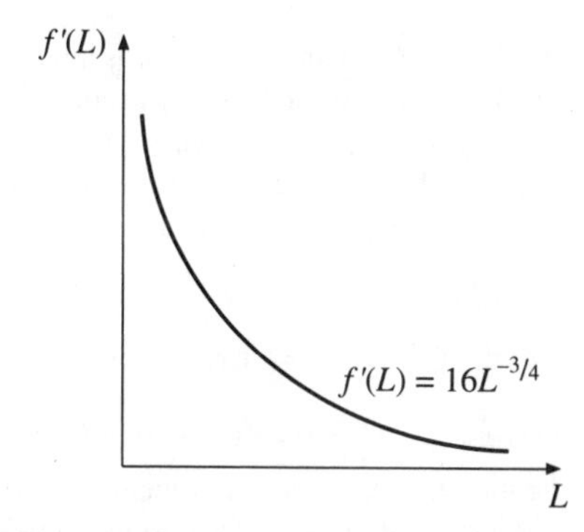

Figure 5.R.3 (a)ii

(b) $f(L) = 10L + 2L^{1/2} \Rightarrow f'(L) = 10 + L^{-1/2} \Rightarrow f''(L) = -(\frac{1}{2})L^{-3/2} < 0$. The second derivative, which gives the slope of the derivative function, is negative. This means that the rate at which output rises with respect to more input being used is decreasing in L. This is consistent with the so-called law of diminishing marginal product of labor and is illustrated by Figures 5.R.3 (b)i-ii.

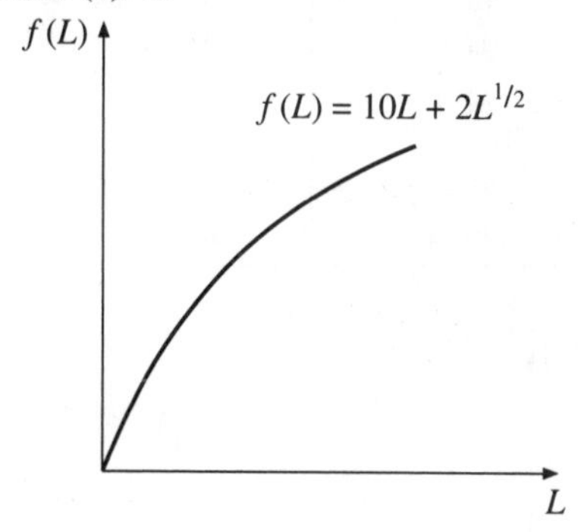

Figure 5.R.3 (b)i

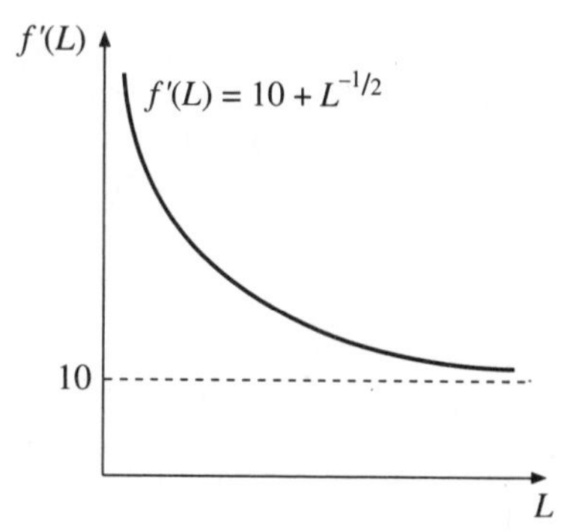

Figure 5.R.3 (b)ii

(c) $f(L) = 5L^3 \Rightarrow f'(L) = 15L^2 \Rightarrow f''(L) = 30L > 0$. The second derivative, which gives the slope of the derivative function, is positive. This means that the rate at which output rises with respect to more input being used is increasing in L. This is inconsistent with the so-called law of diminishing marginal product of labor and is illustrated by Figures 5.R.3 (c)i-ii.

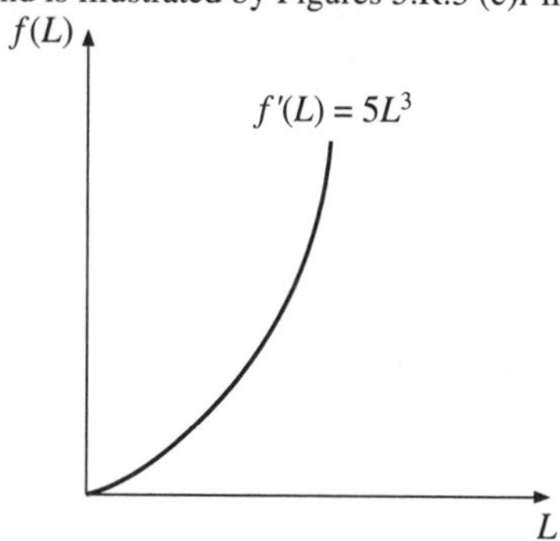

Figure 5.R.3 (c)i

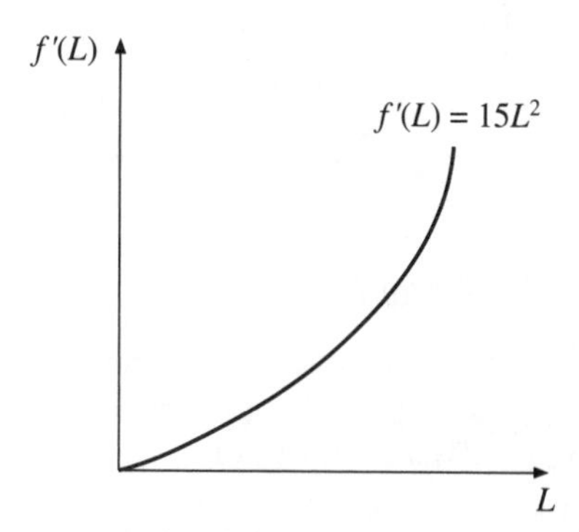

Figure 5.R.3 (c)ii

(d) $f(L) = -L^3 + 12L^2 + 3L \Rightarrow f'(L) = -3L^2 + 24L + 3 \Rightarrow f''(L) = -6L + 24$. It follows that

$$f''(L) = 24 - 6L \begin{cases} > 0 & L < 4 \\ = 0 & L = 4 \\ < 0 & L > 4 \end{cases}$$

For $L < 4$ the second derivative is positive, which means that the rate at which output rises with respect to more input being used is increasing. For $L > 4$, the second derivative of $f(L)$ is negative, which means that the rate at which output rises with respect to more input being used is decreasing. Thus, this production function is consistent *eventually* (i.e., for sufficiently large L) with the so-called law of diminishing marginal product of labor. These results are illustrated in Figures 5.R.3 (d)i-ii.

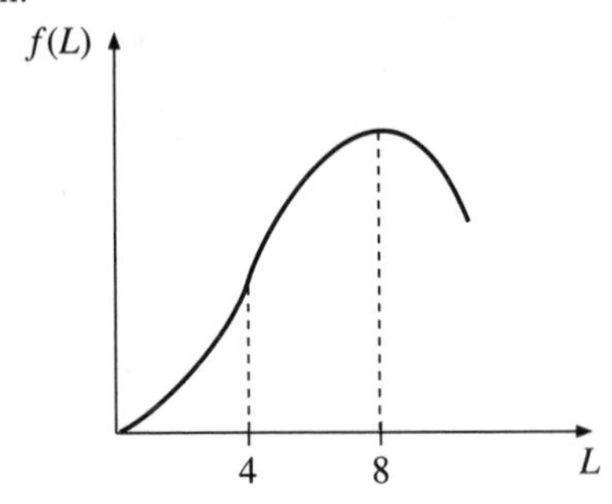

Figure 5.R.3 (d)i

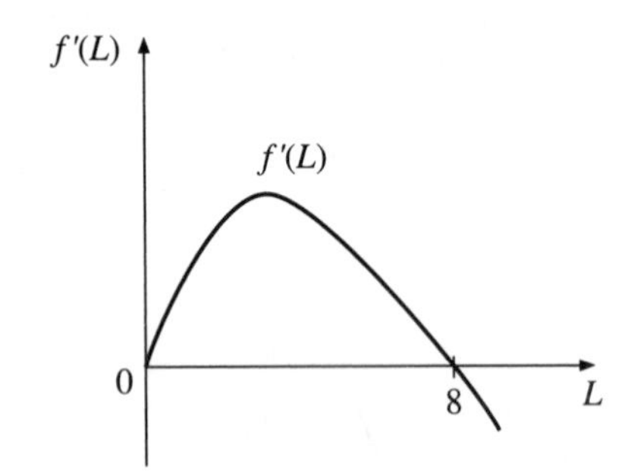

Figure 5.R.3 (d)ii

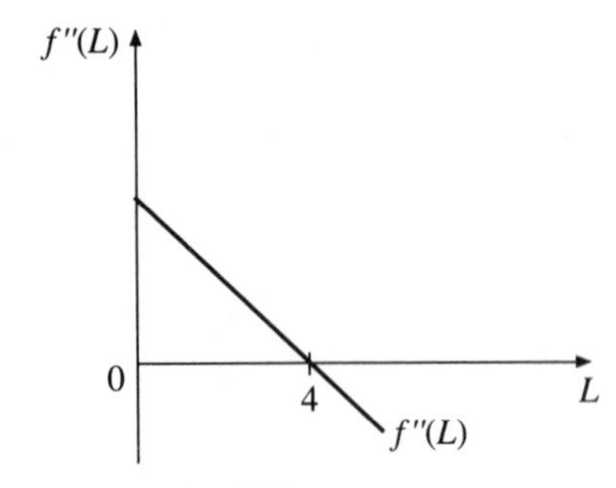

Figure 5.R.3 (d)iii

5. $\epsilon = -\frac{dy}{dp}\frac{p}{y}$ and so for the demand function $y = 200 - 5p$ we get $\epsilon = (-) - 5\frac{p}{y} = 5\frac{p}{y}$. Upon substituting for y gives us $\epsilon = 5\frac{p}{200-5p}$. Thus, we have that

$$\begin{aligned}
\epsilon < 1 \quad &\Leftrightarrow \quad 5\frac{p}{200-5p} < 1 \Leftrightarrow 5p < 200 - 5p \\
&\Leftrightarrow \quad 10p < 200 \Leftrightarrow p < 20 \\
\epsilon = 1 \quad &\Leftrightarrow \quad 5\frac{p}{200-5p} = 1 \Leftrightarrow 5p = 200 - 5p \\
&\Leftrightarrow \quad 10p = 200 \Leftrightarrow p = 20 \\
\epsilon > 1 \quad &\Leftrightarrow \quad 5\frac{p}{200-5p} > 1 \Leftrightarrow 5p > 200 - 5p \\
&\Leftrightarrow \quad 10p > 200 \Leftrightarrow p > 20
\end{aligned}$$

Noting that $y = 100$ when $p = 1$, we can illustrate the above results on the following graph.

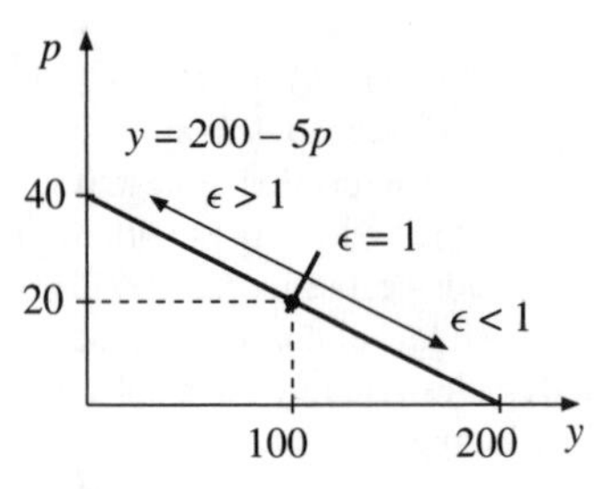

Figure 5.R.5

7. Since $q(L) = aL^b$, $b > 0$ is the production function, its inverse is $L(q) = (q/a)^{1/b}$ and so the cost function is $C(q) = c_0 + w(q/a)^{1/b}$. The derivative of the production function is

$$\frac{dq}{dL} = abL^{b-1} = \frac{ab}{L^{1-b}}$$

If $b < 1$ then dq/dL is decreasing in L. If $b > 1$ then dq/dL is increasing in L. If $b = 1$ then dq/dL is constant (i.e., it is neither decreasing nor increasing in L). The derivative of the cost function is

$$\frac{dC}{dq} = \left(w/a^{1/b}\right)(1/b)q^{(1/b)-1} = \frac{(w/a^{1/b})(1/b)}{q^{1-(1/b)}}$$

If $b < 1$ then $(1/b) - 1 > 0$ and so dC/dq is increasing in q. If $b > 1$ then $(1/b) - 1 < 0$ and so dC/dq is decreasing in q. If $b = 1$ then $(1/b) - 1 = 0$ (i.e., $dC/dq = (w/a)q^0 = w/a$) and so dC/dq is neither increasing nor decreasing in q. Thus, the relationships stated in the question are satisfied.

9. $C'(y) = 3y^2 - 24y + 50$ and $C''(y) = 6y - 24$.

(strictly convex): $\quad C''(y) > 0 \Rightarrow 6y - 24 > 0 \Rightarrow y > 4$

(strictly concave): $\quad C''(y) < 0 \Rightarrow 6y - 24 < 0 \Rightarrow y < 4$

Thus, on the interval $(0, 4)$ the cost function is stricly concave and on the interval $(4, +\infty)$ the cost function is strictly convex.

y	$C(y)$	$C'(y)$	$C''(y)$
0	20	50	-24
1	59	29	-18
2	80	14	-12
3	89	5	-6
4	92	2	0
5	95	5	6
6	104	14	12
7	125	29	18
8	164	50	24
9	227	77	30
10	320	110	36

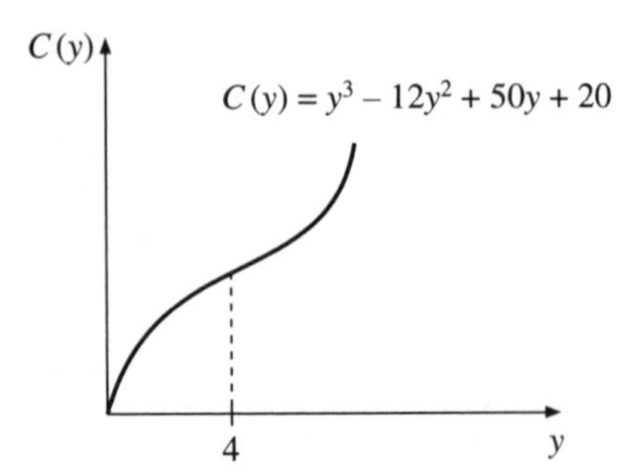

Figure 5.R.9

CHAPTER 6

Section 6.1 (page 216)

1. (a) $y = x^3 - 3x^2 + 1$ and $dy/dx = 3x^2 - 6x = 0 = x(3x - 6) = 0$ which implies $x_1^* = 0, x_2^* = 2$.

At $x = 0$, $y = 1$. At $x = 0.1$, $y = 0.97$. At $x = -0.1$, $y = 0.97$. Thus $x_1^* = 0$ yields a local maximum of the function.

At $x = 2$, $y = -3$. At $x = 2.1$, $y = -2.97$. At $x = 1.9$, $y = -2.97$. Thus $x_2^* = 2$ yields a local minimum of the function.

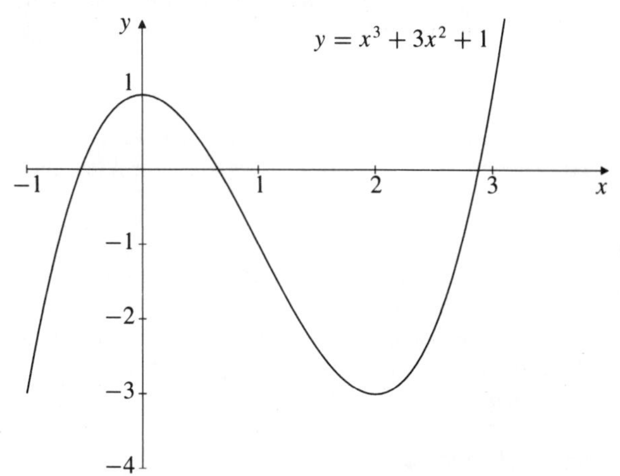

Figure 6.1.1 (a)

(b) $y = x^4 - 4x^3 + 16x^2 - 2$ and $dy/dx = 4x^3 - 12x^2 + 16 = 0$. So $x_1^* = -1$ and $x_2^* = 2$ are stationary values.

At $x = -1$, $y = -13$. At $x = -1.1$, $y = -12.8119$, while at $x = -0.9$, $y = -12.8279$. Thus, $x_1^* = -1$ yields a local minimum of the function.

At $x = 2$, $y = 14$. At $x = 1$, $y = 11$, while at $x = 3$, $y = 19$. Thus, $x_2^* = 2$ yields a local minimum of the function.

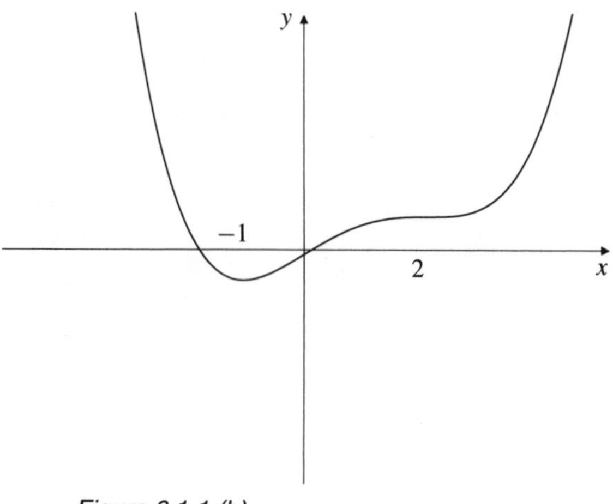

Figure 6.1.1 (b)

(c) $y = 3x^3 - 3x - 2$ and $dy/dx = 9x^2 - 3 = 0 \Rightarrow x^2 = 1/3$ which implies $x_1^* = \sqrt{1/3} \doteq 0.577$, $x_2^* = -\sqrt{1/3} \doteq -0.577$.

At $x = \sqrt{1/3}$, $y = -3.15470$. At $x = 0.58$, $y = -3.15466$. At $x = 0.57$, $y = -3.15442$. Thus $x_1^* = \sqrt{1/3}$ yields a local minimum of the function.

At $x = -\sqrt{1/3}$, $y = -0.845299$. At $x = -0.57$, $y = -0.845579$. At $x = -0.58$, $y = -0.845336$. Thus $x_2^* = -\sqrt{1/3}$ yields a local maximum of the function.

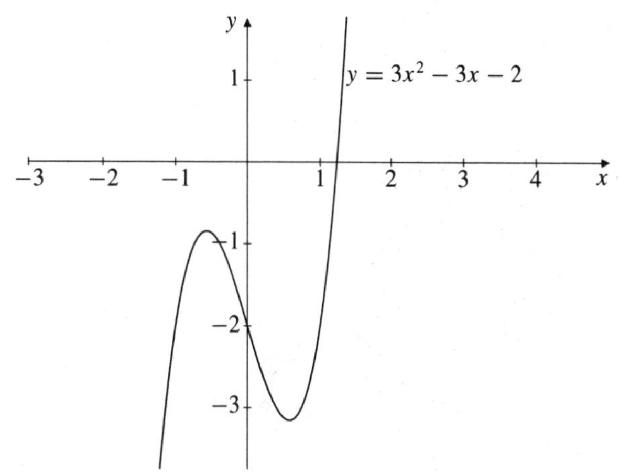

Figure 6.1.1 (c)

(d) $y = 3x^4 - 10x^3 + 6x^2 + 1$ and $dy/dx = 12x^3 - 30x^2 + 12x = x(12x^2 - 30x + 12) = 0$. Hence, $x_1^* = 0$ is a stationary value. Furthermore the roots of the quadratic in brackets, $x_2^* = 0.5$ and $x_3^* = 2$, are stationary values.

At $x = 0$, $y = 1$. At $x = 0.1$, $y = 1.05$. At $x = -0.1$, $y = 1.07$. Thus $x_1^* = 0$ yields a local minimum of the function.

At $x = 0.5$, $y = 1.44$. At $x = 0.6$, $y = 1.39$. At $x = 0.4$, $y = 1.40$. Thus $x_2^* = 0.5$ yields a local maximum of the function.

At $x = 2$, $y = -7$. At $x = 2.1$, $y = -6.8$. At $x = 1.9$, $y = -6.8$. Thus $x_3^* = 2$ yields a local minimum.

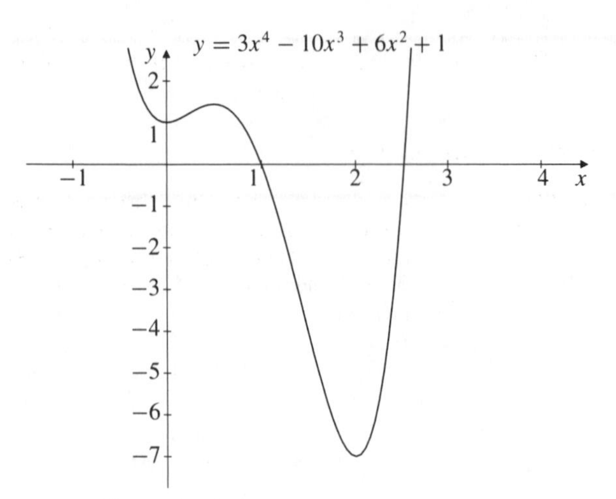

Figure 6.1.1 (d)

(e) $y = 2x/(x^2+1)$ and

$$\frac{dy}{dx} = \frac{2-2x^2}{x^4+2x^2+1} = 0$$

This expression can only be zero if the numerator equals zero. Since the denominator is always positive, the fraction will be defined at the stationary values of the numerator. Thus we obtain the stationary values by setting $2 - 2x^2 = 0$. Hence $x_1^* = 1, x_2^* = -1$.

At $x = 1$, $y = 1$. At $x = 1.1$, $y = 0.995$. At $x = 0.9$, $y = 0.994$. Thus $x_1^* = 1$ yields a local maximum of the function.

At $x = -1$, $y = -1$. At $x = -0.9$, $y = -0.994$. At $x = -1.1$, $y = -0.995$. Thus $x_2^* = -1$ yields a local minimum of the function.

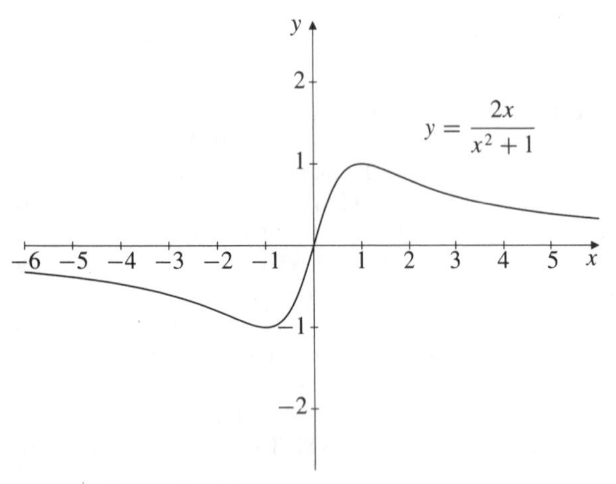

Figure 6.1.1 (e)

3. Using the cost function $C = 0.04x^2 + 3x + 80$ we obtain the profit function $\pi(x) = px - (0.04x^2 + 3x + 80)$. The first order condition for profit maximization is

$$\pi'(x) = p - (0.08x + 3) = 0$$

or $p = 0.08x + 3$ which is simply price equals marginal cost. This equation must give us a maximum since $\pi''(x) = -0.08 < 0$. Thus it follows that $x^* = 12.5p - 37.5$.

Now we must take into account that for the range of prices where price is lower than average variable cost (AVC) the firm will produce zero output. Hence we need to solve for the minimum of the AVC function which is given by

$$\frac{\text{Total variable cost}}{x} = \frac{0.04x^2 + 3x}{x} = 0.04x + 3$$

Obviously AVC is minimized at $x = 0$. The resulting minimized value of AVC is \$3. Thus the firm will shut down for any price below \$3, while at a price at or above \$3 it will produce the output according to the first-order condition. This implies that its supply function is

$$x = \begin{cases} 12.5p - 37.5 & \text{for } p \geq \$3 \\ 0 & \text{for } p < \$3 \end{cases}$$

5. In this problem it is useful to work with the linear inverse demand function $p(x) = a - bx$, $a > 0, b > 0$, since we want to solve for output. This function corresponds to the normal linear demand function

$$x = \frac{a}{b} - \frac{1}{b}p$$

Sales revenue $R(x)$ is simply $p(x)x$ and thus $R(x) = ax - bx^2$. The first-order condition for sales revenue maximization is

$$R'(x) = a - 2bx = 0$$

This gives us a maximum because $R''(x) = -2b < 0$. Thus, we obtain for the sales maximum output $x_{\text{sm}} = a/2b$. The output that the monopolist would produce if it sold its output at a zero price is found by setting $p(x) = 0$. This yields $x_{\text{zp}} = a/b = 2x_{\text{sm}}$.

Section 6.2 (page 225)

1. **(a)** $f''(x) = 6x - 6$, $f''(x_1 = 0) = -6$, $f''(x_2 = 2) = 6$.

(b) $f''(x) = 12x^2 - 24x + 9$, $f''(x_1 = 0) = 9$, $f''(x_2 = 1.5) = 0$.

(c) $f''(x) = 18x$, $f''(x_1 = \sqrt{1/3}) = 10.44$, $f''(x_2 = -\sqrt{1/3}) = -10.44$.

(d) $f''(x) = 36x^2 - 60x + 12$, $f''(x_1 = 0) = 12$, $f''(x_2 = 0.5) = -9$, $f''(x_3 = 2) = 36$.

(e)

$$f''(x) = \frac{4x^3 - 12x}{(x^2+1)^2}$$

$f''(x_1 = 1) = -8$, $f''(-1) = 8$.

3. First, we must have $b \leq a^2/80$ in order to obtain a solution for the first-order condition marginal revenue equal to marginal cost. This can be seen in Figure 6.2.3 (a) If this condition is not fulfilled then the marginal revenue curve does not intersect the marginal cost curve. Increasing output just leads to an increase in loss.

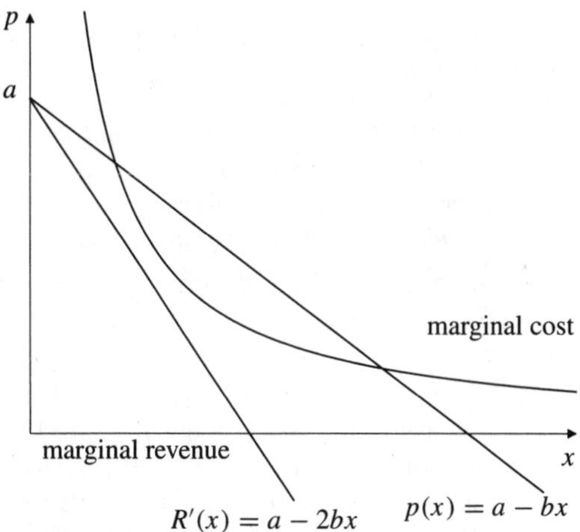

Figure 6.2.3 (a)

Second, the second derivative of the profit function cannot be positive at the profit maximizing output. The significance of this condition becomes clear in the case where the marginal revenue curve has two points of intersection with the marginal cost curve [see Figure 6.2.3 (b)]. The first point of intersection x_1 maximizes losses since to its left marginal cost is higher than marginal revenue. The value of the second derivative of the profit function at x_1 is positive. The second point of intersection x_2, however, maximizes profits because marginal revenue is above marginal cost to the left of x_2. The value of second derivative of the profit function at x_2 is negative.

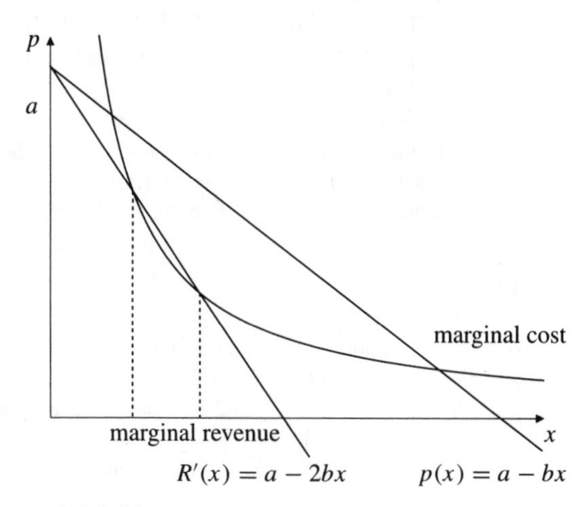

Figure 6.2.3 (b)

5. (a) Writing the inverse of the production function as $L(x) = L$, we have the cost function $C(x) = wL(x)$. Therefore, we obtain for price equal to marginal cost $p = C'(x) = w(dL/dx)$.

Multiplying both sides of the equation by (dx/dL) and noting that $(dx/dL) = (df/dL)$ because L is the inverse function of f, yields $p(df/dL) = w$ which is marginal value product equals input price.

(b) Substituting p for marginal revenue $p + p'(x)x$, the argument is exactly as in (a).

(c) The production function must be strictly concave, i.e., $f'' < 0$. This implies rising marginal costs and a falling marginal value product. Only then will the conditions above describe a profit maximum.

Section 6.3 (page 233)

1. (a) $\max y = 3 + 2x$ subject to $0 \leq x \leq 10$ This is a linear function with a positive slope as depicted in Figure 6.3.1 (a). Thus an interior maximum cannot occur, and the solution is $x^* = 10$ with $y^* = 23$.

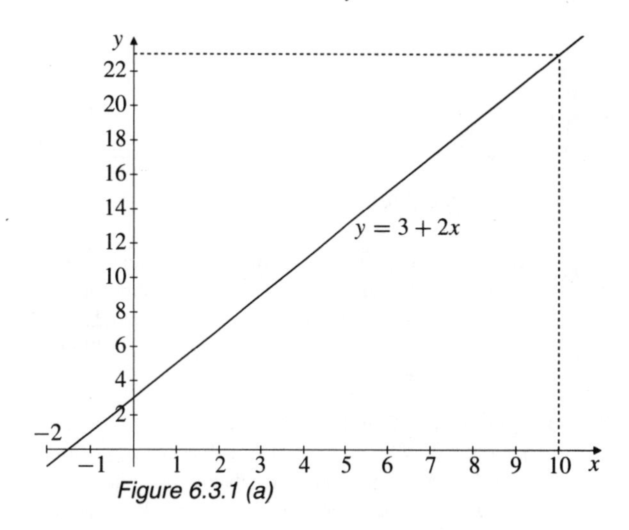

Figure 6.3.1 (a)

(b) $\max y = 1 + 10x^2$ subject to $5 \leq x \leq 20$. This function has the first derivative $dy/dx = 20x$. Clearly, it has positive slope throughout the interval. The maximum must occur at the upper bound as shown in Figure 6.3.1 (b). Therefore $x^* = 20$ with $y^* = 4001$.

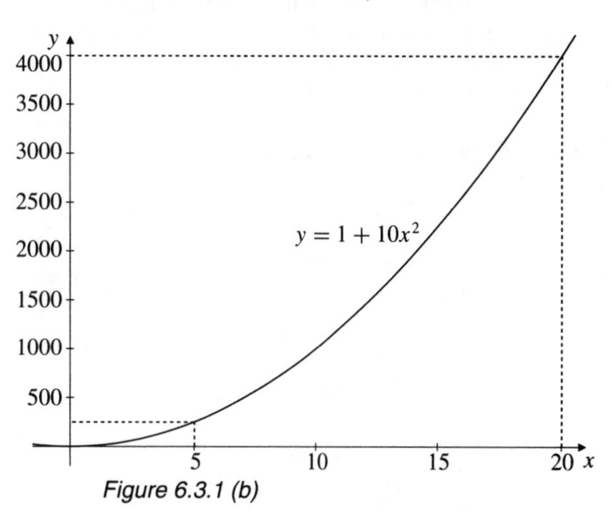

Figure 6.3.1 (b)

(c) $\min y = 5 - x^2$ subject to $0 \leq x \leq 10$. The first derivative of this function is $dy/dx = -2x$. Hence the function has zero or negative slope throughout the interval. At the lower bound the slope is negative. Thus the minimum occurs at $x^* = 10$ with $y^* = -95$ [see Figure 6.3.1 (c)].

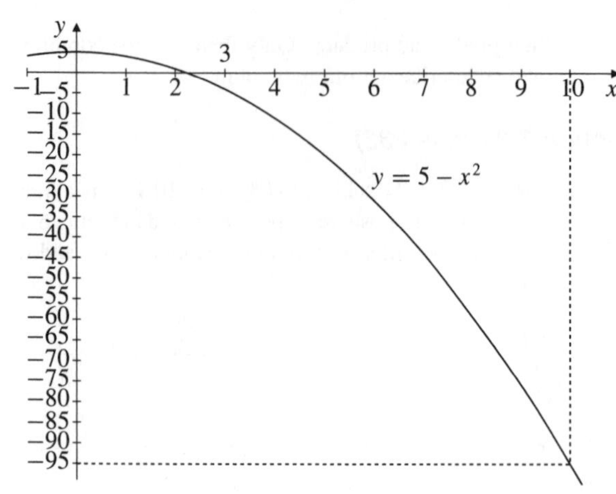

Figure 6.3.1 (c)

These solutions differ from the case in which x is unconstrained because the first derivatives are not equal to zero at the optimum. In (a) we have $dy/dx = 2$, in (b) $dy/dx = 400$, and in (c) $dy/dx = -20$ at the respective optima.

3. (a) A farm's problem is to $\max \pi(x) = R(x) - C(x) = px - C(x) = x - 0.2x = 0.8x$ subject to $x \leq 100$.

Clearly, $x^* = 100$. The shadow price is $\pi'(x) = 0.8$: If the output quota would be relaxed by one unit, a farm would make an extra profit of 0.8 which is the difference between marginal revenue, here price, and marginal costs at the optimum.

(b) Another farmer will be willing to pay as much for the output quota as the extra profits that the additional output quota will allow him to make. Since marginal costs are identical and constant for all farmers this is just the profit each farmer makes, i.e., $\pi(100) = 80$. Thus, the farmer can ask \$80 for his output quota.

Review Exercises (page 235)

1.

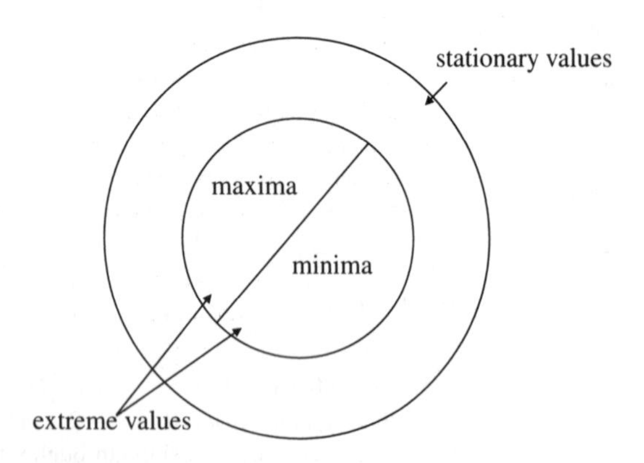

Figure 6.R.1

3. (a) $f(x) = 0.5x^3 - 3x^2 + 6x + 10$. Setting the first derivative equal to zero yields $f'(x) = 1.5x^2 - 6x + 6 = 0$. Dividing the equation by 1.5 we obtain $x^2 - 4x + 4 = (x-2)^2 = 0$. Thus $x^* = 2$ is the only stationary value of $f(x)$.

The second derivative is $f''(x) = 3x - 6$ and $f''(x^* = 2) = 0$ Therefore, we cannot judge from the second derivative whether x^* is a maximum, minimum, or point of inflection. However, $f'''(x) = 3$. Therefore, we will use the third derivative as the final term in the Taylor expansion to examine the nature of the function around $x^* = 2$.

$$f(\hat{x}) = f(x^*) + \frac{f'''(\zeta)(\hat{x} - x^*)^3}{6}$$

for ζ between x^* and $\hat{x}$. Since $f'''(x) = 3$ for all x, we can write this equation as $f(\hat{x}) = f(2) + 0.5(\hat{x} - 2)^3$. For $\hat{x} > 2$, we get $f(\hat{x}) > f(2)$, while for $\hat{x} < 2$, we get $f(\hat{x}) < f(2)$. Therefore, $x^* = 2$ delivers a local minimum.

(b) $f(x) = x^3 - 3x^2 + 5$. Setting the first derivative equal to zero, we obtain $f'(x) = 3x^2 - 6x = 0$. Dividing the equation by three yields $x^2 - 2x = x(x-2) = 0$. Thus the stationary values are $x_1^* = 0$ and $x_2^* = 2$.

The second derivative is $f''(x) = 6x - 6$.

$f''(x_1^* = 0) = -6 < 0 \Rightarrow x_1^*$ is a local maximum.

$f''(x_2^* = 2) = 6 > 0 \Rightarrow x_2^*$ is a local minimum.

(c) $f(x) = x^4 - 4x^3 + 16x$. Setting the first derivative equal to zero yields $f'(x) = 4x^3 - 12x^2 + 16 = 0$. Dividing the equation by 5, we obtain $x^3 - 3x^2 + 4 = (x+1)(x-2)^2 = 0$. Thus the stationary values are $x_1^* = -1$ and $x_2^* = 2$.

The second derivative is $f''(x) = 12x^2 - 24x$.

$f''(x_1^* = -1) = 36 > 0 \Leftrightarrow x_1^*$ is a local minimum.

$f''(x_2^* = 2) = 0$

We could now use a Taylor expansion to determine the nature of x_2^*. However, looking at $f'(x)$ shows that $f'(x)$ must be positive for any x in a small interval around x_2^*, i.e., f has positive slope in the interval around x_2^*. Thus, x_2^* must be a point of inflection.

(d) $f(x) = x + 1/x$. Setting the first derivative equal to zero yields $f'(x) = 1 - 1/x^2 = 0 \Rightarrow x^2 = 1$. Thus, $x_1^* = -1$ and $x_2^* = 1$ are the stationary values.

The second derivative is $f''(x) = 2/x^3$.

$f''(x_1^* = -1) = -2 < 0 \Rightarrow f$ has a local maximum at $x_1^* = -1$

$f''(x_2^* = 1) = 2 > 0 \Rightarrow f$ has a local minimum at $x_2^* = 1$

(e) $f(x) = x^3 - 3x - 1$. Setting the first derivative equal to zero yields $f'(x) = 3x^2 - 3 = 0 \Rightarrow x^2 = 1$. Hence, $x_1^* = -1$ and $x_2^* = 1$ are the stationary values.

The second derivative is $f''(x) = 6x$.

$f''(x_1^* = -1) = -6 < 0 \Rightarrow f$ has a local maximum at $x_1^* = -1$

$f''(x_2^* = 1) = 6 > 0 \Rightarrow f$ has a local minimum at $x_2^* = 1$.

(f) $f(x) = 3x^4 - 10x^3 + 6x^2 + 5$. Setting the first derivative equal to zero yields $f'(x) = 12x^3 - 30x^2 + 12x = 0$ which implies $2x^3 - 5x^2 + 2x = 0$ (after division by 6). So, $x(2x^2 - 5x + 2) = 0$. Thus, $x_1^* = 0$ is a stationary value. The quadratic in brackets has two roots $x_2^* = 0.5$ and $x_3^* = 2$ which are further stationary values.

The second derivative is $f''(x) = 36x^2 - 60x + 12$.

$f''(x_1^* = 0) = 12 > 0 \Rightarrow f$ has a local minimum at $x_1^* = 0$.

$f''(x_2^* = 0.5) = -9 < 0 \Rightarrow f$ has a local maximum at $x_2^* = 0.5$.

$f''(x_3^* = 2) = 36 > 0 \Rightarrow f$ has a local minimum at $x_3^* = 2$.

(g) $f(x) = (1 - x^2)/(1 + x^2)$. Using the quotient rule, the first derivative is

$$\begin{aligned} f'(x) &= \frac{-2x(1+x^2) - (1-x^2)2x}{(1+x^2)^2} \\ &= \frac{-2x}{(1+x^2)} - \frac{2x(1-x^2)}{(1+x^2)^2} \\ &= \frac{-2x}{(1+x^2)}\left[1 + \frac{1-x^2}{1+x^2}\right] \end{aligned}$$

The term in the square brackets is always positive since $(1-x^2)/(1+x^2) > -1$. Therefore, $f'(x)$ can only be zero when $(-2x)/(1+x^2) = 0$, i.e., the only stationary value is $x^* = 0$.

Before we proceed with the second derivative, note that in this case it is obvious that x^* must be a local maximum of f. Look at $f'(x)$. The term in square brackets is always positive. The denominator of $(-2x)/(1+x^2)$ is always positive. Hence, $f' > 0$ for $x < x^* = 0$ and $f' < 0$ for $x > x^*$. Thus, f must have a local maximum at x^*.

This is a sufficient answer for the problem. For those who used the second derivative, we will also present the solution of this approach. Here it is useful to use this formulation of $f'(x)$:

$$f'(x) = \frac{-2x}{(1+x^2)} - \frac{2x(1-x^2)}{(1+x^2)^2}$$

We then obtain

$$\begin{aligned} f''(x) &= \frac{-2(1+x^2) - (-2x)2x}{(1+x^2)^2} \\ &\quad - \frac{\left[2(1-x^2) + 2x(-2x)\right](1+x^2)^2}{(1+x^2)^4} \\ &\quad - \frac{2x(1-x^2)2(1+x^2)^2 2x}{(1+x^2)^4} \end{aligned}$$

If we evaluate this at $x^* = 0$, we have

$$f''(x^* = 0) = \frac{-2-0}{1} - \frac{[2+0]-0}{1} = -1$$

Thus, f has a local maximum at x^*.

(h) $f(x) = (3 - x^2)^{1/2}$. Using the Chain Rule we obtain for the first derivative

$$\begin{aligned} f'(x) &= \frac{1}{2}(3-x^2)^{-1/2}(-2x) \\ &= \frac{-x}{(3-x^2)^{1/2}} \end{aligned}$$

Thus, $x^* = 0$ is the only stationary value. As in problem (g) one can already judge from $f'(x)$ the nature of x^*. The denominator is always positive. Thus $f' > 0$ for $x < x^* = 0$ and $f' < 0$ for $x > x^*$. Thus f must have a local maximum at x^*.

This result is also obtained via the second derivative:

$$\begin{aligned} f''(x) &= \frac{-(3-x^2)^{1/2}}{(3-x^2)} \\ &\quad - \frac{(-x)(-1/2)(3-x^2)^{-1/2}(-2x)}{(3-x^2)} \\ &= \frac{-(3-x^2)^{1/2} + x^2(3-x^2)^{-1/2}}{(3-x^2)} \end{aligned}$$

and so

$$f''(x^* = 0) = \frac{-3^{1/2} + 0}{3} = -3^{-1/2} = -\frac{1}{\sqrt{3}} < 0$$

which implies that f has a local maximum at $x^* = 0$.

(i) $f(x) = (2 - x)/(x^2 + x - 2)$. The first derivative is

$$\begin{aligned} f'(x) &= \frac{-(x^2+x-2) - (2-x)(2x+1)}{(x^2+x-2)^2} \\ &= -\frac{1}{(x^2+x-2)}\left[1 + \frac{(2-x)(2x+1)}{(x^2+x-2)}\right] \end{aligned}$$

The first factor can never be zero. Thus, there will only be stationary values if the second factor becomes zero, i.e.,

$$1 + \frac{(2-x)(2x+1)}{(x^2+x-2)} = 0$$

This can be transformed into the quadratic $x^2 - 4x = x(x-4) = 0$. Thus, $x_1^* = 0$ and $x_2^* = 4$ are the stationary values of f.

To determine the nature of these stationary values we could now use the second derivative. However, in the case of this function this is much too complicated. Therefore, we offer two alternative ways.

First, we can calculate the values of the function at and around the stationary values.

At $x = 0$, $f(x) = -1$. At $x = 0.2$, $f(x) = -1.02$. At $x = -0.2$, $f(x) = -1.02$. Thus, $x_1^* = 0$ is a local maximum of $f(x)$.

At $x = 4$, $f(x) = -0.1111$. At $x = 4.2$, $f(x) = -0.1109$. At $x = 3.8$, $f(x) = -0.1108$. Thus, $x_2^* = 4$ is a local minimum of $f(x)$.

The second way is more elegant. Take a look at the function $f(x)$. The denominator $x^2 + x - 2 = (x-1)(x+2)$ has the roots -2 and 1. At these roots the function is not defined. As x approaches these roots, $f(x)$ must approach minus or plus infinity. We will now determine these signs which will allow us to sketch the function.

The numerator $2 - x$ is positive for $x < 2$ and negative for $x > 2$. The denominator is positive for $x < -2$, negative for $-2 < x < 1$ and positive for $x > 1$. Thus

$$f(x) \text{ is } \begin{cases} \text{positive for } x < -2 \\ \text{negative for } -2 < x < 1 \\ \text{positive for } 1 < x < 2 \\ \text{negative for } x > 2 \end{cases}$$

This information allows us to determine the nature of $x_1^* = 0$. Because $f(x)$ must be minus infinity for x slightly larger than -2 and for x slightly smaller than 1, x_1^* must be a local maximum.

Finally, we find out the nature of x_2^* with an additional characteristic of $f(x)$: $f(x)$ must approach zero as x approaches infinity since x is in the second power in the denominator and only in the first power in the numerator. Thus $f'(x)$ which must be negative at $x = 2$ must change signs at $x_2^* = 4$. Thus x_4^* must be a local maximum.

To confirm these results Figure 6.R.3(i) shows a plot of $f(x)$.

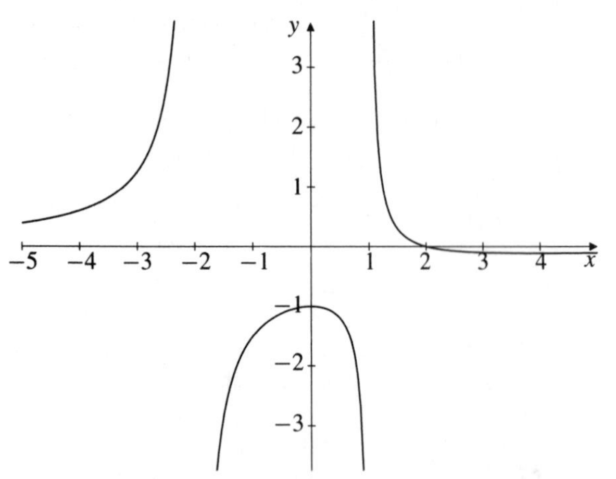

Figure 6.R.3(i)

(j) $f(x) = x^{0.5}e^{-0.1x}$. The first derivative is

$$\begin{aligned} f'(x) &= 0.5x^{-0.5}e^{-0.1x} + x^{0.5}e^{-0.1x}(-0.1) \\ &= e^{-0.1x}(0.5x^{-0.5} - 0.1x^{0.5}) \\ &= e^{-0.1x}x^{-0.5}(0.5 - 0.1x) \end{aligned}$$

$f'(x)$ can only be zero when the third factor is zero since $e^{-0.1x} > 0$ for all x and $x^{-0.5} \neq 0$ for all x. Thus

$f'(x) = 0 \Rightarrow 0.5 - 0.1x = 0 \Rightarrow x = 5.$

and $x^* = 5$ is the only stationary value of $f(x)$.

Using the product rule and $f'(x) = e^{-0.1x}(0.5x^{-0.5} - 0.1x^{0.5})$ we find the second derivative:

$$\begin{aligned} f''(x) &= e^{-0.1x}(-0.25x^{-1.5} - 0.05x^{-0.5}) \\ &\quad + e^{-0.1x}(-0.1)(0.5x^{-0.5} - 0.1x^{0.5}) \\ &= e^{-0.1x}(-0.25x^{-1.5} - 0.1x^{-0.5} + 0.01x^{0.5}) \end{aligned}$$

$f''(x^* = 5) = -0.03 < 0$. Thus $f(x)$ has a local maximum at $x^* = 5$.

5. The first-order condition of the problem

$$\max_x \pi(x) = px - C(x) = px - 0.5x^3 + 3x^2 - 6x - 50$$

can be written as $1.5x^2 - 6x + 6 - p = 0$ and we can use the standard formula to obtain the solutions

$$x_{1,2}^* = 2 \pm \frac{\sqrt{6p}}{3}$$

The second-order condition is $-3x + 6 < 0 \Rightarrow x > 2$. Thus the profit-maximizing output is $x^* = 2 + \sqrt{6p}/3$

For $p = 1$, $x^* = 2 + \sqrt{6}/3$ with the corresponding profit $\pi(x) = -51.5$. For $p = 2$, $x^* = 2 + \sqrt{12}/3$ with $\pi(x) = -48.5$. Thus, in both cases the firm will produce zero output.

7. The effect of the royalty is to raise constant unit cost c to \$0.75. Therefore, the firm's problem is to

$$\begin{aligned} \max_x \pi(x) &= p(x)x - cx = (5 - 0.5x)x - 0.75x \\ &= 4.25x - 0.5x^2 \end{aligned}$$

The first-order condition is $\pi'(x) = 4.25 - x = 0$. Thus $x^* = 4.25$. This is indeed a maximum since $\pi''(x) = -1 < 0$.

The largest bid the firm will make is equal to the profit the firm makes at the optimum which is $\pi(4.25) = 9.5625$. This compares to 10.125 when there is no royalty. Quite intuitively, the largest bid is smaller when a royalty has to be paid.

9. The student's problem is to maximize her grade average:

$$\max_{t_1,t_2}(g_1 + g_2)/2$$

where $g_1 = 20 + 20\sqrt{t_1}$, $g_2 = -80 + 3t_2$ and $t_1 + t_2 = 60$. Substituting g_1 and g_2 in the objective function and replacing t_2 by $60 - t_1$ the problem can be written as

$$\max_{t_1} f(t_1) = 120 + 20\sqrt{t_1} - 3t_1$$

The first-order condition is

$$f'(t_1) = \frac{10}{\sqrt{t_1}} - 3 = 0 \Rightarrow t_1^* = \frac{100}{9} = 11\frac{1}{9}$$

This is a maximum since

$$f''(t_1^*) = -5(t_1^*)^{-1.5} = -0.135 < 0$$

Hence $t_2^* = 60 - t_1^* = 48\frac{8}{9}$

Observe that at the optimum $dg_1/dt_1 = 10/\sqrt{t_1^*} = 3 = dg_2/dt_2$, i.e., one hour more (less) spent on studying for subject 1 yields the same absolute improvement (decline) in g_1 as the decline (improvement) in g_2 by one hour less (more) spent on subject 2. If this were not the case, the student could improve her grade average by devoting more of her time to one of the subjects.

This is essentially an economic problem because the resources necessary to achieve an objective, here time, are scarce and because it must be decided how these resources are best employed.

CHAPTER 7

Section 7.1 (page 249)

1. (a) The first equation implies that $x = 2$ and substituting into the second equation gives $y = 4$. See Figure 7.1.1 (a).

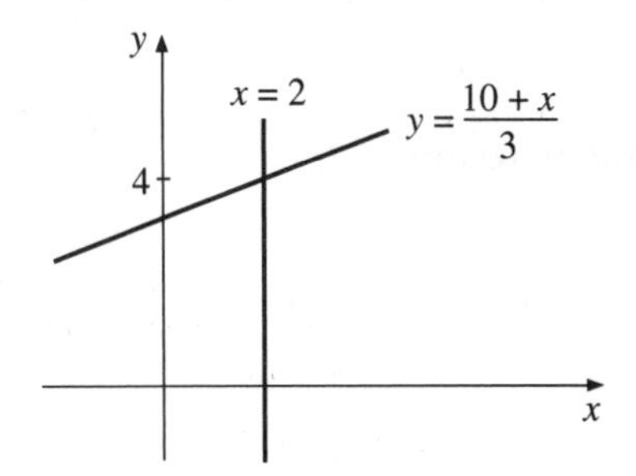

Figure 7.1.1 (a)

(b) The first equation implies that $y = 1 - x$ and the second equation also implies that $y = 1-x$. There are infinitely many solutions because both equations are equivalent. In terms of Figure 7.1.1 (b), the graphs of both equations coincide.

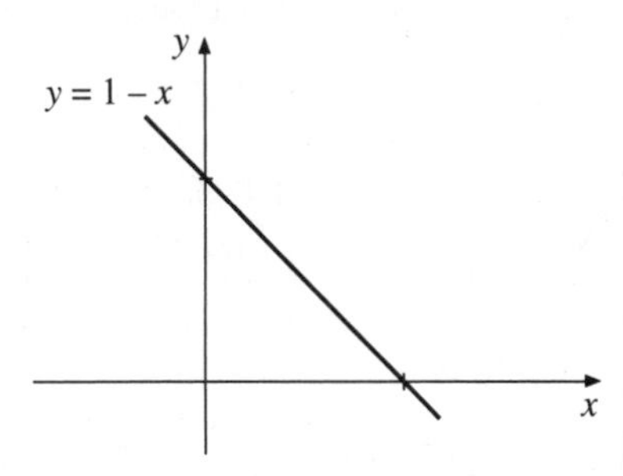

Figure 7.1.1 (b)

(c) The first equation implies that $y = 2.5 + 0.25x$, while the second equation implies that $y = 5 + 0.25x$, and there are no values of y and x that can solve these equations simultaneously, so there is no solution. In terms of Figure 7.1.1 (c) the graphs of these equations are parallel.

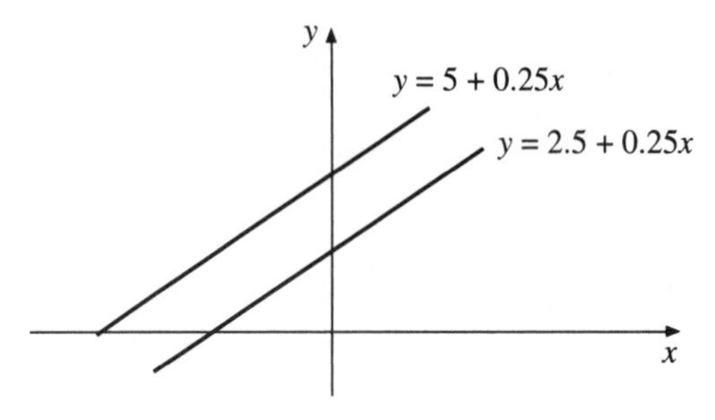

Figure 7.1.1 (c)

(d) Since the first equation gives $y = 10 - x$ and the second equation gives $y = 2x$, we have $10 - x = 2x$, which implies $x = 10/3$. Substituting this value of x into either equation gives $y = 20/3$. This is shown in Figure 7.1.1 (d).

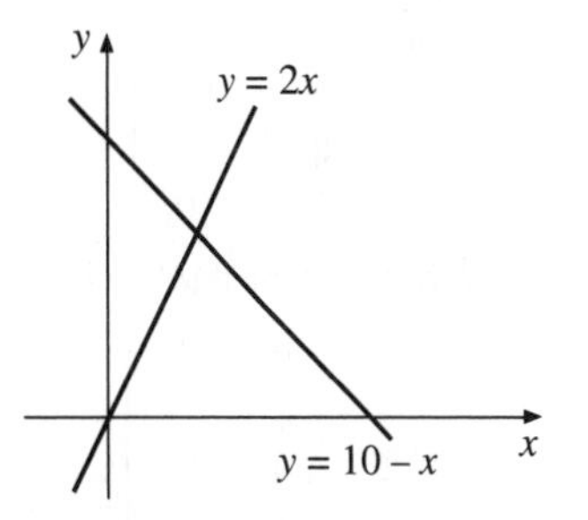

Figure 7.1.1 (d)

3. (a) These are parallel if $c = 4$, since in this case $y = -10 + 2x$ (from the first equation) and $y = 2.5 + 2x$ (from the second equation). There is no solution for $c = 4$.

(b) They have a solution for any other value of c, given by $x = 25/(4 - c)$ and $y = 10(1 + c)/(4 - c)$.

5. $\beta_{21} > 0$ implies that an increase in the price of good 1 increases the supply of good 2. If good 1 is a input into the production of good 2, and if the price of this input is related to its quality or its productivity, then more of good 2 can be produced.

7. (a) If GNP is 7.5, then the first (IS) equation gives $R = 10$. Substituting for this level of GNP and interest rate into the second equation and solving for M gives $M = 25$.

(b) Since the interest rate which makes the goods and money market jointly in equilibrium is 10, the world interest rate is inconsistent with this equilibrium. In the full model, something must change, which will either be the exchange rate or the money supply depending on whether there is a flexible or fixed exchange rate.

Section 7.2 (page 266)

1. (a) $x = 5, y = -3/5, z = 13/5$.

(b) Linearly dependent: Equation 1 = (Equation 2 + Equation 3) × 2.

(c) $x_1 = 11/2, x_2 = -3/2, x_3 = -7/2$.

(d) $x_1 = 2, x_2 = 1, x_3 = 0, x_4 = -2$.

(e) Linearly dependent: Equation 1 = Equation 3 + Equation 4.

3. (a)

$$\begin{bmatrix} 2 & -1 & 0 & 0 & 0 \\ 0 & 1 & 1 & 2 & 100 \\ 1 & 2 & 2 & 0 & 60 \\ -1 & 0 & 1 & -1 & -10 \end{bmatrix}$$

Add the third row to the fourth to get

$$\begin{bmatrix} 2 & -1 & 0 & 0 & 0 \\ 0 & 1 & 1 & 2 & 100 \\ 1 & 2 & 2 & 0 & 60 \\ 0 & 2 & 3 & -1 & 50 \end{bmatrix}$$

Subtract 2 times the third row from the first to get

$$\begin{bmatrix} 2 & -1 & 0 & 0 & 0 \\ 0 & 1 & 1 & 2 & 100 \\ 0 & -5 & -4 & 0 & -120 \\ 0 & 2 & 3 & -1 & 50 \end{bmatrix}$$

Divide the first row by 2 and the third row by -5 to get

$$\begin{bmatrix} 1 & -1/2 & 0 & 0 & 0 \\ 0 & 1 & 1 & 2 & 100 \\ 0 & 1 & 4/5 & 0 & 24 \\ 0 & 2 & 3 & -1 & 50 \end{bmatrix}$$

Subtract twice the second row from the last row, and subtract the third row from the second row:

$$\begin{bmatrix} 1 & -1/2 & 0 & 0 & 0 \\ 0 & 1 & 1 & 2 & 100 \\ 0 & 0 & 1/5 & 2 & 76 \\ 0 & 0 & 1 & -5 & -150 \end{bmatrix}$$

Multiply the third row by 5 and subtract the result from the last row

$$\begin{bmatrix} 1 & -1/2 & 0 & 0 & 0 \\ 0 & 1 & 1 & 2 & 100 \\ 0 & 0 & 1 & 10 & 380 \\ 0 & 0 & 0 & 15 & 530 \end{bmatrix}$$

Divide the last row by 15 and then subtract multiples of rows to yield the reduced row-echelon form

$$\begin{bmatrix} 1 & 0 & 0 & 0 & 4/3 \\ 0 & 1 & 0 & 0 & 8/3 \\ 0 & 0 & 1 & 0 & 80/3 \\ 0 & 0 & 0 & 1 & 106/3 \end{bmatrix}$$

so that $x_1 = 4/3$, $x_2 = 8/3$, $x_3 = 80/3$, $x_4 = 106/3$.

(b)

$$\begin{bmatrix} 1 & 1 & 1 & 0 \\ 0 & 1 & 0 & 20 \\ -1 & 0 & 2 & 10 \end{bmatrix}$$

Add the first row to the last row:

$$\begin{bmatrix} 1 & 1 & 1 & 0 \\ 0 & 1 & 0 & 20 \\ 0 & 1 & 3 & 10 \end{bmatrix}$$

Subtract the last row from the second row:

$$\begin{bmatrix} 1 & 1 & 1 & 0 \\ 0 & 1 & 0 & 20 \\ 0 & 0 & -3 & 10 \end{bmatrix}$$

Divide the last row by -3:

$$\begin{bmatrix} 1 & 1 & 1 & 0 \\ 0 & 1 & 0 & 20 \\ 0 & 0 & 1 & -10/3 \end{bmatrix}$$

Subtract the second row and the third row from the first row to obtain the reduced row-echelon form:

$$\begin{bmatrix} 1 & 0 & 0 & -50/3 \\ 0 & 1 & 0 & 20 \\ 0 & 0 & 1 & -10/3 \end{bmatrix}$$

so that $x_1 = -50/3$, $x_2 = 20$, $x_3 = -10/3$.

5. $p_1 = 8$, $p_2 = 5$, $p_3 = 2$, $p_4 = 1$.

Review Exercises (page 268)

1. (d) and (e)

3. (a) With $x = 0$, the system reduces to

$$\begin{aligned} y - z &= 10 \\ 4y + 2z &= 4 \end{aligned}$$

with solution $y = 4$ and $z = -6$.

(b) $x = y = z = 0$ is the only solution.

(c) $x = -88/19$, $y = 30/19$ $z = 58/19$.

5. When $M = 15$ $Y = 100$ and $R = 10$. If M exceeds this amount there is an excess demand for output as shown by point B in Figure 7.R.5. When M is less than 15, there is an excess supply, with IS and LM intersecting to the left of $Y = 100$.

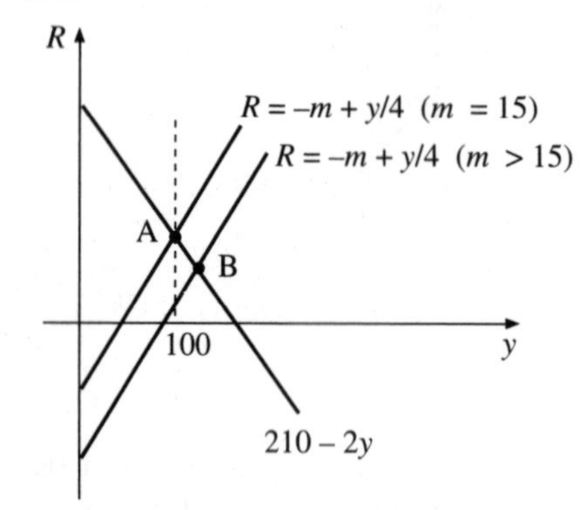

Figure 7.R.5

CHAPTER 8

Section 8.1 (page 274)

1. For

$$\begin{bmatrix} 1 & 2 \\ x-y & 2 \end{bmatrix} = \begin{bmatrix} 1 & y \\ 0 & 2 \end{bmatrix}$$

We need $x - y = 0$ and $y = 2$. Therefore, $x - 2 = 0 \longrightarrow x = 2$.

3. For

$$\begin{bmatrix} 3 & 4 & x \\ 2 & 5 & 7 \end{bmatrix} = \begin{bmatrix} 3 & 4 & y \\ 2 & 5 & 7 \end{bmatrix}$$

We need $x = y$.

5. For

$$\begin{bmatrix} 1 & 0 & 0 \\ 0 & 1 & 1 \\ 0 & 1 & 0 \end{bmatrix} = \begin{bmatrix} 1 & y & z \\ 0 & y & z \\ y & 1 & z \end{bmatrix}$$

There are no values of y and z that can make the above matrices equal since it is implied that $y = z = 0$ and $y = z = 1$, which is not consistent.

Section 8.2 (page 287)

1. If

$$A = \begin{bmatrix} 1 & 0 & 0 \\ 0 & 1 & 0 \\ 0 & 0 & 1 \end{bmatrix}$$

then $3A$ is obtained by multiplying all elements of A by 3. Hence

$$3A = \begin{bmatrix} 3 & 0 & 0 \\ 0 & 3 & 0 \\ 0 & 0 & 3 \end{bmatrix}$$

3. For

$$\mathbf{a} = [1\ 2\ 0] \quad \text{and} \quad \mathbf{b} = \begin{bmatrix} -1 \\ 0 \\ 1 \end{bmatrix}$$

$\mathbf{ab} = 1(-1) + 2(0) + 0(1) = -1$ and

$$\begin{aligned} \mathbf{ba} &= \begin{bmatrix} -1 \\ 0 \\ 1 \end{bmatrix} [1\ 2\ 0] = \begin{bmatrix} -1(1) & -1(2) & -1(0) \\ 0(1) & 0(2) & 0(0) \\ 1(1) & 1(2) & 1(0) \end{bmatrix} \\ &= \begin{bmatrix} -1 & -2 & 0 \\ 0 & 0 & 0 \\ 1 & 2 & 0 \end{bmatrix} \end{aligned}$$

5. For output quantities

$$\mathbf{q} = \begin{bmatrix} 15,000 \\ 27,000 \end{bmatrix}$$

a row price vector $\mathbf{p} = [10\ 12]$, inputs given by

$$\mathbf{z} = \begin{bmatrix} 11,000 \\ 15,000 \\ 15,000 \end{bmatrix}$$

and input prices given by the row vector $\mathbf{w} = [10\ 10\ 8]$, the firm's profit is given by

$$\begin{aligned} \mathbf{pq} - \mathbf{wz} &= \begin{bmatrix} 10 & 12 \end{bmatrix} \begin{bmatrix} 15,000 \\ 27,000 \end{bmatrix} \\ &\quad - \begin{bmatrix} 10 & 10 & 8 \end{bmatrix} \begin{bmatrix} 11,000 \\ 15,000 \\ 15,000 \end{bmatrix} \\ &= 10(15,000) + 12(27,000) \\ &\quad -[10(11,000) + 10(15,000) \\ &\quad +8(15,000)] \\ &= 150,000 + 324,000 \\ &\quad -(110,000 + 150,000 + 120,000) \\ &= 94,000 \end{aligned}$$

Section 8.3 (page 291)

1. (a) The transpose of the matrix

$$I_3 = \begin{bmatrix} 1 & 0 & 0 \\ 0 & 1 & 0 \\ 0 & 0 & 1 \end{bmatrix}$$

is the matrix I_3 itself.

(b) The transpose of the matrix

$$A = \begin{bmatrix} 1 & 2 & 3 \\ 2 & 1 & 0 \\ 3 & 0 & 1 \end{bmatrix}$$

is the matrix A itself.

3. (a) For

$$A = \begin{bmatrix} 1 & 0 & 0 \\ 0 & 0 & 1 \end{bmatrix} \quad \text{and} \quad B = \begin{bmatrix} 4 & 3 \\ 1 & 1 \\ 0 & 2 \end{bmatrix}$$

We have

$$AB = \begin{bmatrix} 1 & 0 & 0 \\ 0 & 0 & 1 \end{bmatrix} \begin{bmatrix} 4 & 3 \\ 1 & 1 \\ 0 & 2 \end{bmatrix} = \begin{bmatrix} 4 & 3 \\ 0 & 2 \end{bmatrix}$$

$$(AB)^T = \begin{bmatrix} 4 & 0 \\ 3 & 2 \end{bmatrix}$$

Also

$$B^T A^T = \begin{bmatrix} 4 & 1 & 0 \\ 3 & 0 & 2 \end{bmatrix} \begin{bmatrix} 1 & 0 \\ 0 & 0 \\ 0 & 1 \end{bmatrix} = \begin{bmatrix} 4 & 0 \\ 3 & 2 \end{bmatrix}$$

Hence, $(AB)^T = B^T A^T$.

(b) For

$$A = \begin{bmatrix} 1 & 0 & 0 \\ 0 & 0 & 1 \end{bmatrix} \quad \text{and} \quad B = \begin{bmatrix} 1 & 2 & 3 \\ 2 & 1 & 0 \\ 3 & 0 & 1 \end{bmatrix}$$

We have

$$AB = \begin{bmatrix} 1 & 0 & 0 \\ 0 & 0 & 1 \end{bmatrix} \begin{bmatrix} 1 & 2 & 3 \\ 2 & 1 & 0 \\ 3 & 0 & 1 \end{bmatrix} = \begin{bmatrix} 1 & 2 & 3 \\ 3 & 0 & 1 \end{bmatrix}$$

so

$$(AB)^T = \begin{bmatrix} 1 & 3 \\ 2 & 0 \\ 3 & 1 \end{bmatrix}$$

Also

$$B^T A^T = \begin{bmatrix} 1 & 2 & 3 \\ 2 & 1 & 0 \\ 3 & 0 & 1 \end{bmatrix} \begin{bmatrix} 1 & 0 \\ 0 & 0 \\ 0 & 1 \end{bmatrix} = \begin{bmatrix} 1 & 3 \\ 2 & 0 \\ 3 & 1 \end{bmatrix}$$

Hence, $(AB)^T = B^T A^T$.

5. If A is 3×5 and AB is 3×7, then B is 5×7, since $A_{(3\times5)} B_{(5\times7)} = AB_{(3\times7)}$.

7. If BA is 2×6, then B must have 2 rows since $B_{(2\times q)} A_{(q\times6)} = BA_{(2\times6)}$ where q is the common inner dimension of B and A.

Section 8.4 (page 296)

1. If

$$I_3 = \begin{bmatrix} 1 & 0 & 0 \\ 0 & 1 & 0 \\ 0 & 0 & 1 \end{bmatrix}$$

then it is easy to show that

$$\begin{bmatrix} 1 & 0 & 0 \\ 0 & 1 & 0 \\ 0 & 0 & 1 \end{bmatrix} \begin{bmatrix} 1 & 0 & 0 \\ 0 & 1 & 0 \\ 0 & 0 & 1 \end{bmatrix} = \begin{bmatrix} 1 & 0 & 0 \\ 0 & 1 & 0 \\ 0 & 0 & 1 \end{bmatrix}$$

3. For the matrix

$$A = \begin{bmatrix} x & -x \\ x-1 & 1-x \end{bmatrix}$$

we have to check that $AA = A$. That is

$$\begin{bmatrix} x & -x \\ x-1 & 1-x \end{bmatrix} \begin{bmatrix} x & -x \\ x-1 & 1-x \end{bmatrix}$$

$$= \begin{bmatrix} x^2 - x^2 + x & -x^2 - x + x^2 \\ x^2 - x + x - x^2 + x - 1 & -x^2 + x + 1 - 2x + x^2 \end{bmatrix}$$

$$= \begin{bmatrix} x & -x \\ x-1 & 1-x \end{bmatrix}$$

5. (a) For

$$A = \begin{bmatrix} 1/6 & -1/3 & 1/6 \\ -1/3 & 2/3 & -1/3 \\ 1/6 & -1/3 & 1/6 \end{bmatrix}$$

We can verify that A is idempotent, since

$$\begin{bmatrix} 1/6 & -1/3 & 1/6 \\ -1/3 & 2/3 & -1/3 \\ 1/6 & -1/3 & 1/6 \end{bmatrix} \times$$

$$\begin{bmatrix} 1/6 & -1/3 & 1/6 \\ -1/3 & 2/3 & -1/3 \\ 1/6 & -1/3 & 1/6 \end{bmatrix}$$

$$= \begin{bmatrix} 1/6 & -1/3 & 1/6 \\ -1/3 & 2/3 & -1/3 \\ 1/6 & -1/3 & 1/6 \end{bmatrix}$$

Therefore, $AA = A$ and $(AA)A = A$ and trace(AAA) = trace(AA) = trace$(A) = 1$.

(b) For

$$A = 1/11 \begin{bmatrix} 6 & -2 & -5 & 1 \\ -2 & 8 & -2 & -4 \\ -5 & -2 & 6 & 1 \\ 1 & -4 & 1 & 2 \end{bmatrix}$$

We can verify that A is idempotent (see Exercise 2). Therefore, $AAA = AA = A$ and trace(AAA) = trace(AA) = trace$(A) = 2$.

Review Exercises (page 297)

1. (a) AB is 2×2.

(b) $A^T B$ is a scalar.

(c) $A^T BA$ is a scalar.

(d) $AA^T B$ is 5×5.

3. For

$$A = \begin{bmatrix} 1 & -3 \\ -2 & 4 \end{bmatrix} \quad \text{and} \quad \mathbf{x} = \begin{bmatrix} 5 \\ 3 \end{bmatrix}$$

$$A\mathbf{x} = \begin{bmatrix} 1 & -3 \\ -2 & 4 \end{bmatrix} \begin{bmatrix} 5 \\ 3 \end{bmatrix} = \begin{bmatrix} -4 \\ 2 \end{bmatrix}$$

$$(A\mathbf{x})^T = \begin{bmatrix} -4 & 2 \end{bmatrix}$$

$$\mathbf{x}^T A^T = \begin{bmatrix} 5 & 3 \end{bmatrix} \begin{bmatrix} 1 & -2 \\ -3 & 4 \end{bmatrix} = \begin{bmatrix} -4 & 2 \end{bmatrix}$$

$$\mathbf{x}\mathbf{x}^T = \begin{bmatrix} 5 \\ 3 \end{bmatrix} \begin{bmatrix} 5 & 3 \end{bmatrix} = \begin{bmatrix} 25 & 15 \\ 15 & 9 \end{bmatrix}$$

$$\mathbf{x}^T\mathbf{x} = \begin{bmatrix} 5 & 3 \end{bmatrix} \begin{bmatrix} 5 \\ 3 \end{bmatrix} = 34$$

$A^T_{(2\times 2)}\mathbf{x}^T_{(1\times 2)}$ is not defined, since there is no common inner dimension.

5. For

$$A = \begin{bmatrix} 3 & -4 \\ -5 & 1 \end{bmatrix} \quad \text{and} \quad B = \begin{bmatrix} 7 & 4 \\ 5 & k \end{bmatrix}$$

$$AB = \begin{bmatrix} 3 & -4 \\ -5 & 1 \end{bmatrix}\begin{bmatrix} 7 & 4 \\ 5 & k \end{bmatrix} = \begin{bmatrix} 1 & 12-4k \\ -30 & -20+k \end{bmatrix}$$

$$BA = \begin{bmatrix} 7 & 4 \\ 5 & k \end{bmatrix}\begin{bmatrix} 3 & -4 \\ -5 & 1 \end{bmatrix} = \begin{bmatrix} 1 & -24 \\ 15-5k & -20+k \end{bmatrix}$$

For $AB = BA$ we need $12-4k = -24$ and $15-5k = -30$. We can see that $k = 9$ satisfies both of these.

7. For Exercise 6 of Section 8.2, we have that (to two decimal places accuracy)

$$P^3 = \begin{bmatrix} 0.56 & 0.28 & 0.13 \\ 0.18 & 0.39 & 0.11 \\ 0.26 & 0.33 & 0.76 \end{bmatrix}$$

Then, $P^4 = P^3P$ is given by

$$\begin{aligned} P^4 &= \begin{bmatrix} 0.56 & 0.28 & 0.13 \\ 0.18 & 0.39 & 0.11 \\ 0.26 & 0.33 & 0.76 \end{bmatrix} \times \\ &\quad \begin{bmatrix} 0.80 & 0.15 & 0.05 \\ 0.10 & 0.70 & 0.05 \\ 0.10 & 0.15 & 0.90 \end{bmatrix} \\ &= \begin{bmatrix} 0.49 & 0.30 & 0.16 \\ 0.20 & 0.32 & 0.13 \\ 0.32 & 0.38 & 0.71 \end{bmatrix} \end{aligned}$$

Then, $\mathbf{x}^4 = P^4\mathbf{x}^0$ is obtained as

$$\begin{aligned} \mathbf{x}^4 &= \begin{bmatrix} 0.49 & 0.30 & 0.16 \\ 0.20 & 0.32 & 0.13 \\ 0.32 & 0.38 & 0.71 \end{bmatrix}\begin{bmatrix} 5 \\ 10 \\ 6 \end{bmatrix} \\ &= \begin{bmatrix} 6.41 \\ 4.98 \\ 9.66 \end{bmatrix} \end{aligned}$$

CHAPTER 9

Section 9.1 (page 312)

1. (a) The inverse of

$$\begin{bmatrix} 5 & 0 \\ 0 & 3 \end{bmatrix} \text{ is } \begin{bmatrix} 1/5 & 0 \\ 0 & 1/3 \end{bmatrix}$$

(b) The inverse of

$$\begin{bmatrix} 3 & 2 \\ -1 & 1 \end{bmatrix} \text{ is } \frac{1}{5}\begin{bmatrix} 1 & -2 \\ 1 & 3 \end{bmatrix}$$

where 5 is the determinant.

(c) The inverse does not exist, since the rows and columns are linearly dependent.

3. (a) We need to solve $\mathbf{z} = A\mathbf{y}$

$$\begin{bmatrix} z_1 \\ z_2 \end{bmatrix} = \begin{bmatrix} 2 & 1 \\ 3 & 4 \end{bmatrix}\begin{bmatrix} 5 \\ 10 \end{bmatrix} = \begin{bmatrix} 20 \\ 55 \end{bmatrix}$$

(b) $\mathbf{w}^T A\mathbf{y} = \mathbf{w}^T\mathbf{z}$ is a scalar and represents total costs, in our case

$$\begin{bmatrix} 5 & 10 \end{bmatrix}\begin{bmatrix} 20 \\ 55 \end{bmatrix} = 650$$

5. The IS curve is obtained from $Y = C + I + G + X$

$$\begin{aligned} Y &= 15 + 0.8Y - 0.8(-25 + 0.25Y) \\ &\quad +65 - R + 94 + 40 - 0.1Y \\ &= 15 + 20 + 65 + 94 + 40 + 0.8Y \\ &\quad -0.2Y - 0.1Y - R \\ 0.5Y &= 234 - R \\ Y &= 468 - 2R \end{aligned}$$

The LM is obtained as

$$\begin{aligned} 5Y - 50R &= 1500 \\ Y &= 300 + 10R \end{aligned}$$

Equilibrium Y and R are obtained from solving

$$\begin{aligned} Y &= 468 - 2R \\ Y &= 300 + 10R \end{aligned}$$

or $300 + 10R = 468 - 2R$, so that $R^* = 14$. Then $Y^* = 300 + 10(14) = 440$. The budget deficit in equilibrium is given by

$$G - T = 94 - [-25 + 0.25(440)] = 94 + 25 - 110 = 9$$

The trade surplus is $X = 50 - 10 - 0.1(440) = -4.4$. This is a deficit.

Section 9.2 (page 317)

1. (a) For

$$A = \begin{bmatrix} 3 & 0 & 4 \\ 2 & 3 & 2 \\ 0 & 5 & -1 \end{bmatrix}$$

Expanding along the first column:

$$|A| = 3\begin{vmatrix} 3 & 2 \\ 5 & -1 \end{vmatrix} - 2\begin{vmatrix} 0 & 4 \\ 5 & -1 \end{vmatrix} = -39 - 2(-20) = 1$$

(b) For

$$B = \begin{bmatrix} 2 & -4 & 3 \\ 3 & 1 & 2 \\ 1 & 4 & -1 \end{bmatrix}$$

Expanding along the first column:

$$\begin{aligned} |B| &= 2\begin{vmatrix} 1 & 2 \\ 4 & -1 \end{vmatrix} - 3\begin{vmatrix} -4 & 3 \\ 4 & -1 \end{vmatrix} + \begin{bmatrix} -4 & 3 \\ 1 & 2 \end{bmatrix} \\ &= -18 + 24 - 11 = -5 \end{aligned}$$

(c) For

$$C = \begin{bmatrix} 2 & 3 & -4 \\ 4 & 0 & 5 \\ 5 & 1 & 6 \end{bmatrix}$$

Expanding along the second column:

$$|C| = -3\begin{vmatrix} 4 & 5 \\ 5 & 6 \end{vmatrix} - 1\begin{vmatrix} 2 & -4 \\ 4 & 5 \end{vmatrix} = 3 - 26 = -23$$

(d) For

$$D = \begin{bmatrix} 4 & 3 & 0 \\ 6 & 5 & 2 \\ 9 & 7 & 3 \end{bmatrix}$$

Expanding along the first row:

$$|D| = 4\begin{vmatrix} 5 & 2 \\ 7 & 3 \end{vmatrix} - 3\begin{vmatrix} 6 & 2 \\ 9 & 3 \end{vmatrix} = 4$$

3. We find the determinant of

$$B = \begin{bmatrix} a & b & c \\ d & e & f \\ 3g & 3h & 3i \end{bmatrix}$$

if

$$|A| = \begin{vmatrix} a & b & c \\ d & e & f \\ g & h & i \end{vmatrix} = 7$$

Using Theorem 9.7, $|B| = 3|A| = 21$.

5. We find the determinant of

$$D = \begin{bmatrix} a & b & c \\ 3d+a & 3e+b & 3f+c \\ g & h & i \end{bmatrix}$$

if

$$|A| = \begin{vmatrix} a & b & c \\ d & e & f \\ g & h & i \end{vmatrix} = 3$$

Using Theorem 9.5 and 9.7, $|D| = 3|A| = 9$.

7. $|A^3| = |A||A^2| = |A||A||A|$. But

$$|A| = \begin{vmatrix} 1 & 0 & 1 \\ 1 & 1 & 2 \\ 1 & 2 & 1 \end{vmatrix} = \begin{vmatrix} 1 & 2 \\ 2 & 1 \end{vmatrix} + \begin{vmatrix} 1 & 1 \\ 1 & 2 \end{vmatrix}$$
$$= -3 + 1 = -2$$

Therefore $|A^3| = (-2)^3 = -8$.

Section 9.3 (page 326)

1. (a) For the matrix

$$\begin{bmatrix} 1 & 2 & 3 \\ 0 & 1 & 2 \\ 0 & 0 & 1 \end{bmatrix}$$

The determinant is

$$\begin{vmatrix} 1 & 2 & 3 \\ 0 & 1 & 2 \\ 0 & 0 & 1 \end{vmatrix} = 1\begin{vmatrix} 1 & 2 \\ 0 & 1 \end{vmatrix} = 1$$

The cofactors are given by

$$C_{11} = \begin{vmatrix} 1 & 2 \\ 0 & 1 \end{vmatrix} = 1 \qquad C_{21} = -\begin{vmatrix} 2 & 3 \\ 0 & 1 \end{vmatrix} = -2$$

$$C_{12} = -\begin{vmatrix} 0 & 2 \\ 0 & 1 \end{vmatrix} = 0 \qquad C_{22} = \begin{vmatrix} 1 & 3 \\ 0 & 1 \end{vmatrix} = 1$$

$$C_{31} = \begin{vmatrix} 2 & 3 \\ 1 & 2 \end{vmatrix} = 1 \qquad C_{13} = \begin{vmatrix} 0 & 1 \\ 0 & 0 \end{vmatrix} = 0$$

$$C_{32} = -\begin{vmatrix} 1 & 3 \\ 0 & 2 \end{vmatrix} = -2 \qquad C_{23} = -\begin{vmatrix} 1 & 2 \\ 0 & 1 \end{vmatrix} = 0$$

$$C_{33} = \begin{vmatrix} 1 & 2 \\ 0 & 1 \end{vmatrix} = 1$$

so

$$\text{adj}(C) = \begin{bmatrix} 1 & -2 & 1 \\ 0 & 1 & -2 \\ 0 & 0 & 1 \end{bmatrix}$$

and because the determinant is 1, the adjoint matrix is the inverse.

(b) For the matrix

$$\begin{bmatrix} 1 & 2 & -1 \\ 0 & 1 & 0 \\ -5 & 2 & 3 \end{bmatrix}$$

The determinant is given by

$$\begin{vmatrix} 1 & 2 & -1 \\ 0 & 1 & 0 \\ -5 & 2 & 3 \end{vmatrix} = \begin{vmatrix} 1 & 0 \\ 2 & 3 \end{vmatrix} - 5\begin{vmatrix} 2 & -1 \\ 1 & 0 \end{vmatrix} = 3 - 5 = -2$$

The cofactors are given by

$$C_{11} = \begin{vmatrix} 1 & 0 \\ 2 & 3 \end{vmatrix} = 3 \qquad C_{21} = -\begin{vmatrix} 2 & -1 \\ 2 & 3 \end{vmatrix} = -8$$

$$C_{12} = -\begin{vmatrix} 0 & 0 \\ -5 & 3 \end{vmatrix} = 0 \qquad C_{22} = \begin{vmatrix} 1 & -1 \\ -5 & 3 \end{vmatrix} = -2$$

$$C_{31} = \begin{vmatrix} 2 & -1 \\ 1 & 0 \end{vmatrix} = 1 \qquad C_{13} = \begin{vmatrix} 0 & 1 \\ -5 & 2 \end{vmatrix} = 5$$

$$C_{32} = -\begin{vmatrix} 1 & -1 \\ 0 & 0 \end{vmatrix} = 0 \qquad C_{23} = -\begin{vmatrix} 1 & 2 \\ -5 & 2 \end{vmatrix} = -12$$

$$C_{33} = \begin{vmatrix} 1 & 2 \\ 0 & 1 \end{vmatrix} = 1$$

so

$$\text{adj}(C) = \begin{bmatrix} 3 & -8 & 1 \\ 0 & -2 & 0 \\ 5 & -12 & 1 \end{bmatrix}$$

and

$$C^{-1} = \begin{bmatrix} -3/2 & 4 & -1/2 \\ 0 & 1 & 0 \\ -5/2 & 6 & -1/2 \end{bmatrix}$$

(c) For the matrix

$$\begin{bmatrix} 3 & 0 & 0 \\ 0 & -1 & 0 \\ 0 & 0 & 3 \end{bmatrix}$$

The determinant is given by

$$\begin{vmatrix} 3 & 0 & 0 \\ 0 & -1 & 0 \\ 0 & 0 & 3 \end{vmatrix} = 3\begin{vmatrix} -1 & 0 \\ 0 & 3 \end{vmatrix} = -9$$

The cofactors are given by

$$C_{11} = \begin{vmatrix} -1 & 0 \\ 0 & 3 \end{vmatrix} = -3 \qquad C_{21} = -\begin{vmatrix} 0 & 0 \\ 0 & 3 \end{vmatrix} = 0$$

$$C_{12} = -\begin{vmatrix} 0 & 0 \\ 0 & 3 \end{vmatrix} = 0 \qquad C_{22} = \begin{vmatrix} 3 & 0 \\ 0 & 3 \end{vmatrix} = 9$$

$$C_{31} = \begin{vmatrix} 0 & 0 \\ -1 & 0 \end{vmatrix} = 0 \qquad C_{13} = \begin{vmatrix} 0 & -1 \\ 0 & 0 \end{vmatrix} = 0$$

$$C_{32} = -\begin{vmatrix} 0 & 0 \\ -1 & 0 \end{vmatrix} = 0 \qquad C_{23} = -\begin{vmatrix} 3 & 0 \\ 0 & 0 \end{vmatrix} = 0$$

$$C_{33} = \begin{vmatrix} 3 & 0 \\ 0 & -1 \end{vmatrix} = -3$$

so

$$\text{adj}(C) = \begin{bmatrix} -3 & 0 & 0 \\ 0 & 9 & 0 \\ 0 & 0 & -3 \end{bmatrix}$$

and

$$C^{-1} = \begin{bmatrix} 1/3 & 0 & 0 \\ 0 & -1 & 0 \\ 0 & 0 & 1/3 \end{bmatrix}$$

3. With

$$A = \begin{bmatrix} 1 & 0 & 5 \\ 1 & 1 & 0 \\ 3 & 2 & 6 \end{bmatrix} \quad \text{and} \quad \mathbf{y} = \begin{bmatrix} 5 \\ 5 \\ 10 \end{bmatrix}$$

we need to find $\mathbf{z} = A\mathbf{y}$

$$\begin{bmatrix} z_1 \\ z_2 \\ z_3 \end{bmatrix} = \begin{bmatrix} 1 & 0 & 5 \\ 1 & 1 & 0 \\ 3 & 2 & 6 \end{bmatrix} \begin{bmatrix} 5 \\ 5 \\ 10 \end{bmatrix} = \begin{bmatrix} 55 \\ 10 \\ 85 \end{bmatrix}$$

5. (a)

$$|A| = \begin{vmatrix} 1 & 5 & -6 \\ -1 & -4 & 4 \\ -2 & -7 & 9 \end{vmatrix}$$

Add row 1 to row 3

$$|A| = \begin{vmatrix} 1 & 5 & -6 \\ -1 & -4 & 4 \\ -1 & -2 & 3 \end{vmatrix}$$

Add row 1 to row 2

$$|A| = \begin{vmatrix} 1 & 5 & -6 \\ 0 & 1 & -2 \\ -1 & -2 & 3 \end{vmatrix}$$

Add row 1 to row 3

$$|A| = \begin{bmatrix} 1 & 5 & -6 \\ 0 & 1 & -2 \\ 0 & 3 & -3 \end{bmatrix} = \begin{vmatrix} 1 & -2 \\ 3 & -3 \end{vmatrix} = 3$$

(b)

$$|B| = \begin{vmatrix} 1 & 3 & 0 & 2 \\ -2 & -5 & 7 & 4 \\ 3 & 5 & 2 & 1 \\ -1 & 0 & -9 & -5 \end{vmatrix}$$

Add row 2 to row 3

$$|B| = \begin{vmatrix} 1 & 3 & 0 & 2 \\ -2 & -5 & 7 & 4 \\ 1 & 0 & 9 & 5 \\ -1 & 0 & -9 & -5 \end{vmatrix}$$

Add row 3 to row 4

$$|B| = \begin{vmatrix} 1 & 3 & 0 & 2 \\ -2 & -5 & 7 & 4 \\ 1 & 0 & 9 & 5 \\ 0 & 0 & 0 & 0 \end{vmatrix} = 0$$

(c)

$$|C| = \begin{vmatrix} 1 & -1 & -3 & 0 \\ 0 & 1 & 0 & 4 \\ -1 & 2 & 8 & 5 \\ -1 & -1 & -2 & 3 \end{vmatrix}$$

Add row 1 to row 3

$$|C| = \begin{vmatrix} 1 & -1 & -3 & 0 \\ 0 & 1 & 0 & 4 \\ 0 & 1 & 5 & 5 \\ -1 & -1 & -2 & 3 \end{vmatrix}$$

Add row 1 to row 4

$$|C| = \begin{vmatrix} 1 & -1 & -3 & 0 \\ 0 & 1 & 0 & 4 \\ 0 & 1 & 5 & 5 \\ 0 & -2 & -5 & 3 \end{vmatrix}$$

Add row 2 to row 3

$$|C| = \begin{vmatrix} 1 & -1 & -3 & 0 \\ 0 & 1 & 0 & 4 \\ 0 & 2 & 5 & 9 \\ 0 & -2 & -5 & 3 \end{vmatrix}$$

Add row 3 to row 4

$$|C| = \begin{vmatrix} 1 & -1 & -3 & 0 \\ 0 & 1 & 0 & 4 \\ 0 & 2 & 5 & 9 \\ 0 & 0 & 0 & 12 \end{vmatrix} = \begin{vmatrix} 1 & 0 & 4 \\ 2 & 5 & 9 \\ 0 & 0 & 12 \end{vmatrix}$$

$$= 12 \begin{vmatrix} 1 & 0 \\ 2 & 5 \end{vmatrix} = 60$$

Section 9.4 (page 339)

1.

$$A = \begin{bmatrix} 2 & 1 & 0 \\ -3 & 0 & 1 \\ 0 & 1 & 2 \end{bmatrix} \quad \mathbf{x} = \begin{bmatrix} x_1 \\ x_2 \\ x_3 \end{bmatrix} \quad \mathbf{b} = \begin{bmatrix} 7 \\ -8 \\ -3 \end{bmatrix}$$

where

$$|A| = -\begin{vmatrix} -3 & 1 \\ 0 & 2 \end{vmatrix} - \begin{vmatrix} 2 & 0 \\ -3 & 1 \end{vmatrix} = 6 - 2 = 4$$

Now

$$A_1 = \begin{bmatrix} 7 & 1 & 0 \\ -8 & 0 & 1 \\ -3 & 1 & 2 \end{bmatrix}$$

so

$$|A_1| = -\begin{vmatrix} -8 & 1 \\ -3 & 2 \end{vmatrix} - \begin{vmatrix} 7 & 0 \\ -8 & 1 \end{vmatrix} = 13 - 7 = 6$$

Hence, $x_1 = 6/4 = 1.5$. Similarly

$$A_2 = \begin{bmatrix} 2 & 7 & 0 \\ -3 & -8 & 1 \\ 0 & -3 & 2 \end{bmatrix}$$

so

$$|A_2| = 2\begin{vmatrix} -8 & 1 \\ -3 & 2 \end{vmatrix} + 3\begin{vmatrix} 7 & 0 \\ -3 & 2 \end{vmatrix} = -26 + 42 = 16$$

Hence, $x_2 = 16/4 = 4$. Finally

$$A_3 = \begin{bmatrix} 2 & 1 & 7 \\ -3 & 0 & -8 \\ 0 & 1 & -3 \end{bmatrix}$$

so

$$|A_3| = -\begin{vmatrix} -3 & -8 \\ 0 & -3 \end{vmatrix} - \begin{vmatrix} 2 & 7 \\ -3 & -8 \end{vmatrix} = -9 - 5 = -14$$

Hence, $x_3 = -14/4$.

3.

$$A = \begin{bmatrix} 2 & 4 & -1 \\ 1 & -3 & 2 \\ 6 & 5 & 1 \end{bmatrix} \quad \mathbf{x} = \begin{bmatrix} x_1 \\ x_2 \\ x_3 \end{bmatrix} \quad \mathbf{b} = \begin{bmatrix} 3 \\ -1 \\ 5 \end{bmatrix}$$

so

$$\begin{aligned} |A| &= 2\begin{vmatrix} -3 & 2 \\ 5 & 1 \end{vmatrix} - \begin{vmatrix} 4 & -1 \\ 5 & 1 \end{vmatrix} + 6\begin{vmatrix} 4 & -1 \\ -3 & 2 \end{vmatrix} \\ &= 2(-13) - 9 + 6(5) = -5 \end{aligned}$$

Now

$$A_1 = \begin{bmatrix} 3 & 4 & -1 \\ -1 & -3 & 2 \\ 5 & 5 & 1 \end{bmatrix}$$

so

$$\begin{aligned} |A_1| &= 3\begin{vmatrix} -3 & 2 \\ 5 & 1 \end{vmatrix} + \begin{vmatrix} 4 & -1 \\ 5 & 1 \end{vmatrix} + 5\begin{vmatrix} 4 & -1 \\ -3 & 2 \end{vmatrix} \\ &= 3(-13) + 9 + 5(5) = -5 \end{aligned}$$

Hence, $x_1 = 5/5 = 1$. Similarly

$$A_2 = \begin{bmatrix} 2 & 3 & 1 \\ 1 & -1 & 2 \\ 6 & 5 & 1 \end{bmatrix}$$

so

$$\begin{aligned} |A_2| &= 2\begin{vmatrix} -1 & 2 \\ 5 & 1 \end{vmatrix} - \begin{vmatrix} 3 & -1 \\ 5 & 1 \end{vmatrix} + 6\begin{vmatrix} 3 & -1 \\ -1 & 2 \end{vmatrix} \\ &= 2(-11) - 8 + 6(5) = 0 \end{aligned}$$

Hence, $x_2 = 0$. Finally

$$A_3 = \begin{bmatrix} 2 & 4 & 3 \\ 1 & -3 & -1 \\ 6 & 5 & 5 \end{bmatrix}$$

so

$$\begin{aligned} |A_3| &= 2\begin{vmatrix} -3 & -1 \\ 5 & 5 \end{vmatrix} - \begin{vmatrix} 4 & 3 \\ 5 & 5 \end{vmatrix} + \begin{vmatrix} 4 & 3 \\ -3 & -1 \end{vmatrix} \\ &= 2(-10) - 5 + 6(5) = 5 \end{aligned}$$

Hence, $x_3 = 5/5 = 1$.

5. The IS-LM model is summerized as $A\mathbf{x} = \mathbf{b}$ or

$$\begin{bmatrix} 1 & -1 & -1 & 0 \\ -b(1-t) & 1 & 0 & 0 \\ 0 & 0 & 1 & l \\ k & 0 & 0 & -h \end{bmatrix} \begin{bmatrix} Y \\ C \\ I \\ R \end{bmatrix} = \begin{bmatrix} G \\ a \\ e \\ \bar{M} \end{bmatrix}$$

The determinant of the above matrix in Section 9.4 was found to be $|A| = -h(1 - b(1-t)) - lk$. We have

$$A_1 = \begin{bmatrix} G & -1 & -1 & 0 \\ a & 1 & 0 & 0 \\ e & 0 & 1 & l \\ \bar{M} & 0 & 0 & -h \end{bmatrix}$$

so

$$\begin{aligned} |A_1| &= \begin{vmatrix} a & 0 & 0 \\ l & 1 & l \\ \bar{M} & 0 & -h \end{vmatrix} + \begin{vmatrix} G & -1 & 0 \\ e & 1 & l \\ \bar{M} & 0 & -h \end{vmatrix} \\ &= a\begin{vmatrix} 1 & l \\ 0 & -h \end{vmatrix} + G\begin{vmatrix} 1 & l \\ 0 & -h \end{vmatrix} + \begin{vmatrix} e & l \\ \bar{M} & -h \end{vmatrix} \\ &= -ah - Gh - (eh + \bar{M}l) \end{aligned}$$

Therefore

$$Y = \frac{ah + Gh + eh + \bar{M}l}{h(1 - b(1-t)) + lk}$$

Also

$$A_2 = \begin{bmatrix} 1 & G & -1 & 0 \\ -b(1-t) & a & 0 & 0 \\ 0 & e & 1 & l \\ k & \bar{M} & 0 & -h \end{bmatrix}$$

so, expanding along the third column

$$\begin{aligned} |A_2| &= -\begin{vmatrix} -b(1-t) & a & 0 \\ 0 & e & l \\ k & \bar{M} & -h \end{vmatrix} + \\ &\quad \begin{vmatrix} 1 & G & 0 \\ -b(1-t) & a & 0 \\ k & \bar{M} & -h \end{vmatrix} \\ &= -[-b(1-t)]\begin{vmatrix} e & l \\ \bar{M} & -h \end{vmatrix} - k\begin{vmatrix} a & 0 \\ e & l \end{vmatrix} \\ &\quad -h\begin{vmatrix} 1 & G \\ -b(1-t) & a \end{vmatrix} \\ &= -b(1-t)(eh + \bar{M}l) - kal - h[a + Gb(1-t)] \end{aligned}$$

Therefore

$$C = \frac{b(1-t)(eh + \bar{M}l) + kal + h[a + Gb(1-t)]}{h[1 - b(1-t)] + lk}$$

Review Exercises (page 340)

1. If

$$A = \begin{bmatrix} 12 & 13 & 14 \\ 15 & 16 & 17 \\ 18 & 19 & 20 \end{bmatrix}$$

Subtract column 1 from column 2

$$\begin{bmatrix} 12 & 1 & 14 \\ 15 & 1 & 17 \\ 18 & 1 & 20 \end{bmatrix}$$

Subtract column 1 from column 3

$$\begin{bmatrix} 12 & 1 & 2 \\ 15 & 1 & 2 \\ 18 & 1 & 2 \end{bmatrix}$$

Multiply column 2 by -2 and add the product to column 3

$$\begin{bmatrix} 12 & 1 & 0 \\ 15 & 1 & 0 \\ 18 & 1 & 0 \end{bmatrix}$$

Clearly the above has a zero determinant and hence no inverse.

3.

$$A = \begin{bmatrix} 1 & 5 & 0 \\ 2 & 4 & -1 \\ 0 & -2 & 0 \end{bmatrix}$$

so, expanding along the first column

$$|A| = \begin{vmatrix} 4 & -1 \\ -2 & 0 \end{vmatrix} - 2\begin{vmatrix} 5 & 0 \\ -2 & 0 \end{vmatrix} = -2$$

5. The system $A\mathbf{x} = \mathbf{b}$ is given by

$$\begin{bmatrix} 2 & 1 & 0 \\ -3 & 0 & 1 \\ 0 & 1 & 2 \end{bmatrix}\begin{bmatrix} x_1 \\ x_2 \\ x_3 \end{bmatrix} = \begin{bmatrix} 3 \\ -8 \\ -2 \end{bmatrix}$$

we have

$$|A| = 2\begin{vmatrix} 0 & 1 \\ 1 & 2 \end{vmatrix} + 3\begin{vmatrix} 1 & 0 \\ 1 & 2 \end{vmatrix} = -2 + 6 = 4$$

Since

$$A_1 = \begin{bmatrix} 3 & 1 & 0 \\ -8 & 0 & 1 \\ -2 & 1 & 2 \end{bmatrix}$$

we have

$$|A_1| = -\begin{vmatrix} -8 & 1 \\ -2 & 2 \end{vmatrix} - \begin{vmatrix} 3 & 0 \\ -8 & 1 \end{vmatrix} = 14 - 3 = 11$$

Hence, $x_1 = |A_1|/|A| = 11/4$. Similarly

$$A_2 = \begin{bmatrix} 2 & 3 & 0 \\ -3 & -8 & 1 \\ 0 & -2 & 2 \end{bmatrix}$$

so

$$|A_2| = -\begin{vmatrix} 2 & 3 \\ 0 & -2 \end{vmatrix} + 2\begin{vmatrix} 2 & 3 \\ -3 & -8 \end{vmatrix} + 4 - 14 = -10$$

Hence, $x_2 = |A_2|/|A| = -10/4$. Finally

$$A_3 = \begin{bmatrix} 2 & 1 & 3 \\ -3 & 0 & -8 \\ 0 & 1 & -2 \end{bmatrix}$$

so

$$|A_3| = -\begin{vmatrix} -3 & -8 \\ 0 & -2 \end{vmatrix} - \begin{vmatrix} 2 & 3 \\ -3 & -8 \end{vmatrix} = -6 + 7 = 1$$

Hence, $x_3 = |A_3|/|A| = 1/4$.

7. Since $A^T A = I$, using Theorem 9.1 and 9.8 we have $|A^T A| = |I|$, $|A|^2 = 1$, so $|A| = \pm 1$.

9. $A^3 = AAA$, then $|A^3| = |AAA| = |A||A||A|$ by Theorem 9.8. So $|A^3| = 0$ if and only if $|A| = 0$. Hence, A has no inverse.

CHAPTER 10*

Section 10.1 (page 354)

1. The length of $\mathbf{y}$ is given by

$$\|\mathbf{y}\| = \sqrt{1^2 + 2^2 + (-1)^2} = \sqrt{6}$$

The length of $\mathbf{w}$ is given by

$$\|\mathbf{w}\| = \sqrt{(-1)^2 + (-2)^2 + 1^2} = \sqrt{6}$$

The length of $\mathbf{z}$ is given by

$$\|z\| = \sqrt{0 + 0 + (-1)^2} = 1$$

The length of $\mathbf{v}$ is given by

$$\|\mathbf{v}\| = \sqrt{(1/3)^2 + (1/3)^2 + (1/3)^2} = \frac{1}{\sqrt{3}}$$

3. (a) The vectors $\mathbf{e}_1$, $\mathbf{e}_2$, $\mathbf{e}_3$ and $\mathbf{e}_4$ are linearly independent since

$$\lambda_1\mathbf{e}_1 + \lambda_2\mathbf{e}_2 + \lambda_3\mathbf{e}_3 + \lambda_4\mathbf{e}_4 = \mathbf{0}$$

implies that $\lambda_1 = \lambda_2 = \lambda_3 = \lambda_4 = 0$.

(b) The vectors $\mathbf{v}_1$, $\mathbf{v}_2$, $\mathbf{v}_3$, and $\mathbf{v}_4$ are linearly independent since

$$\lambda_1\mathbf{v}_1 + \lambda_2\mathbf{v}_2 + \lambda_3\mathbf{v}_3 + \lambda_4\mathbf{v}_4 = \mathbf{0}$$

implies that $\lambda_1 = \lambda_2 = \lambda_3 = \lambda_4 = 0$.

5. (a) The pair $\mathbf{y} = [1\,0\,{-1}]^T$ and $\mathbf{w} = [1\,0\,1]^T$ are orthogonal vectors, since

$$\mathbf{y}^T\mathbf{w} = \begin{bmatrix} 1 & 0 & -1 \end{bmatrix}\begin{bmatrix} 1 \\ 0 \\ 1 \end{bmatrix} = 1 - 1 = 0$$

(b) The pair $\mathbf{y} = [0\,0-1]^T$ and $\mathbf{w} = [3\,1\,0]^T$ are orthogonal vectors since

$$\mathbf{y}^T\mathbf{w} = \begin{bmatrix} 0 & 0 & -1 \end{bmatrix}\begin{bmatrix} 3 \\ 1 \\ 0 \end{bmatrix} = 0$$

(c) The pair $\mathbf{y} = [1\;2]^T$ and $\mathbf{w} = [-2\;1]^T$ are orthogonal vectors since

$$\mathbf{y}^T\mathbf{w} = \begin{bmatrix} 1 & 2 \end{bmatrix}\begin{bmatrix} -2 \\ 1 \end{bmatrix} = 0$$

7. (a) Since $\mathcal{V}$ describes the positive quadrant in the xy-plane, then $\mathbf{u} + \mathbf{v}$ will be also in $\mathcal{V}$ if $\mathbf{u}$ and $\mathbf{v}$ are already in $\mathcal{V}$, since $\mathbf{u} + \mathbf{v}$ will be non-negative.

(b) If $\mathbf{u}$ belongs to $\mathcal{V}$, then if λ is a negative number, then $\lambda\mathbf{u}$ will not belong to $\mathcal{V}$ since it is negative.

Because of (b) we conclude that $\mathcal{V}$ is not a vector space.

9. The rank of

$$A = \begin{bmatrix} 1 & 0 & 1 \\ 0 & 1 & 0 \\ 1 & 0 & 0 \end{bmatrix}$$

can be obtained by looking at the linear dependence or independence of its columns (rows). In this case, since

$$\lambda_1\begin{bmatrix} 1 \\ 0 \\ 1 \end{bmatrix} + \lambda_2\begin{bmatrix} 0 \\ 1 \\ 0 \end{bmatrix} + \lambda_3\begin{bmatrix} 1 \\ 0 \\ 0 \end{bmatrix} = \begin{bmatrix} 0 \\ 0 \\ 0 \end{bmatrix}$$

implies that $\lambda_1 = \lambda_2 = \lambda_3 = 0$, the columns of A are linearly independent and its rank is 3.

The rank of

$$B = \begin{bmatrix} 1 & 0 & 1 & 0 \\ 0 & 1 & 0 & 1 \\ 1 & 0 & 0 & 1 \end{bmatrix}$$

can be also seen to be 3. Since among the four columns of B, there are only three linearly independent ones. The fourth column can be obtained as a linear combination of the other three.

Section 10.2 (page 367)

1. For the matrix

$$A = \begin{bmatrix} 2 & 1 \\ 1 & 2 \end{bmatrix}$$

(a) The characteristic equation is given by

$$|A - \lambda I| = \begin{bmatrix} 2-\lambda & 1 \\ 1 & 2-\lambda \end{bmatrix} = (2-\lambda)^2 - 1 = 0$$

The roots are $\lambda_1 = 3$, $\lambda_2 = 1$.

(b) Corresponding to $\lambda_1 = 3$, we have

$$\begin{bmatrix} -1 & 1 \\ 1 & -1 \end{bmatrix}\begin{bmatrix} q_1 \\ q_2 \end{bmatrix} = \begin{bmatrix} 0 \\ 0 \end{bmatrix}$$

So $q_1 = q_2$. By normalization $q_1^2 + q_2^2 = 1$, we obtain $2q_1^2 = 1$. So $q_1 = 1/\sqrt{2}$ (discard the negative root). Hence

$$q_1 = \begin{bmatrix} 1/\sqrt{2} \\ 1/\sqrt{2} \end{bmatrix}$$

Similarly for $\lambda_2 = 1$, we have

$$\begin{bmatrix} 1 & 1 \\ 1 & 1 \end{bmatrix}\begin{bmatrix} q_1 \\ q_2 \end{bmatrix} = \begin{bmatrix} 0 \\ 0 \end{bmatrix}$$

So $q_1 = -q_2$. By normalization $q_1^2 + q_2^2 = 1$ which leads to

$$q_1 = \begin{bmatrix} 1/\sqrt{2} \\ -1/\sqrt{2} \end{bmatrix}$$

(c) The orthogonalization takes place since $Q^T A Q = \Lambda$, where

$$Q = \begin{bmatrix} 1/\sqrt{2} & 1/\sqrt{2} \\ 1/\sqrt{2} & -1/\sqrt{2} \end{bmatrix} \quad Q^T = \begin{bmatrix} 1/\sqrt{2} & 1/\sqrt{2} \\ 1/\sqrt{2} & -1/\sqrt{2} \end{bmatrix}$$

$$\Lambda = \begin{bmatrix} 3 & 0 \\ 0 & 1 \end{bmatrix}$$

It can be seen that

$$\begin{bmatrix} 1/\sqrt{2} & 1/\sqrt{2} \\ 1/\sqrt{2} & -1/\sqrt{2} \end{bmatrix} \begin{bmatrix} 2 & 1 \\ 1 & 2 \end{bmatrix} \begin{bmatrix} 1/\sqrt{2} & 1/\sqrt{2} \\ 1/\sqrt{2} & -1/\sqrt{2} \end{bmatrix} = \begin{bmatrix} 3 & 0 \\ 0 & 1 \end{bmatrix}$$

3. (a) If $P = X(X^T X)^{-1} X^T$,

$$\begin{aligned} PP &= X(X^T X)^{-1} X^T X (X^T X)^{-1} X^T \\ &= X(X^T X)^{-1} X^T = P \end{aligned}$$

Since $(X^T X)(X^T X)^{-1} = I$. Hence, P is idempotent.

(b) If

$$X = \begin{bmatrix} 1 & 2 \\ 1 & 4 \\ 1 & 1 \\ 1 & 3 \end{bmatrix}$$

then P is obtained through matrix multiplication, after the inversion of the 2×2 $(X^T X)$ matrix as

$$P = \begin{bmatrix} 0.3 & 0.1 & 0.4 & 0.2 \\ 0.1 & 0.7 & -0.2 & 0.4 \\ 0.4 & -0.2 & 0.7 & 0.1 \\ 0.2 & 0.4 & 0.1 & 0.3 \end{bmatrix}$$

Finding the eigenvalues of P is very messy, but the problem simplifies down to solving

$$\lambda^2(\lambda - 1)^2 = 0$$

so that the eigenvalues are 0 and 1.

5. Let A be orthogonal such that $A^T A = I$ and B be orthogonal such that $B^T B = I$. Then AB is also orthogonal since

$$(AB)^T AB = B^T A^T AB = B^T B = I$$

Section 10.3 (page 379)

1. (a) For

$$A = \begin{bmatrix} 1 & 1 & 0 \\ 1 & 4 & 2 \\ 0 & 2 & 3 \end{bmatrix}$$

We check the leading principal minor condition, Theorem 10.17, to obtain

$$1, \quad \begin{vmatrix} 1 & 1 \\ 1 & 4 \end{vmatrix},$$

$$\begin{vmatrix} 1 & 1 & 0 \\ 1 & 4 & 2 \\ 0 & 2 & 3 \end{vmatrix} = \begin{vmatrix} 4 & 2 \\ 2 & 3 \end{vmatrix} - \begin{vmatrix} 1 & 0 \\ 2 & 3 \end{vmatrix} = 12 - 4 - 3 = 5$$

Therefore A is positive definite.

(b) For

$$B = \begin{bmatrix} 5 & -6 & -6 \\ -1 & 4 & 2 \\ 3 & -6 & -4 \end{bmatrix}$$

We have found in Exercise 10.2.4 that $\lambda_1 = 2$, $\lambda_2 = 2$, $\lambda_3 = 1$. Since all eigenvalues are positive, by Theorem 10.13 we conclude that B is positive definite.

(c) For

$$C = \begin{bmatrix} 1 & 0 & 1 \\ 0 & 1 & 1 \\ 1 & 1 & 2 \end{bmatrix}$$

We have found in Exercise 10.2.2 that $\lambda_1 = 1$, $\lambda_2 = 0$, $\lambda_3 = 3$. By Theorem 10.14 we conclude that C is positive semidefinite.

3. Let $g(\mathbf{x}) = 5x_1^2 + 3x_2^2 + 2x_3^2 - x_1x_2 + 8x_2x_3$. The coefficients of x_1^2, x_2^2, x_3^2 go on the diagonal of A. To make A symmetric, the coefficient of x_ix_j for $i \neq j$ must be split between the (i, j) and (j, i) entries in A. The coefficient of x_1x_3 is 0. Therefore

$$\begin{aligned} g(\mathbf{x}) &= \mathbf{x}^T B\mathbf{x} \\ &= \begin{bmatrix} x_1 & x_2 & x_3 \end{bmatrix} \begin{bmatrix} 5 & -1/2 & 0 \\ -1/2 & 3 & 4 \\ 0 & 4 & 2 \end{bmatrix} \times \\ &\quad \begin{bmatrix} x_1 \\ x_2 \\ x_3 \end{bmatrix} \end{aligned}$$

5. $g(x) = 3x_1^2 + 2x_2^2 + x_3^2 + 4x_1x_2 + 4x_2x_3$. We write $g(\mathbf{x})$ as $x^T Ax$.

$$\mathbf{x}^T A\mathbf{x} = \begin{bmatrix} x_1 & x_2 & x_3 \end{bmatrix} \begin{bmatrix} 3 & 2 & 0 \\ 2 & 2 & 2 \\ 0 & 2 & 1 \end{bmatrix} \begin{bmatrix} x_1 \\ x_2 \\ x_3 \end{bmatrix}$$

We look at the leading principal minor condition.

$$3, \quad \begin{vmatrix} 3 & 2 \\ 2 & 2 \end{vmatrix} = 2, \quad \begin{vmatrix} 3 & 2 & 0 \\ 2 & 2 & 2 \\ 0 & 2 & 1 \end{vmatrix} = -10$$

Hence, the form is indefinite.

Review Exercises (page 381)

1. (a) For $\mathbf{y}^T = [0\ 1]$, $\mathbf{w}^T = [-1\ 1]$. We can see that

$$\lambda_1 \begin{bmatrix} 0 \\ 1 \end{bmatrix} + \lambda_2 \begin{bmatrix} -1 \\ 1 \end{bmatrix} = \begin{bmatrix} 0 \\ 0 \end{bmatrix}$$

implies $\lambda_1 = 0$, $\lambda_2 = 0$. Hence $\mathbf{y}$ and $\mathbf{w}$ are linearly independent.

(b) For $\mathbf{y}^T = [0\ 1]$, $\mathbf{w}^T = [0\ -1]$. We can see that

$$\lambda_1 \begin{bmatrix} 0 \\ 1 \end{bmatrix} + \lambda_2 \begin{bmatrix} 0 \\ -1 \end{bmatrix} = \begin{bmatrix} 0 \\ 0 \end{bmatrix}$$

implies $\lambda_1 = \lambda_2$ for all values of λ_1 and λ_2. Hence, $\mathbf{y}$ and $\mathbf{w}$ are linearly independent.

(c) For $\mathbf{y}^T = [1\ 0\ -1]$, $\mathbf{w}^T = [1\ 1\ 0]$ and $\mathbf{z}^T = [0\ 3\ 0]$. We can establish that

$$\lambda_1 \begin{bmatrix} 1 \\ 0 \\ -1 \end{bmatrix} + \lambda_2 \begin{bmatrix} 1 \\ 1 \\ 0 \end{bmatrix} + \lambda_3 \begin{bmatrix} 0 \\ 3 \\ 0 \end{bmatrix} = \begin{bmatrix} 0 \\ 0 \\ 0 \end{bmatrix}$$

implies $\lambda_1 = 0$, $\lambda_2 = 0$ and $\lambda_3 = 0$.

3. (a) If A is 9×7, the largest possible rank of A is 7.

(b) If A is 7×9, the largest possible rank of A is 7.

5. Two possible bases for $\mathbb{R}^3$ are

(a)

$$\mathbf{v}_1 = \begin{bmatrix} 2 \\ 0 \\ 0 \end{bmatrix} \quad \mathbf{v}_2 = \begin{bmatrix} 1 \\ 1 \\ 0 \end{bmatrix} \quad \mathbf{v}_3 = \begin{bmatrix} 1 \\ 1 \\ 1 \end{bmatrix}$$

(b)

$$\mathbf{w}_1 = \begin{bmatrix} 3 \\ 0 \\ 0 \end{bmatrix} \quad \mathbf{w}_2 = \begin{bmatrix} 2 \\ 2 \\ 0 \end{bmatrix} \quad \mathbf{w}_3 = \begin{bmatrix} 2 \\ 2 \\ 2 \end{bmatrix}$$

In fact any three linearly independent vectors in $\mathbb{R}^3$ constitute a basis.

7. (a) The eigenvalues of

$$A = \begin{bmatrix} 2 & 3 \\ 3 & -6 \end{bmatrix}$$

can be found from solving the characteristic equation

$$\begin{vmatrix} 2-\lambda & 3 \\ 3 & -6-\lambda \end{vmatrix} = \lambda^2 + 4\lambda - 21 = 0$$

The roots are $\lambda_1 = 3$ and $\lambda_2 = -7$.

(b) For

$$B = \begin{bmatrix} 2 & 7 \\ 7 & 2 \end{bmatrix}$$

The characteristic equation for B is given by

$$\begin{vmatrix} 2-\lambda & 7 \\ 7 & 2-\lambda \end{vmatrix} = (2-\lambda)^2 - 49 = 0$$

The roots are $\lambda_1 = 9$ and $\lambda_2 = -5$.

9. The orthogonal decomposition of

$$A = \begin{bmatrix} 7 & 2 \\ 2 & 4 \end{bmatrix}$$

is obtained by first solving for the eigenvalues

$$\begin{vmatrix} 7-\lambda & 2 \\ 2 & 4-\lambda \end{vmatrix} = (7-\lambda)(4-\lambda) - 4 = 0$$

The roots are $\lambda_1 = 8$, $\lambda_2 = 3$. The eigenvector corresponding to $\lambda_1 = 8$ is given by

$$\begin{bmatrix} -1 & 2 \\ 2 & -4 \end{bmatrix} \begin{bmatrix} q_1 \\ q_2 \end{bmatrix} = \begin{bmatrix} 0 \\ 0 \end{bmatrix}$$

which implies $-q_1 + 2q_2 = 0$ and $2q_1 - 4q_2 = 0$, so $q_1 = 2q_2$. Using the normalization condition yields $q_1^2 + q_2^2 = 1$, so $q_1 = 2/\sqrt{5}, q_2 = 1/\sqrt{5}$. Hence

$$\mathbf{q}_1 = \begin{bmatrix} 2/\sqrt{5} \\ 1/\sqrt{5} \end{bmatrix}$$

The eigenvector corresponding to $\lambda_2 = 3$ is given by

$$\begin{bmatrix} 4 & 2 \\ 2 & 1 \end{bmatrix} \begin{bmatrix} q_1 \\ q_2 \end{bmatrix} = \begin{bmatrix} 0 \\ 0 \end{bmatrix}$$

which implies $4q_1 + 2q_2 = 0$ and $2q_1 + q_2 = 0$, so $q_2 = -2q_1$. Using the normalization condition yields $q_1 = 1/\sqrt{5}$, $q_2 = -2/\sqrt{5}$, Hence

$$\mathbf{q}_2 = \begin{bmatrix} 1/\sqrt{5} \\ -2/\sqrt{5} \end{bmatrix}$$

Hence

$$Q = \begin{bmatrix} 2/\sqrt{5} & 1/\sqrt{5} \\ 1/\sqrt{5} & -2/\sqrt{5} \end{bmatrix}$$

and $Q^T A Q = \Lambda$.

CHAPTER 11

Section 11.1 (page 395)

1.

$$\frac{\partial f(x_1, x_2)}{\partial x_1}$$

$$= \lim_{\Delta x_1 \to 0} \frac{f(x_1 + \Delta x_1, x_2) - f(x_1, x_2)}{\Delta x_1}$$

$$= \lim_{\Delta x_1 \to 0} \frac{[3(x_1 + \Delta x_1) + 5x_2] - [3x_1 + 5x_2]}{\Delta x_1}$$

$$= \lim_{\Delta x_1 \to 0} \frac{3\Delta x_1}{\Delta x_1} = 3$$

$$\frac{\partial f(x_1, x_2)}{\partial x_2}$$

$$= \lim_{\Delta x_2 \to 0} \frac{f(x_1, x_2 + \Delta x_2) - f(x_1, x_2)}{\Delta x_2}$$

$$= \lim_{\Delta x_2 \to 0} \frac{[3x_1 + 5(x_2 + \Delta x_2)] - [3x_1 + 5x_2]}{\Delta x_2}$$

$$= \lim_{\Delta x_2 \to 0} \frac{5\Delta x_2}{\Delta x_2} = 5$$

3.

$$\frac{\partial R(x_1, x_2)}{\partial x_2}$$

$$= \lim_{\Delta x_2 \to 0} \frac{R(x_1, x_2 + \Delta x_2) - R(x_1, x_2)}{\Delta x_2}$$

$$= \lim_{\Delta x_2 \to 0} \frac{[p_1 x_1 + p_2(x_2 + \Delta x_2)] - [p_1 x_1 + p_2 x_2]}{\Delta x_2}$$

$$= \lim_{\Delta x_2 \to 0} \frac{p_2 \Delta x_2}{\Delta x_2} = p_2$$

This partial derivative is the rate at which the firm's revenue increases with respect to an increase in x_2, or, in other words, is the increase in revenue resulting from a one unit increase in the sales of good 2 (i.e., x_2). Since p_2 is the unit price of good 2, this result makes intuitive economic sense.

5.

$$\frac{\partial f(x_1, x_2)}{\partial x_2} = \lim_{\Delta x_2 \to 0} \frac{f(x_1, x_2 + \Delta x_2) - f(x_1, x_2)}{\Delta x_2}$$

$$= \lim_{\Delta x_2 \to 0} \frac{[x_1^2(x_2 + \Delta x_2)] - [x_1^2 x_2]}{\Delta x_2}$$

$$= \lim_{\Delta x_2 \to 0} \frac{[x_1^2 x_2 + x_1^2 \Delta x_2] - x_1^2 x_2}{\Delta x_2}$$

$$= \lim_{\Delta x_2 \to 0} \frac{x_1^2 \Delta x_2}{\Delta x_2} = x_1^2$$

7. Given $y = 10x_1^{1/2}x_2^{1/3}x_3^{1/4}$ we have

$$\frac{\partial y}{\partial x_1} = (1/2)10x_1^{-1/2}x_2^{1/3}x_3^{1/4}$$

$$= 5x_1^{-1/2}x_2^{1/3}x_3^{1/4} = \frac{5x_2^{1/3}x_3^{1/4}}{x_1^{1/2}}$$

$$\frac{\partial y}{\partial x_2} = (1/3)10x_1^{1/2}x_2^{-2/3}x_3^{1/4}$$

$$= (10/3)x_1^{1/2}x_2^{-2/3}x_3^{1/4} = \frac{10x_1^{1/2}x_3^{1/4}}{3x_2^{2/3}}$$

$$\frac{\partial y}{\partial x_3} = (1/4)10x_1^{1/2}x_2^{1/3}x_3^{-3/4}$$

$$= 2.5x_1^{1/2}x_2^{1/3}x_3^{-3/4} = \frac{2.5x_1^{1/2}x_2^{1/3}}{x_3^{3/4}}$$

9. The marginal product for input 1 is

$$\frac{\partial y}{\partial x_1} = 12(-2)[0.4x_1^{-1/2} + 0.6x_2^{-1/2}]^{-3} \times [(-1/2)0.4x_1^{-3/2}]$$

$$= (4.8)x_1^{-3/2}[0.4x_1^{-1/2} + 0.6x_2^{-1/2}]^{-3}$$

The marginal product for input 2 is

$$\frac{\partial y}{\partial x_2} = 12(-2)[0.4x_1^{-1/2} + 0.6x_2^{-1/2}]^{-3}[(-1/2)0.6x_1^{-3/2}]$$

$$= (7.2)x_1^{-3/2}[0.4x_1^{-1/2} + 0.6x_2^{-1/2}]^{-3}$$

11. Using

$$\frac{dY}{dt} = f_t + f_K \frac{dK}{dt}$$

we get

$$\frac{dY}{dt} = \underbrace{0.1(1+t)^{-1/2}K}_{f_t} + \underbrace{0.2(1+t)^{1/2}}_{f_K}\underbrace{(0.05K_0 e^{0.05t})}_{dK/dt}$$

and so

$$\frac{dY}{dt} = 0.1(1+t)^{-1/2}K_0 e^{0.05t} + 0.01(1+t)^{1/2}K_0 e^{0.05t}$$

The first part of this expression, f_t, gives the rate at which output would rise if the capital stock were artificially held fixed. That is, the increase in Y would be the result of using the existing capital stock more effectively. The second part, $f_K(dK/dt)$, measures the increase in output due to the fact that the capital stock is increasing over time.

Section 11.2 (page 402)

1.

$$\nabla f = \begin{bmatrix} a_1 \\ a_2 \end{bmatrix} \qquad \nabla_2 F = \begin{bmatrix} 0 & 0 \\ 0 & 0 \end{bmatrix}$$

Being a linear function, $y = a_1 x_1 + a_2 x_2$, the rate at which y changes as x_i, $i = 1, 2$, changes is constant. Therefore, the first derivatives are constants and all second-order partial derivatives are zero.

3.

$$\nabla f = \begin{bmatrix} 3x_1^2x_2^4 \\ 4x_1^3x_2^3 \end{bmatrix} \qquad \nabla_2 F = \begin{bmatrix} 6x_1x_2^4 & 12x_1^2x_2^3 \\ 12x_1^2x_2^3 & 12x_1^3x_2^2 \end{bmatrix}$$

5.

$$\nabla f = \begin{bmatrix} 2x_1 \\ 2x_2 \end{bmatrix} \qquad \nabla_2 F = \begin{bmatrix} 2 & 0 \\ 0 & 2 \end{bmatrix}$$

7.

$$\begin{aligned}
\frac{\partial y}{\partial x_1} &= 25x_1^{-1/2}x_2^{2/3} > 0 \\
\frac{\partial y}{\partial x_2} &= \frac{100}{3}x_1^{1/2}x_2^{-1/3} > 0 \\
\frac{\partial^2 y}{\partial x_1^2} &= \frac{-25}{2}x_1^{-3/2}x_2^{2/3} < 0 \\
\frac{\partial^2 y}{\partial x_2^2} &= -\frac{100}{9}x_1^{1/2}x_2^{-4/3} < 0 \\
\frac{\partial^2 y}{\partial x_1 \partial x_2} &= \frac{\partial^2 y}{\partial x_2 \partial x_1} = \frac{50}{3}x_1^{-1/2}x_2^{-1/3} > 0
\end{aligned}$$

The results $\partial y/\partial x_1 > 0$ and $\partial y/\partial x_2 > 0$ means that the marginal product of each input is positive (i.e., increasing x_i leads to an increase in output, $i = 1, 2$). The result $\partial^2 y/\partial x_1^2 < 0$ means that the marginal product of input 1, although positive, falls as x_1 (only) rises. Similarly for $\partial^2 y/\partial x_2$. The cross-partial derivatives are positive indicating that the marginal product of either input is increased by more use of the other input (i.e., inputs are complementary).

9.

$$\begin{aligned}
f_{12} &= f_{21} = (6x_1 + 3x_1^2x_3)e^{3x_2+x_1x_3} - 6x_2^2/x_1^2 \\
f_{13} &= f_{31} = (3x_1^2 + x_1^3x_3)e^{3x_2+x_1x_3}
\end{aligned}$$

Section 11.3 (page 420)

1. (a) $du = 5dx_1 + 3dx_2$

(b) See Figure 11.3.1.

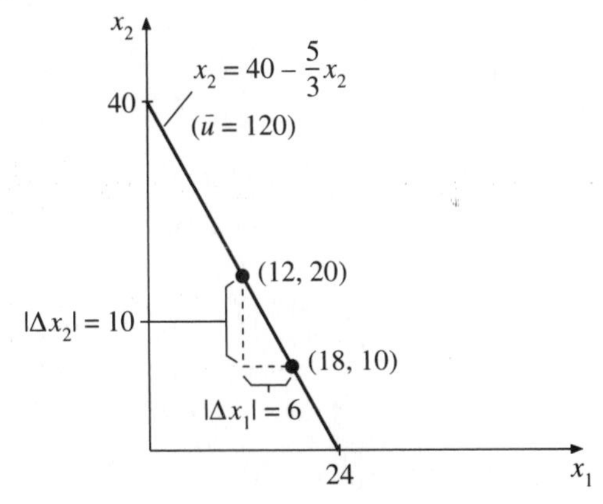

Figure 11.3.1

(c) Since $u = 5x_1 + 3x_2$ is a linear function, MRS = 5/3 (we do not need to take the limit as $\Delta x_1 \to 0$).

3. $dy = 0 \Rightarrow x_2dx_1 + x_1dx_2 = 0$, or MRTS $= -dx_2/dx_1 = x_2/x_1$. The equation of an isoquant is $x_2 = \bar{y}/x_1$ and so $dx_2/dx_1 = -\bar{y}/x_1^2 < 0$ and $d^2x_2/dx_1^2 = 2\bar{y}/x_1^3 > 0$. So the isoquants are negatively-sloped and strictly convex.

5. (a) $du = x_2dx_1 + x_1dx_2 = 0$, so $dx_2/dx_1 = -x_2/x_1$

(b) $d\hat{u} = 2x_1x_2^2dx_1 + 2x_1^2x_2dx_2 = 0$ so $dx_2/dx_1 = -x_2/x_1$.

(c)

$$d\tilde{u} = BKx_1^{K-1}x_2^Kdx_1 + BKx_1^Kx_2^{K-1}dx_2$$

so

$$dx_2/dx_1 = -\frac{BKx_1^{K-1}x_2^K}{BKx_1^Kx_2^{K-1}} = -x_2/x_1$$

7.

$$\begin{aligned}
dy &= \left(\left(-\frac{1}{3}\right)[0.3x_1^{-3} + 0.7x_2^{-3}]^{-4/3}((-3)0.3x_1^{-4})\right)dx_1 \\
&\quad + \left(\left(-\frac{1}{3}\right)[0.3x_1^{-3} + 0.7x_2^{-3}]^{-4/3}((-3)0.7x_2^{-4})\right)dx_2
\end{aligned}$$

and so

$$\begin{aligned}
dy &= (0.3x_1^{-4}[0.3x_1^{-3} + 0.7x_2^{-3}]^{-4/3})dx_1 \\
&\quad + (0.7x_2^{-4}[0.3x_1^{-3} + 0.7x_2^{-3}]^{-4/3})dx_2
\end{aligned}$$

Thus, $dy = 0$ implies

$$\begin{aligned}
\frac{dx_2}{dx_1} &= -\frac{0.3x_1^{-4}[0.3x_1^{-3} + 0.7x_2^{-3}]^{-4/3}}{0.7x_2^{-4}[0.3x_1^{-3} + 0.7x_2^{-3}]^{-4/3}} \\
&= -\frac{3x_2^4}{7x_1^4}
\end{aligned}$$

Thus, the MRTS $= 3x_2^4/7x_1^4$ gets smaller as one moves along an isoquant from left to right (and downwards) because as x_1 rises and x_2 falls the denominator gets bigger and the numerator gets smaller. Thus, isoquants are strictly convex to the origin.

9. $u(8, 1) = 8^2 \times 1 = 64$, $u(4, 4) = 4^2 \times 4 = 64$ and $u(2, 16) = 2^2 \times 16 = 64$. Thus, the points A, B, C lie on the same indifference curve.

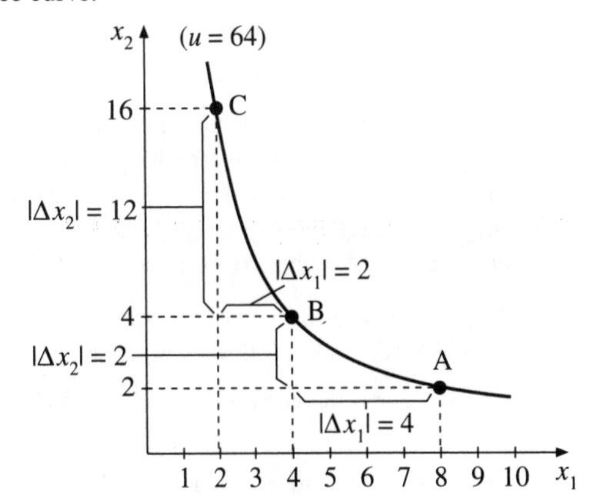

Figure 11.3.9

Between B and C, $|\Delta x_2|/|\Delta x_1| = 12/2 = 6$, and between A and B, $|\Delta x_2|/|\Delta x_1| = 4/2 = 2$. In moving from A to B, the consumer is indifferent to consuming more x_2 in exchange for less x_1 at the rate of 2 units of x_2 for every 1 unit of x_1. In moving from B to C, the consumer requires 6 units of x_2 in exchange for each unit of x_1 to remain indifferent. The intuition for this is that in moving from B to C, as compared to moving from A to B, the consumer begins with more of x_2 and less of x_1 and so is less willing to give up good 1 for good 2, and so requires greater compensation of good 2 in exchange for each unit of good 1 given up.

Section 11.4 (page 434)

1. $f_1 = 2(x_1+x_2)$, $f_2 = 2(x_1+x_2)$, $f_{11} = 2$, $f_{22} = 2$, $f_{12} = 2$. Thus, we have that

$$d^2y = 2dx_1^2 + 4dx_1dx_2 + 2dx_2^2 = 2(dx_1+dx_2)^2 \geq 0$$

for any pair, dx_1, dx_2.

3. $f_1 = (1/2)x_1^{-1/2}x_2^{1/3}$, $f_2 = (1/3)x_1^{1/2}x_2^{-2/3}$ and so $f_{11} = -(1/4)x_1^{-3/2}x_2^{1/3}$, $f_{12} = f_{21} = (1/6)x_1^{-1/2}x_2^{-2/3}$, $f_{22} = -(2/9)x_1^{1/2}x_2^{-5/3}$. Thus, we get

$$\begin{aligned} H &= \begin{bmatrix} f_{11} & f_{12} \\ f_{21} & f_{22} \end{bmatrix} \\ &= \begin{bmatrix} -(1/4)x_1^{-3/2}x_2^{1/3} & (1/6)x_1^{-1/2}x_2^{-2/3} \\ (1/6)x_1^{-1/2}x_2^{-2/3} & -(2/9)x_1^{1/2}x_2^{-5/3} \end{bmatrix} \end{aligned}$$

and so, $|H_1| = -(1/4)x_1^{-1/2}x_2^{1/3} < 0$.

$$\begin{aligned} |H_2| &= (-(1/4)x_1^{-3/2}x_2^{1/3})(-(2/9)x_1^{1/2}x_2^{-5/3}) \\ &\quad -((1/6)x_1^{-1/2}x_2^{-2/3})^2 \\ &= (2/36)x_1^{-1}x_2^{-4/3} - (1/36)x_1^{-1}x_2^{-4/3} \\ &= (1/36)x_1^{-1}x_2^{-4/3} > 0 \quad \text{for } x_1, x_2 > 0 \end{aligned}$$

Thus, H is negative definite (see Theorem 11.9) and so f is strictly concave.

5. $f_1 = (1/2)(x_1+x_2)^{-1/2}$, $f_2 = (1/2)(x_1+x_2)^{-1/2}$ and so $f_{11} = -(1/4)(x_1+x_2)^{-3/2}$, $f_{12} = f_{21} = -(1/4)(x_1+x_2)^{-3/2}$, $f_{22} = -(1/4)(x_1+x_2)^{-3/2}$. So

$$\begin{aligned} d^2y &= f_{11}dx_1^2 + 2f_{12}dx_1dx_2 + f_{22}dx_2^2 \\ &= -(1/4)(x_1+x_2)^{-3/2}dx_1^2 \\ &\quad -(1/2)(x_1+x_2)^{-3/2}dx_1dx_2 \\ &\quad -(1/4)(x_1+x_2)^{-3/2}dx_2^2 \\ &= -(1/4)(x_1+x_2)^{-3/2}(dx_1+dx_2)^2 \leq 0 \end{aligned}$$

and so f is a concave function. For any (x_1, x_2)-values satisfying $x_1+x_2 = k$, k a constant, we get $dx_1+dx_2 = 0$ and so $(dx_1+dx_2)^2 = 0$ and so $d^2y = 0$. Such (x_1, x_2)-values generate horizontal line segments (i.e., $y = (x_1+x_2)^{1/2} = k^{1/2}$) on the graph of f and so this function is not *strictly* concave.

7. To prove this result we need to show that for an additively separable function it follows that, for all $i = 1, 2, \ldots, n$

(i) if $f_{ii} < 0$ then $d^2y < 0$ for all dx_i values (not all zero) and

(ii) $d^2y < 0$ for all dx_i values (not all zero) only if it is the case that $f_{ii} < 0$ for all i.

We begin with result (i). Since $f_{ij} = 0$ for $i \neq j$ it follows that $d^2y = \sum_{i=1}^n f_{ii}dx_i^2$. Since $dx_i^2 \geq 0$ for any value dx_i, each term in the summation will be nonpositive if $f_{ii} < 0$. Since for at least one dx_i we must have $dx_i \neq 0 \Leftrightarrow dx_i^2 > 0$ it follows that, $f_{ii} < 0$ implies $d^2y < 0$. This proves result (i).

Now consider result (ii). Suppose $d^2y < 0$ for all combinations of dx_i values. This must include, for example, the case where $dx_1 \neq 0$, but all other dx_i values are equal to zero, (i.e., $dx_2 = dx_3 = \ldots = dx_n = 0$). In this case, we would have

$$d^2y = \sum_{i=1}^n f_{ii}dx_i^2 = f_{11}dx_1^2$$

and so $d^2y < 0$ requires that $f_{11} < 0$. We could pick any particular i, from $i = 1, 2, \ldots, n$ and repeat this exercise. Thus, for d^2y to be non-negative for all possible combinations of dx_i values requires that each $f_{ii} < 0$, which proves result (ii).

Section 11.5 (page 453)

(There are several algebraic steps required in each of the parts of questions 1 and 2. We show these in complete detail for question 1(a). The others are determined by following similar steps.)

1. To answer this question refer to Theorem 11.12 and Theorem 11.9.

(a) The first-order partial derivative of f are

$$f_1 = \frac{1}{2}x_1^{-1/2}x_2^{1/4} \text{ and } f_2 = \frac{1}{4}x_1^{1/2}x_2^{-3/4}$$

and so the second-order partials are

$$f_{11} = -\frac{1}{4}x_1^{-3/2}x_2^{1/4}, \quad f_{12} = \frac{1}{8}x_1^{-1/2}x_2^{-3/4},$$

$$f_{22} = -\frac{3}{16}x_1^{1/2}x_2^{-7/4}$$

Upon substituting these expressions into the bordered Hessian,

$$|\bar{H}_2| = |\bar{H}| = -f_1^2 f_{22} + 2f_1f_2f_{12} - f_2^2 f_{11}$$

we obtain the result that

$$\begin{aligned} |\bar{H}| &= -\left(\frac{1}{2}x_1^{-1/2}x_2^{1/4}\right)^2\left(-\frac{3}{16}x_1^{1/2}x_2^{-7/4}\right) \\ &\quad +2\left(\frac{1}{2}x_1^{-1/2}x_2^{1/4}\right)\left(\frac{1}{4}x_1^{1/2}x_2^{-3/4}\right)\left(\frac{1}{8}x_1^{-1/2}x_2^{-3/4}\right) \\ &\quad -\left(\frac{1}{4}x_1^{1/2}x_2^{-3/4}\right)^2\left(-\frac{1}{4}x_1^{-3/2}x_2^{1/4}\right) \\ &= -\left(\frac{1}{4}x_1^{-1}x_2^{1/2}\right)\left(-\frac{3}{16}x_1^{1/2}x_2^{-7/4}\right) \\ &\quad +2\left(\frac{1}{2}x_1^{-1/2}x_2^{1/4}\right)\left(\frac{1}{4}x_1^{1/2}x_2^{-3/4}\right)\left(\frac{1}{8}x_1^{-1/2}x_2^{-3/4}\right) \\ &\quad -\left(\frac{1}{16}x_1^{1}x_2^{-3/2}\right)\left(-\frac{1}{4}x_1^{-3/2}x_2^{1/4}\right) \\ &= \frac{3}{64}x_1^{-1/2}x_2^{-5/4} + \frac{1}{32}x_1^{-1/2}x_2^{-5/4} \\ &\quad +\frac{1}{64}x_1^{-1/2}x_2^{-5/4} \\ &= \frac{3}{32}x_1^{-1/2}x_2^{-5/4} > 0 \end{aligned}$$

Thus, f is quasiconcave.

To check if f is also strictly concave, note that

$$f_{11} = -\frac{1}{4}x_1^{-3/2}x_2^{1/4} < 0$$

and

$$f_{11}f_{22} > f_{12}^2$$

$$\left(-\frac{1}{4}x_1^{-3/2}x_2^{1/4}\right)\left(-\frac{3}{16}x_1^{1/2}x_2^{-7/4}\right) > \left(\frac{1}{8}x_1^{-1/2}x_2^{-3/4}\right)^2$$

$$\frac{3}{64}x_1^{-1}x_2^{-3/2} > \frac{1}{64}x_1^{-1}x_2^{-3/2}$$

which is the case. Thus, f is strictly concave.

(b) Computing first- and second-order partial derivatives gives us

$$\begin{aligned} f_1 &= \frac{1}{3}x^{-2/3}x_2^{2/3} \\ f_2 &= \frac{2}{3}x^{1/3}x_2^{-1/3} \\ f_{11} &= -\frac{2}{9}x^{-5/3}x_2^{2/3} \\ f_{12} &= \frac{2}{9}x^{-2/3}x_2^{-1/3} \\ f_{22} &= -\frac{2}{9}x^{1/3}x_2^{-4/3} \end{aligned}$$

Upon substituting these into appropriate expressions, we find that f is quasiconcave.

f is quasiconcave since

$$|\bar{H}| = -f_1^2f_{22} + 2f_1f_2f_{12} - f_2^2f_{11} = \frac{18}{81}x_1^{-1} > 0$$

f is not strictly concave because $f_{11}f_{22} = f_{12}^2$. However, $|H_1^*| = f_{11}$, f_{22} are both ≤ 0 and $|H_2^*| = |H| = f_{11}f_{22} - f_{12}^2 \leq 0$ and so f is (weakly) concave.

(c) Computing first- and second-order partial derivatives gives us

$$\begin{aligned} f_1 &= 2x_1x_2^3 \\ f_2 &= 3x_1^2x_2^2 \\ f_{11} &= 2x_3^3 \\ f_{12} &= 6x_1x_2^2 \\ f_{22} &= 6x_1^2x_2 \end{aligned}$$

Upon substituting these into the appropriate expression we find that f is quasiconcave since

$$|\bar{H}| = -f_1^2f_{22} + 2f_1f_2f_{12} - f_2^2f_{11} = 30x_1^4x_2^7 > 0$$

f is neither strictly concave nor (weakly) concave because $f_{11}f_{22} < f_{12}^2$.

3. From Theorem 11.12 we see that we need to show that $|\bar{H}_2| > 0$ and $|\bar{H}_3| < 0$ in order to show that the function f is quasiconcave. To show $|\bar{H}_2| > 0$ use the same procedure as in Question 1. To show that $|\bar{H}_3| < 0$ requires substantially more work. We provide an outline of the method below for the function in part (a). The solution to the function in part (b) follows a similar procedure.

First compute all of the first- and second-order partial derivatives of f:

$$\begin{aligned} f_1 &= \frac{1}{4}x_1^{-3/4}x_2^{1/3}x_3^{1/4} \\ f_2 &= \frac{1}{3}x_1^{1/4}x_2^{-2/3}x_3^{1/4} \\ f_3 &= \frac{1}{4}x_1^{1/4}x_2^{1/3}x_3^{-3/4} \\ f_{11} &= -\frac{3}{16}x_1^{-7/4}x_2^{1/3}x_3^{1/4} \\ f_{22} &= -\frac{2}{9}x_1^{1/4}x_2^{-5/3}x_3^{1/4} \\ f_{33} &= -\frac{3}{16}x_1^{1/4}x_2^{1/3}x_3^{-7/4} \\ f_{12} = f_{21} &= \frac{1}{12}x_1^{-3/4}x_2^{-2/3}x_3^{1/4} \\ f_{13} = f_{31} &= \frac{1}{16}x_1^{-3/4}x_2^{1/3}x_3^{-3/4} \\ f_{23} = f_{32} &= \frac{1}{12}x_1x_2x_3 \end{aligned}$$

We next substitute these values into the following expression:

$$\begin{aligned} |\bar{H}_3| &= |\bar{H}| = \begin{vmatrix} 0 & f_1 & f_2 & f_3 \\ f_1 & f_{11} & f_{12} & f_{13} \\ f_2 & f_{21} & f_{22} & f_{23} \\ f_3 & f_{31} & f_{32} & f_{33} \end{vmatrix} \\ &= -f_1\begin{vmatrix} f_1 & f_{12} & f_{13} \\ f_2 & f_{22} & f_{23} \\ f_3 & f_{32} & f_{33} \end{vmatrix} \\ &\quad +f_2\begin{vmatrix} f_1 & f_{11} & f_{13} \\ f_2 & f_{21} & f_{23} \\ f_3 & f_{31} & f_{33} \end{vmatrix} \\ &\quad -f_3\begin{vmatrix} f_1 & f_{11} & f_{12} \\ f_2 & f_{21} & f_{22} \\ f_3 & f_{31} & f_{32} \end{vmatrix} \end{aligned}$$

Upon making these substitutions into the first term, for example, we find that

$$-f_1\begin{vmatrix} f_1 & f_{12} & f_{13} \\ f_2 & f_{22} & f_{23} \\ f_3 & f_{32} & f_{33} \end{vmatrix} =$$

$$-\left(\frac{1}{4}x_1^{-3/4}x_2^{1/3}x_3^{1/4}\right)\left(\frac{12}{576}x_1^{-1/4}x_2^{-1/2}x_3^{-5/4}\right) =$$

$$-\frac{12}{2304}x_1^{-1}x_2^{-1/6}x_3^{-1}$$

By carrying out the same steps for each part of the expression for $|\bar{H}|$ we can show that $|\bar{H}| < 0$.

To show that this function is also strictly concave refer to Theorem 11.9. We need to show that $|H_1| = f_{11} < 0$

$$\begin{aligned} |H_2| &= \begin{vmatrix} f_{11} & f_{12} \\ f_{21} & f_{22} \end{vmatrix} \\ &= f_{11}f_{22} - f_{12}^2 > 0 \quad \text{or } f_{11}f_{22} > f_{12}^2 \end{aligned}$$

and

$$\begin{aligned} |H_3| &= |H| = \begin{vmatrix} f_{11} & f_{12} & f_{13} \\ f_{21} & f_{22} & f_{23} \\ f_{31} & f_{32} & f_{33} \end{vmatrix} \\ &= f_{11}(f_{22}f_{33} - f_{23}^2) - f_{12}(f_{21}f_{33} - f_{23}f_{31}) \\ &\quad +f_{13}(f_{21}f_{32} - f_{22}f_{31}) < 0 \end{aligned}$$

Now

$$f_{11} = -\frac{3}{16}x_1^{-7/4}x_2^{1/3}x_3^{1/4} < 0$$

and so the first condition is satisfied

To see that $f_{11}f_{22} > f_{12}^2$ holds, note that

$$\begin{aligned} f_{11}f_{22} &= \left(-\frac{3}{16}x_1^{-7/4}x_2^{1/3}x_3^{1/4}\right) \times \\ & \left(-\frac{2}{9}x_1^{1/4}x_2^{-5/3}x_3^{1/4}\right) \\ &= \frac{6}{144}x_1^{-3/2}x_2^{-4/3}x_3^{1/2} \end{aligned}$$

which is greater than

$$\begin{aligned} f_{12}^2 &= \left(\frac{1}{12}x_1^{-3/4}x_2^{-2/3}x_3^{1/4}\right)^2 \\ &= \frac{1}{144}x_1^{-3/2}x_2^{-4/3}x_3^{1/2} \end{aligned}$$

To see that the third condition, $|H_3| = |H| < 0$ holds, make the substitutions required and collect terms to get:

$$\begin{aligned} |H| &= f_{11}(f_{22}f_{33} - f_{23}^2) - f_{12}(f_{21}f_{33} - f_{23}f_{31}) \\ & +f_{13}(f_{21}f_{32} - f_{22}f_{31}) \\ &= -\frac{15}{2304}x_1^{-5/4}x_2^{-1}x_3^{-5/4} + \frac{5}{2304}x_1^{-5/4}x_2^{-1}x_3^{-5/4} \\ & -\frac{3}{2304}x_1^{-5/4}x_2^{-1}x_3^{-5/4} \\ &= -\frac{7}{2304}x_1^{-5/4}x_2^{-1}x_3^{-5/4} < 0 \end{aligned}$$

which implies that f is strictly concave.

(b) For this function repeat the same exercise as above to show this function is quasiconcave. However, this function is not concave. To see this, also compute the determinants as was done in part (a) and compare your results to the conditions given in Theorem 11.9 and you will see that these conditions are not satisfied.

5. $f_1 = \alpha A x_1^{\alpha-1}x_2^{\beta}$, $f_2 = \beta A x_1^{\alpha}x_2^{\beta-1}$. Along a ray from the origin $x_2 = kx_1$, so

$$\text{MRTS} = \frac{f_1}{f_2} = \frac{\alpha x_2}{\beta x_1} = \frac{\alpha k x_1}{\beta x_1} = \frac{\alpha k}{\beta}$$

which is a constant. Using $x_2 = kx_1$ in f_1 and f_2 shows that marginal products are independent of x_1 only if $\alpha + \beta = 1$. To see this note that upon setting $x_2 = kx_1$ we obtain

$$f_1 = \alpha A k^{\beta} x_1^{\alpha+\beta-1}$$

and

$$f_2 = \beta A k^{\beta-1} x_1^{\alpha+\beta-1}$$

7. (a) $x_1^{1/2}x_2^{1/2}$ is a homogeneous function

$$(sx_1)^{1/2}(sx_2)^{1/2} = sx_1^{1/2}x_2^{1/2}$$

(i.e., it is homogeneous of degree one). Thus, since f is a monotonic transformation of x_1x_2, it follows that f is homothetic. To see that f is not homogeneous, note that $f(1,1) = k+1$, $f(2,2) = k+2$, and $f(4,4) = k+4$. If f is homogeneous it must be the case that $f(2,2) = 2^t f(1,1)$ and $f(4,4) = 2^t f(2,2)$ for some value t. This requires that $(k+2) = 2^t(k+1)$ and $(k+4) = 2^t(k+2)$. Taking ratios this means that $2^t = (k+2)/(k+1)$ and $2^t = (k+4)/(k+2)$, which in turn implies that $(k+2)/(k+1) = (k+4)/(k+2)$ or $(k+2)(k+2) = (k+1)(k+4)$. These results, however, are not compatible as

$$(k+2)(k+2) = k^2 + 4k + 4$$

$$(k+1)(k+4) = k^2 + 5k + 4$$

and it is not possible for

$$k^2 + 4k + 4 = k^2 + 5k + 4 \text{ or } 4k = 5k$$

unless $k = 0$. Thus, f cannot be a homogeneous function.

(b) $x_1^2x_2$ is a homogeneous function, since

$$(sx_1)^2(sx_2) = s^3x_1^2x_2$$

(i.e., it is homogeneous of degree three). Thus, since f is a monotonic transformation of x_1x_2, it follows that f is homothetic. To see that f is not homogeneous, note that $f(1,1) = e$, $f(2,2) = e^8$, and $f(4,4) = e^{64}$. If f is homogeneous it must be the case that $f(2,2) = 2^t f(1,1)$ and $f(4,4) = 2^t f(2,2)$ for some value t. This requires that $e^8 = 2^t e$ and $e^{64} = 2^t e^8$. Taking ratios this means that $2^t = e^8/e = e^7$ and $2^t = e^{64}/e^8 = e^{56}$, which are clearly incompatible statements. Thus, f cannot be a homogeneous function.

Section 11.6 (page 458)

1. $\Delta y = f(3,4) - f(1,1) = -15 - 8 = -23$, $dy = 2f_1(1,1) + 3f_2(1,1) = -10$, thus $dy > \Delta y$.

Review Exercises (page 459)

1. $f_1 = \alpha A x_1^{\alpha-1}x_2^{\beta}$, $f_2 = \beta A x_1^{\alpha}x_2^{\beta-1}$

3. $f_1 = a$, $f_{11} = 0$, $f_{12} = 0$, $f_{13} = 0$, $f_2 = \beta x_2^{\beta-1}x_3^{\gamma}$, $f_{21} = 0$, $f_{22} = \beta(\beta-1)x_2^{\beta-2}x_3^{\gamma}$, $f_{23} = \gamma\beta x_2^{\beta-1}x_3^{\gamma-1}$, $f_3 = \gamma x_2^{\beta}x_3^{\gamma-1}$, $f_{31} = 0$, $f_{32} = \gamma\beta x_2^{\beta-1}x_3^{\gamma-1}$, $f_{33} = \gamma(\gamma-1)x_2^{\beta}x_3^{\gamma-2}$. Notice that Young's Theorem applies: $f_{12} = f_{21}$, $f_{13} = f_{31}$, $f_{23} = f_{32}$.

5. The first- and second-order partial derivatives for this function are

$$f_1 = \frac{1}{4}x_1^{-3/4}x_2^{1/2}$$

and

$$f_2 = \frac{1}{2}x_1^{1/4}x_2^{-1/2}$$

and

$$f_{11} = -\frac{3}{16}x_1^{-7/4}x_2^{1/2}$$

$$f_{12} = f_{21} = \frac{1}{8}x_1^{-3/4}x_2^{-1/2}$$

$$f_{22} = -\frac{1}{4}x_1^{1/4}x_2^{-3/2}$$

Thus

$$|H_1| = f_{11} = -\frac{3}{16}x_1^{-7/4}x_2^{1/2} < 0$$

$$\begin{aligned} |H_2| &= f_{11}f_{22} - f_{12}^2 \\ &= \left(-\frac{3}{16}x_1^{-7/4}x_2^{1/2}\right)\left(-\frac{1}{4}x_1^{1/4}x_2^{-3/2}\right) \end{aligned}$$

$$-\left(\frac{1}{8}x_1^{-3/4}x_2^{-1/2}\right)^2$$

$$= \frac{1}{32}x_1^{-3/2}x_2^{-1} > 0$$

7. See Figures 11.R.7 (a) and 11.R.7 (b).

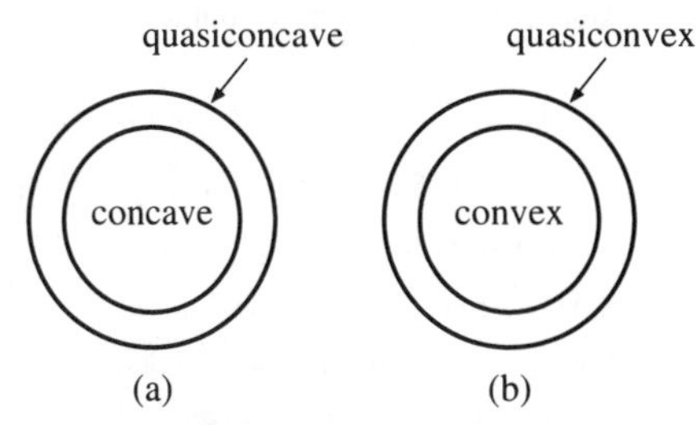

Figure 11.R.7 (a) and (b)

9. (a) The function $f(x_1, x_2, x_3) = x_1^{a_1}x_2^{a_2}x_3^{a_3}$ is homogeneous of degree $a_1 + a_2 + a_3$. To see this note that

$$\begin{aligned}
f(sx_1, sx_2, sx_3) &= (sx_1)^{a_1}(sx_2)^{a_2}(sx_3)^{a_3}\\
&= s^{a_1}x_1^{a_1}s^{a_2}x_2^{a_2}s^{a_3}x_3^{a_3}\\
&= s^{a_1}s^{a_2}s^{a_3}(x_1^{a_1}x_2^{a_2}x_3^{a_3})\\
&= s^{a_1+a_2+a_3}(x_1^{a_1}x_2^{a_2}x_3^{a_3})\\
&= s^{a_1+a_2+a_3}f(x_1, x_2, x_3)
\end{aligned}$$

Since any function which is homogeneous is also homothetic, it follows that this function is both homogeneous and homothetic.

(b) The function $g(x_1, x_2, x_3) = x_1^{1/2}x_2^{1/4}x_3^{1/4}$ is homogeneous of degree $1/2 + 1/4 + 1/4 = 1$. To see this note that

$$\begin{aligned}
g(sx_1, sx_2, sx_3) &= (sx_1)^{1/2}(sx_2)^{1/4}(sx_3)^{1/4}\\
&= s^{1/2}x_1^{1/2}s^{1/4}x_2^{1/4}s^{1/4}x_3^{1/4}\\
&= s^{1/2}s^{1/4}s^{1/4}(x_1^{1/2}x_2^{1/4}x_3^{1/4})\\
&= s^{1/2+1/4+1/4}(x_1^{1/2}x_2^{1/4}x_3^{1/4})\\
&= s^1 g(x_1, x_2, x_3)
\end{aligned}$$

Since the function $f(x_1, x_2, x_3) = 1 + g(x_1, x_2, x_3)$ is a monotonic transformation of $g(x_1, x_2, x_3)$, where $g(x_1, x_2, x_3)$ is a homogeneous function, it follows that f is a homothetic function. However, f is not homogeneous. To see this pick the points $(1, 1, 1)$, $(2, 2, 2)$, and $(4, 4, 4)$. If f is homogeneous of degree k, then it follows that

$$f(2, 2, 2) = 2^k f(1, 1, 1)$$

$$f(4, 4, 4) = 2^k f(2, 2, 2)$$

To see that this is not possible, note that

$$\begin{aligned}
f(1, 1, 1) &= 1 + 1^{1/2}1^{1/4}1^{1/4} = 2\\
f(2, 2, 2) &= 1 + 2^{1/2}2^{1/4}2^{1/4}\\
&= 1 + 2^{1/2+1/4+1/4} = 1 + 2^1 = 1 + 2 = 3\\
f(4, 4, 4) &= 1 + 4^{1/2}4^{1/4}4^{1/4}\\
&= 1 + 4^{1/2+1/4+1/4} = 1 + 4^1 = 1 + 4 = 5
\end{aligned}$$

Substituting these results into the above conditions we get

$$f(2, 2, 2) = 2^k f(1, 1, 1) \Rightarrow 3 = 2^k \times 2$$

and

$$f(4, 4, 4) = 2^k f(2, 2, 2) \Rightarrow 5 = 2^k \times 3$$

The first result implies $2^k = 3/2$ while the second implies $2^k = 5/3$. This is a contradiction and so the function f is not homogeneous.

(c) The function $f(x_1, x_2) = x_1^{1/2}x_2^{1/3} + x_2^{3/2}$ is neither homogeneous nor homothetic. To see that it is not homogeneous one can simply note that upon scaling up all x_i values by the factor s, we obtain

$$f(sx_1, sx_2) = s^{5/6}x_1^{1/2}x_2^{1/3} + s^{3/2}x_2^{3/2}$$

which cannot be written as $s^k f(x_1, x_2)$. However, it is not so obvious how to show that f cannot be written as a monotonic transformation of some function which is homogeneous and so this approach is not so helpful in showing that f is not homothetic. Instead, note that if f is a homothetic function, then it must satisfy the property that the slope of each level curve is equal along any ray from the origin. Thus, f_1/f_2 must be equal at points $(k\bar{x}_1, k\bar{x}_2)$ and $(\bar{x}_1, \bar{x}_2)$. To see that this doesn't hold for the function $f(x_1, x_2) = x_1^{1/2}x_2^{1/3} + x_2^{3/2}$, compute the first-order partial derivatives:

$$f_1(x_1, x_2) = \frac{1}{2}x_1^{-1/2}x_2^{1/3} = \frac{\frac{1}{2}x_2^{1/3}}{x_1^{1/2}}$$

$$\begin{aligned}
f_2(x_1, x_2) &= \frac{1}{3}x_1^{1/2}x_2^{-2/3} + \frac{3}{2}x_2^{1/2}\\
&= \frac{\frac{1}{3}x_1^{1/2}}{x_2^{2/3}} + \frac{3}{2}x_2^{1/2}
\end{aligned}$$

If the function f is homothetic, it follows that the ratio f_1/f_2 must be equal, for example, at the points (1,1) and (2,2). To see that this is not so, note that

$f_1(1, 1) = 1/2$ and $f_1(2, 2) = 1/2^{7/6} \doteq 0.445$

$f_2(1, 1) = 11/6$ and $f_2(2, 2) = \dfrac{1}{3 \times 2^{1/6}} + \dfrac{3}{2^{1/2}} \doteq 2.42$

Thus, at the point (1,1) we get $f_1/f_2 = (1/2)/(11/6) \doteq 0.27$ while at the point (2,2) we get $f_1/f_2 \doteq 0.445/2.42 \doteq 0.18$. Since these two values are not equal, we know that f is not homothetic.

11. (a)

$$\begin{aligned}
dy &= ((-\frac{1}{r})A[\delta x_1^{-r} + (1-\delta)x_2^{-r}]^{-1-1/r}\\
&\quad \times((-r)\delta x_1^{-r-1}))dx_1\\
&\quad +((-\frac{1}{r})A[\delta x_1^{-r} + (1-\delta)x_2^{-r}]^{-1-1/r}((-r)\\
&\quad \times(1-\delta)x_2^{-r-1}))dx_2
\end{aligned}$$

and so

$$\begin{aligned}
dy &= (\delta x_1^{-r-1}A[\delta x_1 + (1-\delta)x_2]^{-1-1/r})dx_1 +\\
&\quad ((1-\delta)x_2^{-r-1}A[\delta x_1^{-r} + (1-\delta)x_2^{-r}]^{-1-1/r})dx_2
\end{aligned}$$

Thus, $dy = 0$ implies

$$\begin{aligned}
\frac{dx_2}{dx_1} &= -\frac{\delta x_1^{-r-1}A[\delta x_1^{-r} + (1-\delta)x_2^{-r}]^{-1-1/r}}{(1-\delta)x_2^{-r-1}A[\delta x_1^{-r} + (1-\delta)x_2^{-r}]^{-1-1/r}}\\
&= -\frac{\delta x_2^{r+1}}{(1-\delta)x_1^{r+1}}
\end{aligned}$$

Thus, the

$$\text{MRTS} = \frac{\delta x_2^{r+1}}{(1-\delta)x_1^{r+1}}, \quad r+1>0$$

gets smaller as one moves along an isoquant from left to right (and downwards) because as x_1 rises and x_2 falls the denominator gets bigger and the numerator gets smaller. Thus, isoquants are strictly convex to the origin.

(d) To show f is homogeneous of degree one, we multiply each input by the same factor s to obtain

$$\begin{aligned} f(sx_1, sx_2) &= A[\delta(sx_1)^{-r} + (1-\delta)(sx_2)^{-r}]^{-1/r} \\ &= A[\delta s^{-r}x_1^{-r} + (1-\delta)s^{-r}x_2^{-r}]^{-1/r} \\ &= A[(s^{-r})(\delta x_1^{-r} + (1-\delta)x_2^{-r})]^{-1/r} \\ &= A(s^{-r})^{-1/r}[\delta x_1^{-r} + (1-\delta)x_2^{-r}]^{-1/r} \\ &= As^1[\delta x_1^{-r} + (1-\delta)x_2^{-r}]^{-1/r} \\ &= sA[\delta x_1^{-r} + (1-\delta)x_2^{-r}]^{-1/r} \\ &= sf(x_1, x_2) \end{aligned}$$

(g) Since f is homogeneous of degree $k=1$, Euler's Theorem, in this case, specifies that

$$f_1x_1 + f_2x_2 = f(x_1, x_2)$$

Upon making the substitutions for f_1 and f_2 on the left side of this equation, we obtain

$$\begin{aligned} & f_1x_1 + f_2x_2 \\ =\ & (\delta x_1^{-r-1}A[\delta x_1^{-r} + (1-\delta)x_2^{-r}]^{-1-1/r})x_1 \\ & +((1-\delta)x_2^{-r-1}A[\delta x_1^{-r} + (1-\delta)x_2^{-r}]^{-1-1/r})x_2 \\ =\ & \delta x_1^{-r}A[\delta x_1^{-r} + (1-\delta)x_2^{-r}]^{-1-1/r} \\ & +(1-\delta)x_2^{-r}A[\delta x_1^{-r} + (1-\delta)x_2^{-r}]^{-1-1/r} \\ =\ & [\delta x_1^{-r} + (1-\delta)x_2^{-r}]A[\delta x_1^{-r} + (1-\delta)x_2^{-r}]^{-1-1/r} \\ =\ & A[\delta x_1^{-r} + (1-\delta)x_2^{-r}]^{-1/r} \end{aligned}$$

and so $f_1x_1 + f_2x_2 = f(x_1, x_2)$.

CHAPTER 12

Section 12.1 (page 472)

1. (a) $y = 3x_1^2 + 2x_2^2 + 5$. The first-order conditions are $6x_1 = 0, 4x_2 = 0$. These can only be satisfied at $x_1^* = x_2^* = 0$. Therefore, $(0, 0)$ is the only stationary point.

(b) $y = 2x_1^2 - 4x_2^2 + 1$. The first-order conditions are $4x_1 = 0, -8x_2 = 0$. These can only be satisfied at $x_1^* = x_2^* = 0$. Therefore, $(0, 0)$ is the only stationary point.

(c) $y = 2x_1 + x_2 - 3x_1^2 - 4x_2^2 + x_1x_2$. The first-order conditions are $2 - 6x_1 + x_2 = 0$, $1 - 8x_2 + x_1 = 0$, which can be written as

$$\begin{bmatrix} 6 & -1 \\ -1 & 8 \end{bmatrix} \begin{bmatrix} x_1 \\ x_2 \end{bmatrix} = \begin{bmatrix} 2 \\ 1 \end{bmatrix}$$

Solving by Cramer's rule yields

$$x_1^* = 17/47 = 0.36 \quad x_2^* = 8/47 = 0.17$$

Thus, $(17/47, 8/47)$ is the only stationary point.

(d) $y = 0.5x_1 - 2x_1^2 + 4x_2 - 3x_2^2 + 2x_1x_2$. The first-order conditions are $0.5 - 4x_1 + 2x_2 = 0$, $4 - 6x_2 + 2x_1 = 0$, which can be written as

$$\begin{bmatrix} 4 & -2 \\ -2 & 6 \end{bmatrix} \begin{bmatrix} x_1 \\ x_2 \end{bmatrix} = \begin{bmatrix} 0.5 \\ 4 \end{bmatrix}$$

Solving by Cramer's rule yields $x_1^* = 11/20 = 0.55$, $x_2^* = 17/20 = 0.85$. Thus, $(11/20, 17/20)$ is the only stationary point.

(e) $y = 2x_1^3 - 3x_1x_2 + x_1^2 - 2x_2^2$. The first order conditions are $6x_1^2 - 3x_2 + 2x_1 = 0$ and $-3x_1 - 4x_2 = 0$. From the second equation we have

$$x_2 = -\frac{3}{4}x_1$$

Substituting into the first yields

$$6x_1^2 + \frac{9}{4}x_1 + \frac{8}{4}x_1 = 0 \Rightarrow x_1^2 = -\frac{17}{24}x_1$$

The solutions for x_1 are $x_{11}^* = 0$ and $x_{12}^* = -17/24 = -0.71$. This implies $x_{21}^* = 0$ and $x_{22}^* = 17/32 = 0.53$. Thus, (0.0) and $(-17/24, 17/32)$ are the stationary values of the function.

(f) $y = x_1x_2^{0.5}(10 - x_1 - x_2)^{0.4}$. First, we will find the stationary values of this function. This becomes extremely tedious if you try to solve it directly. Instead we rewrite the function as

$$\begin{aligned} y &= (x_1^{2.5})^{0.4}(x_2^{1.25})^{0.4}(10 - x_1 - x_2)^{0.4} \\ &= (10x_1^{2.5}x_2^{1.25} - x_1^{3.5}x_2^{1.25} - x_1^{2.5}x_2^{2.25})^{0.4} \end{aligned}$$

Now we define

$$\begin{aligned} \tilde{f} &\equiv y^{2.5} \\ &= 10x_1^{2.5}x_2^{1.25} - x_1^{3.5}x_2^{1.25} - x_1^{2.5}x_2^{2.25} \end{aligned}$$

The stationary values of this function are the stationary values of f and vice versa. The first-order conditions are

$$\begin{aligned} \tilde{f}_1 &= 25x_1^{1.5}x_2^{1.25} - 3.5x_1^{2.5}x_2^{1.25} - 2.5x_1^{1.5}x_2^{2.25} \\ &= x_1^{1.5}x_2^{1.25}(25 - 3.5x_1 - 2.5x_2) = 0 \\ \tilde{f}_2 &= 12.5x_1^{2.5}x_2^{0.25} - 1.25x_1^{3.5}x_2^{0.25} - 2.25x_1^{2.5}x_2^{1.25} \\ &= x_1^{2.5}x_2^{0.25}(12.5 - 1.25x_1 - 2.25x_2) = 0 \end{aligned}$$

Obviously, if $x_1 = 0$ or $x_2 = 0$, the first-order conditions are fulfilled. Thus, all points $(0, x_2)$ and $(x_1, 0)$ are stationary values.

A further stationary value is found by determining the value of x_1 and x_2 for which the terms in brackets of $\tilde{f}_1$ and $\tilde{f}_2$ are zero. This problem can be written as

$$\begin{bmatrix} -3.5 & -2.5 \\ -1.25 & -2.25 \end{bmatrix} \begin{bmatrix} x_1 \\ x_2 \end{bmatrix} = \begin{bmatrix} -25 \\ -12.5 \end{bmatrix}$$

Using Cramer's rule we obtain the stationary value $(5\frac{5}{19}, 2\frac{12}{19})$.

(g) $y = (x_1^2 + x_2^4 + x_3^6)^2$. The first-order conditions are

$$\begin{aligned} f_1 &= 2(x_1^2 + x_2^4 + x_3^6)2x_1 = 0 \\ f_2 &= 2(x_1^2 + x_2^4 + x_3^6)4x_2^3 = 0 \\ f_3 &= 2(x_1^2 + x_2^4 + x_3^6)6x_3^5 = 0 \end{aligned}$$

$x_1^* = 0$, $x_2^* = 0$, $x_3^* = 0$ is a solution of this system of equations. The terms in brackets are positive if either x_1 or x_2 or x_3 are not zero. Thus, $(0, 0, 0)$ is the only stationary value of the function.

(h) $y = 4x_1^2 + 2x_2^2 + x_3^2 - x_1 + 5x_3$. The first-order conditions are

$$\begin{aligned} 8x_1 - 1 &= 0 \Rightarrow x_1^* = 1/8 \\ 4x_2 &= 0 \Rightarrow x_2^* = 0 \\ 2x_3 + 5 &= 0 \Rightarrow x_3^* = -5/2 \end{aligned}$$

Thus, $(\frac{1}{8}, 0, -\frac{5}{2})$ is the only stationary value of the function.

(i) $y = x_1^3 + x_2^3 - 3x_1x_2$. The first-order conditions are

$$\begin{aligned} 3x_1^2 - 3x_2 &= 0 \\ 3x_2^2 - 3x_1 &= 0 \end{aligned}$$

From the second equation we obtain $x_1 = x_2^2$ and substituting into the first yields

$$3x_2^4 - 3x_2 = 0 \Rightarrow x_2^4 - x_2 = x_2(x_2^3 - 1) = 0$$

Thus, $x_{21}^* = 0$ is a solution. Since $x_2^3 - 1 = 0 \Rightarrow x_2 = \sqrt[3]{1} = 1$, we have $x_{22}^* = 1$ as the second solution. The corresponding values for x_1 are $x_{11}^* = 0$ and $x_{12}^* = 1$. Therefore $(0, 0)$ and $(1, 1)$ are the stationary values of the function.

(j)

$$
\begin{aligned}
y &= 2(x_1 - x_2)^2 - x_1^4 - x_2^4 \\
&= 2x_1^2 - 4x_1x_2 + 2x_2^2 - x_1^4 - x_2^4
\end{aligned}
$$

The first-order conditions are

$$
\begin{aligned}
4x_1 - 4x_2 - 4x_1^3 &= 0 \Rightarrow x_1 - x_2 - x_1^3 = 0 \\
-4x_1 + 4x_2 - 4x_2^3 &= 0 \Rightarrow -x_1 + x_2 - x_2^3 = 0
\end{aligned}
$$

From the second equation we have $x_1 = x_2 - x_2^3$. Substituting into the first we obtain

$$x_2 - x_2^3 - x_2 = (x_2 - x_2^3)^3$$

or

$$-x_2^3 = (x_2 - x_2^3)^3$$

Taking the third root we obtain

$$-x_2 = x_2 - x_2^3$$

or

$$x_2^3 - 2x_2 = x_2(x_2 - 2) = 0$$

This yields the three solutions $x_{21}^* = 0$, $x_{22}^* = \sqrt{2} = 1.41$ and $x_{23}^* = -\sqrt{2} = -1.41$. The corresponding values of x_1 are $x_{11}^* = 0$, $x_{12}^* = -\sqrt{2}$ and $x_{13}^* = \sqrt{2}$.

Thus, $(0, 0)$, $(\sqrt{2}, -\sqrt{2})$, and $(-\sqrt{2}, \sqrt{2})$ are the stationary values of the function.

3. The firm's profit function is

$$
\begin{aligned}
\pi(q_1, q_2) &= p_1q_1 + p_2q_2 - C \\
&= p_1q_1 + (100 - q_2)q_2 - (q_1 + q_2)^2 \\
&= p_1q_1 + 100q_2 - q_2^2 - q_1^2 - 2q_1q_2 - q_2^2 \\
&= p_1q_1 - q_1^2 - 2q_1q_2 + 100q_2 - 2q_2^2
\end{aligned}
$$

The firm maximizes profits. The first-order conditions are

$$
\begin{aligned}
\pi_1(q_1^*, q_2^*) &= p_1 - 2q_1 - 2q_2 = 0 \\
\pi_2(q_1^*, q_2^*) &= -2q_1 + 100 - 4q_2 = 0
\end{aligned}
$$

which can be written as

$$
\begin{bmatrix} 2 & 2 \\ 2 & 4 \end{bmatrix} \begin{bmatrix} q_1 \\ q_2 \end{bmatrix} = \begin{bmatrix} p_1 \\ 100 \end{bmatrix}
$$

Using Cramer's rule we obtain $q_1^* = p_1 - 50$, $q_2^* = 50 - 0.5p_1$, $p_2^* = 50 + 0.5p_1$.

Case (i) $p_1 = \$60$: $q_1^* = 10$, $q_2^* = 20$, $p_2 = \$80$. The firm will sell the quantity in the market in which it has a monopoly that equates marginal revenue to price in the competitive market. This can be seen if we set $R_2'(q_2) = 100 - 2q_2 = 60$. Then we obtain the optimal output $q_2^* = 20$.

Case (ii) $p_1 = \$10$: Since quantities must be non-negative the solution is $q_1^* = 0$, $q_2^* = 25$ and $p_2^* = \$75$.

Now the firm produces only for the market in which it has a monopoly. The reason is that the price in the competitive market has fallen below the marginal revenue the firm obtains at its monopoly market optimum which is \$50. Therefore the competitive market is of no interest to the firm.

5. The firm's profit functions are

$$
\begin{aligned}
\pi_i(q_1, q_2) &= pq_i - c_iq_i \\
&= 10q_i - 0.1(q_1 + q_2)q_i - c_iq_i \quad i = 1, 2
\end{aligned}
$$

where c_i is constant marginal cost of firm i with $c_1 = 0.25$ and $c_2 = 0.5$.

Cournot duopoly:

If we maximize firm 1's profit with the other firm's output as given, we obtain for the two firms the first-order conditions

$$
\begin{aligned}
\frac{\partial \pi_1}{\partial q_1} &= 10 - 0.2q_1 - 0.1q_2 - 0.25 = 0 \\
\frac{\partial \pi_2}{\partial q_2} &= 10 - 0.2q_2 - 0.1q_1 - 0.50 = 0
\end{aligned}
$$

which can be written as

$$
\begin{bmatrix} 0.2 & 0.1 \\ 0.1 & 0.2 \end{bmatrix} \begin{bmatrix} q_1 \\ q_2 \end{bmatrix} = \begin{bmatrix} 9.75 \\ 9.5 \end{bmatrix}
$$

Using Cramer's rule yields $q_1^* = 33.33$, $q_2^* = 30.83$. Substituting into the demand and profit functions we obtain the further equilibrium solutions

$$
\begin{aligned}
p^* &= 3.58 \\
\pi_1(q_1^*, q_2^*) &= 111.12 \\
\pi_2(q_1^*, q_2^*) &= 95.07 \\
\pi_1(q_1^*, q_2^*) + \pi_2(q_1^*, q_2^*) &= 206.19
\end{aligned}
$$

Thus, the firm with the higher unit costs makes lower profits and produces less than the firm with the lower unit costs. This second characteristic becomes clear when we consider the production decision of the firms. They maximize profits given the other firm's output. This is similar to the monopolist's problem but now the demand function the firm faces is residual demand. The first-order condition, however, is still marginal revenue equal to marginal cost. Obviously, the higher is marginal cost, the lower will be the optimal output because marginal revenue depends negatively on output. Thus, given any output of the other firm, the higher marginal costs, the lower the optimal quantity of a firm. This effect causes the firm with higher marginal cost to produce less in the Cournot equilibrium.

Joint-profit maximization:

When the firms maximize joint profits, only firm i with the lower unit costs will produce, i.e. $q_2^* = 0$. Hence, the problem is identical to the monopoly case with firm 1 as the monopolist. Thus, joint-profit maximization implies

$$
\begin{aligned}
\max_{q_1} \pi_i(q_1, q_2) &= 10q_1 - 0.1q_1^2 - 0.25q_1 \\
&= 9.75q_1 - 0.1q_1^2
\end{aligned}
$$

The first-order condition is

$$\frac{\partial \pi_1}{\partial q_1} = 9.75 - 0.2q_1 = 0$$

so $q_1^* = 48.75$. This implies $p^* = 5.13$ and

$$\pi_1(q_1^*, q_2^*) = \pi_1(q_1^*, q_2^*) + \pi_2(q_1^*, q_2^*) = 237.66$$

which compares to 206.19 in the Cournot duopoly case. Thus joint profits increase. There are two reasons for this. First, the less efficient producer does not produce. Second, there is no competition between the firms.

7. The monopolist maximizes the profit function

$$\begin{aligned}\pi(q_1, q_2) &= p(q_1+q_2) - C_1 - C_2 \\ &= [100-(q_1+q_2)](q_1+q_2) - 2q_1^2 - 3q_2^2 \\ &= 100q_1 + 100q_2 - 2q_1q_2 - 3q_1^2 - 4q_2^2\end{aligned}$$

The first-order conditions are

$$\frac{\partial \pi}{\partial q_1} = 100 - 2q_2 - 6q_1 = 0$$

$$\frac{\partial \pi}{\partial q_2} = 100 - 2q_1 - 8q_2 = 0$$

which can be written as

$$\begin{bmatrix} 6 & 2 \\ 2 & 8 \end{bmatrix}\begin{bmatrix} q_1 \\ q_2 \end{bmatrix} = \begin{bmatrix} 100 \\ 100 \end{bmatrix}$$

Using Cramer's rule yields $q_1^* = 13.64$, $q_2^* = 9.09$. Thus, $p^* = 77.27$. From this we can calculate marginal costs of the two plants at the optimum. They are

for plant 1: $4q_1^* = 54.55$
for plant 2: $6q_2^* = 54.55$

and so they are equalized. This is a general characteristic of this problem. The intuition behind this result is straightforward. At the optimum, marginal revenues of the products of two plants are the same since their products are identical. If now marginal costs are unequal, the monopolist could reduce its costs and thus increase profits by switching the production of one unit of output to the plant with the lower marginal cost.

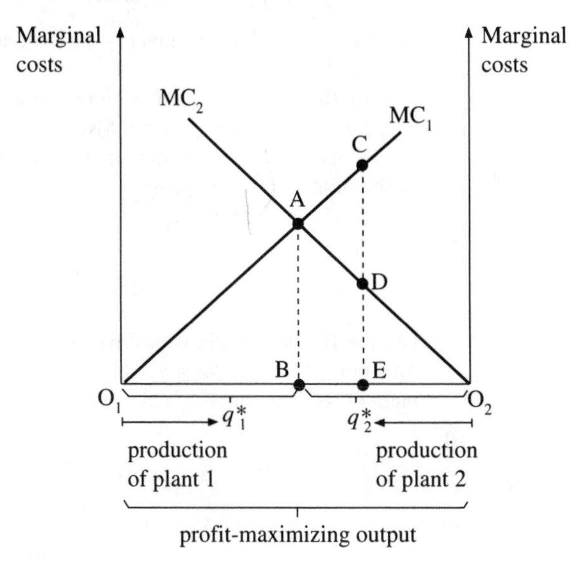

Figure 12.1.7

Figure 12.1.7 illustrates this result. The length of the x-axis corresponds to profit-maximizing output. From the left we measure output of plant 1, from the right output of plant 2. The marginal cost functions are shown. Note that the area under the marginal cost functions is the total cost of the plants. The division of output B describes the optimum. Total cost equals the areas of the triangles 0_1AB (plant 1) and BA0_2 (plant 2), i.e., the area of the triangle 0_1A0_2. To see that this must be the optimum, consider any other division of output among the plants, for example the one described by E. Total costs are now the sum of the areas of the triangle 0_1CE (plant 1) and ED0_2 (plant 2). Compared to the optimum, costs are now higher by the area of the triangle ACD.

Section 12.2 (page 481)

1. (a) $y = 3x_1^2 + 2x_2^2 + 5$

$$f_1 = 6x_1 \qquad f_2 = 4x_2$$
$$f_{11} = 6 \qquad f_{12} = 0$$
$$f_{21} = 0 \qquad f_{22} = 4 \quad \text{for any } x \in \mathbb{R}^n$$

$$F = \begin{vmatrix} 6 & 0 \\ 0 & 4 \end{vmatrix} = 24 > 0$$

for any $x \in \mathbb{R}^n$. Thus, the Hessian Matrix is positive definite. Applying Theorem 12.3 shows that the stationary point $(0, 0)$ yields a minimum of the function.

(b) $y = 2x_1^2 - 4x_2^2 + 1$

$$f_1 = 4x_1 \qquad f_2 = -8x_2$$
$$f_{11} = 4 \qquad f_{12} = 0$$
$$f_{21} = 0 \qquad f_{22} = -8 \text{ for any } x \in \mathbb{R}^n$$

$$F = \begin{vmatrix} 4 & 0 \\ 0 & -8 \end{vmatrix} = -32 < 0$$

for any $x \in \mathbb{R}^n$. Neither Theorem 12.3 nor 12.5 can be applied in this case. In fact, the function has a saddle point at the stationary point $(0, 0)$ as can be seen by noting that $f_{11} > 0$ means that the function reaches a minimum in the x_1-direction while $f_{22} < 0$ means that it reaches a maximum in the x_2-direction.

(c) $y = 2x_1 + x_2 - 3x_1^2 - 4x_2^2 - x_1x_2$

$$f_1 = 2 - 6x_1 + x_2$$
$$f_2 = 1 - 8x_2 + x_1$$
$$f_{11} = -6 \qquad f_{12} = f_{21} = 1 \quad f_{22} = -8$$

for any $x \in \mathbb{R}^n$.

$$F = \begin{vmatrix} -6 & 1 \\ 1 & -8 \end{vmatrix} = 47 > 0$$

for any $x \in \mathbb{R}^n$. Hence, the Hessian matrix is negative definite. Thus, by Theorem 12.3, we have that the stationary point $(\frac{17}{47}, \frac{8}{47})$ yields a maximum.

(d) $y = 0.5x_1 - 2x_1^2 + 4x_2 - 3x_2^2 + 2x_1x_2$

$$f_1 = 0.5 - 4x_1 + 2x_2$$
$$f_2 = 4 - 6x_2 + 2x_1 = 0$$
$$f_{11} = -4 \qquad f_{12} = f_{21} = 2 \quad f_{22} = -6$$

for any $x \in \mathbb{R}^n$.

$$F = \begin{vmatrix} -4 & 2 \\ 2 & -6 \end{vmatrix} = 20 > 0$$

for any $x \in \mathbb{R}^n$. Thus, the Hessian matrix is negative definite. Applying Theorem 12.3 shows that the stationary point $(\frac{11}{20}, \frac{17}{20})$ gives us a maximum of the function.

(e) $y = 2x_1^3 - 3x_1x_2 + x_1^2 - 2x_2^2$

$$f_1 = 6x_1^2 - 3x_2 + 2x_1$$

$$f_2 = 3x_1 - 4x_2$$
$$f_{11} = 12x_1 + 2 \quad f_{12} = f_{21} = -3 \quad f_{22} = -4$$

for any $x \in \mathbb{R}^n$. The function has the stationary points $(0, 0)$ and $(-\frac{17}{24}, \frac{17}{32})$. At the first of these $f_{11} = 2 > 0$ and

$$F = \begin{vmatrix} 2 & -3 \\ -3 & -4 \end{vmatrix} = -17 < 0$$

Neither Theorem 12.3 nor 12.5 can be applied in this case. In fact, the function has a saddle point at $(0, 0)$ as can be seen by noting that $f_{11} > 0$ means that the function reaches a minimum in the x_1-direction while $f_{22} < 0$ means that it reaches a maximum in the x_2-direction.

At $(-\frac{17}{24}, \frac{17}{32})$ $f_{11} = -17/2 + 2 = -13/2 < 0$ and

$$F = \begin{vmatrix} -13/2 & -3 \\ 3 & -4 \end{vmatrix} = 35 > 0$$

i.e., the Hessian matrix is negative definite. Applying Theorem 12.3 tells us that this point yields a maximum of the function.

(f) $y = x_1 x_2^{0.5}(10 - x_1 - x^2)^{0.4}$

First, we examine the stationary values $(0, x_2)$. We then have $y = 0$. Note that $x_2^{0.5}$ and $(10 - x_1 - x_2)^{0.4}$ must be non-negative. Thus, we have $y \le 0$ for $x_1 < 0$ and $y \ge 0$ for $x_1 > 0$. Since variation of x_2 leaves $y = 0$, we have neither an extremum nor a saddle point.

Second, we take a look at the stationary values $(x_1, 0)$. Note that we must have $x_2 \ge 0$. Otherwise $x_2^{0.5}$ is not defined. Thus, we only need to examine $x_2 > 0$. Now $(10 - x_1 - x_2)^{0.4}$ must be non- negative.

(i) If this term is positive, i.e., $x_1 + x_2 < 10$, we then have

$$y \begin{cases} < 0 & \text{for } x_1 < 0 \\ > 0 & \text{for } x_1 > 0 \end{cases}$$

Thus, $(x_1, 0)$ yields a maximum for $x_1 < 0$ and a minimum for $x_1 > 0$. If $x_1 = 0$, we have neither an extremum nor a saddle point.

(ii) If $(10 - x_1 - x_2)^{0.4}$ is zero, i.e., $x_1 + x_2 = 10$, we can only increase x_2 if we simultaneously decrease x_1. If we decrease x_1 by more than we increase x_2, the term $(10 - x_1 - x_2)^{0.4}$ turns positive and the result must be as in (i).

Finally, we determine the nature of $(5\frac{5}{19}, 2\frac{12}{19})$. Here we will use the first-order conditions of the function $\tilde{f}$. The results are the same as if we would examine the first-order conditions of the original function f because $\tilde{f}$ is just a monotonic transformation of f.

$$\tilde{f}_1 = x_1^{1.5} x_2^{1.25}(25 - 3.5x_1 - 2.5x_2)$$
$$\tilde{f}_2 = x_1^{2.5} x_2^{0.25}(12.5 - 1.25x_1 - 2.25x_2)$$

The first two terms of the first-order conditions are positive around the stationary value. Thus, we can easily see that the first derivatives turn negative if we increase x_1 and/or x_2 and positive if we decrease x_1 and/or x_2. Thus, $(5\frac{5}{19}, 2\frac{12}{19})$ must be a maximum.

(g) $y = (x_1^2 + x_2^4 + x_3^6)^2$. $(0, 0, 0)$ must be a global minimum since $f(0, 0, 0) = 0$ and $f > 0$ for all other (x_1, x_2, x_3). Thus, it is not necessary to compute the Hessian matrix.

(h) $y = 4x_1^2 + 2x_2^2 + x_3^2 - x_1 + 5x_3$

$$f_1 = 8x_1 - 1$$
$$f_2 = 4x_2$$
$$f_3 = 2x_3 + 5$$
$$f_{11} = 8 \quad f_{12} = 0$$
$$f_{13} = 0 \quad f_{21} = 0$$
$$f_{22} = 4 \quad f_{23} = 0$$
$$f_{31} = 0 \quad f_{32} = 0$$
$$f_{33} = 2 \quad \text{for any } x \in \mathbb{R}^n$$

$$F = \begin{vmatrix} 8 & 0 & 0 \\ 0 & 4 & 0 \\ 0 & 0 & 2 \end{vmatrix} = 32 > 0$$

for any $x \in \mathbb{R}^n$. Thus, the Hessian matrix is positive definite. By Theorem 12.5 the stationary point $(\frac{1}{8}, 0, -\frac{5}{2})$ yields a minimum.

(i) $y = x_1^3 + x_2^3 - 3x_1x_2$

$$f_1 = 3x_1^2 - 3x_2$$
$$f_2 = 3x_2^2 - 3x_1$$
$$f_{11} = 6x_1 \quad f_{12} = -3$$
$$f_{21} = -3 \quad f_{22} = 6x_2$$

for any $x \in \mathbb{R}^n$. The function has the stationary points $(0, 0)$ and $(1, 1)$.

At $(0, 0)$, $f_{11} = 0$. Therefore neither Theorem 12.3 nor 12.5 can be applied. Also the second-order conditions f_{11} and f_{22} do not allow us to identify the nature of this stationary point.

At $(1, 1)$, $f_{11} = 6$ and

$$F = \begin{vmatrix} 6 & -3 \\ -3 & 6 \end{vmatrix} = 27 > 0$$

i.e., the Hessian matrix is positive definite. Applying Theorem 12.5 the stationary value $(1, 1)$ yields a minimum of the function.

(j)

$$y = 2(x_1 - x_2)^2 - x_1^4 - x_2^4$$
$$= 2x_1^2 - 4x_1x_2 + 2x_2^2 - x_1^4 - x_2^4$$

and so

$$f_1 = 4x_1 - 4x_2 - 4x_1^3$$
$$f_2 = -4x_1 + 4x_2 - 4x_2^3$$
$$f_{11} = 4 - 12x_1^2$$
$$f_{12} = f_{21} = -4$$
$$f_{22} = 4 - 12x_2^2$$

for any $x \in \mathbb{R}^n$. The function has the three stationary points $(0, 0)$, $(\sqrt{2}, -\sqrt{2})$ and $(-\sqrt{2}, \sqrt{2})$.

At $(0, 0)$, $f_{11} = 4$

$$F = \begin{vmatrix} 4 & -4 \\ -4 & 4 \end{vmatrix} = 0$$

Neither Theorem 12.3 nor 12.5 can be applied in this case. Also the second-order conditions f_{11} and f_{22} do not allow us to identify the nature of this stationary point.

At $(\sqrt{2}, -\sqrt{2})$, $f_{11} = -20$

$$F = \begin{vmatrix} -20 & -4 \\ -4 & -20 \end{vmatrix} = 384 > 0$$

i.e., the Hessian matrix is negative definite. Applying Theorem 12.3 shows that $(\sqrt{2}, -\sqrt{2})$ yields a maximum.

At $(-\sqrt{2}, \sqrt{2})$, $f_{11} = -20$ and

$$F = \begin{vmatrix} -20 & -4 \\ -4 & -20 \end{vmatrix} = 384 > 0$$

i.e., the Hessian matrix is negative definite. By Theorem 12.3 we obtain that $(-\sqrt{2}, \sqrt{2})$ yields a maximum.

3. In the Cournot duopoly example, each firm maximizes its profits, given by

$$\pi_i = pq_i = 100q_i - (q_1 + q_2)q_i \quad i = 1, 2$$

The first-order conditions are

$$\frac{\partial \pi_i}{\partial q_i} = 100 - 2q_i - q_j = 0 \quad i \neq j$$

which implies $q_i^* = 50 - \frac{q_j}{2}$.

The second-order conditions are

$$\frac{\partial^2 \pi_i}{\partial q_i} = -2 < 0$$

Thus, q_1^* must be profit-maximizing output.

5. The firm's objective is to

$$\max_{K,L} \pi(L, K) = 64L^{0.25}K^{0.5} - 2L - 4K$$

The first-order conditions are

$$\frac{\partial \pi}{\partial L} = 16L^{-0.75}K^{0.5} - 2 = 0$$
$$\frac{\partial \pi}{\partial K} = 32L^{0.25}K^{-0.5} - 4 = 0$$

Rearranging to

$$16L^{-0.75}K^{0.5} = 2$$
$$32L^{0.25}K^{-0.5} = 4$$

and dividing the first by the second equation yields

$$\frac{1}{2}\frac{K}{L} = \frac{1}{2} \Rightarrow K = L$$

Substituting L for K in the first equation we obtain

$$16K^{-0.25} = 2 \Rightarrow 8 = K^{0.25} \Rightarrow K = 4096$$

Thus, the profit-maximizing use of labor and capital is $K = L = 4096$.

We confirm this result by using the second-order conditions:

$$\frac{\partial^2 \pi}{\partial L^2} = -12L^{-1.75}K^{0.5}$$
$$\frac{\partial^2 \pi}{\partial L \partial K} = 8L^{-0.75}K^{-0.5}$$
$$\frac{\partial^2 \pi}{\partial K \partial L} = 8L^{-0.75}K^{-0.5}$$
$$\frac{\partial^2 \pi}{\partial K^2} = -16L^{-0.75}K^{-0.5}$$

At the optimum we find that the Hessian matrix of second-order partials is

$$H^* = \frac{1}{8192}\begin{bmatrix} -3 & 2 \\ 2 & -4 \end{bmatrix}$$

with $F = \frac{8}{8192} > 0$. Thus, H^* is negative definite. By Theorem 12.3 $K = L = 4096$ yields a maximum.

7. The problem is to maximize $y = -(x_1^4 + x_2^4)$. The first-order conditions are

$$f_1 = -4x_1^3 = 0 \Rightarrow x_1^* = 0$$
$$f_2 = -4x_2^3 = 0 \Rightarrow x_2^* = 0$$

Thus, $(0, 0)$ is the only stationary value of the function. Looking at the function we immediately see that this must be maximum. At $(0, 0)$, $y = 0$ while for all other points (x_1, x_2) we have $y < 0$. However, a problem arises if we want to apply Theorem 12.3. With the second-order conditions

$$f_{11} = -12x_1^2$$
$$f_{12} = f_{21} = 0$$
$$f_{22} = -12x_2^2$$

we have at $(0, 0)$

$$F = \begin{vmatrix} 0 & 0 \\ 0 & 0 \end{vmatrix} = 0$$

Hence, Theorem 12.3 cannot be applied. Fortunately, there is a trick which allows us to apply Theorem 12.3 anyway. Define $z_1 = x_1^2$ and $z_2 = x_2^2$ and substitute into the original function so you obtain $y = -(z_1^2 + z_2^2)$. This function will have a negative definite Hessian confirming that we have found a maximum.

Section 12.3 (page 491)

1. (a)

$$\min y = 3x_1^2 + 2x_2^2 + 5$$
$$\text{s.t. } 0 \leq x_1 \leq 10 \quad 2 \leq x_2 \leq 10$$

This function is quadratic and increasing in x_1 and in x_2 in the intervals to which the variables are restricted. Thus, we can guess immediately that the solution is at the lower bounds of x_1 and x_2, i.e.,

$$x_1^* = 0 \quad x_2^* = 2$$

This point satisfies the necessary conditions for a minimum since

$$f_1(0, 2) = 0 \geq 0 \qquad (0 - 0)0 = 0$$
$$f_2(0, 2) = 8 \geq 0 \qquad (2 - 2)8 = 0$$

at $(0, 2)$.

(b)

$$\max y = 2x_1 + x_2 - 3x_1^2 - 4x_2^2 + x_1x_2$$
$$\text{s.t. } 0 \leq x_1 \leq 1 \quad 0 \leq x_2 \leq 1$$
$$f_1 = 2 - 6x_1 + x_2$$
$$f_2 = 1 - 8x_2 + x_1$$

f_1 is positive at $x_1 = 0$ and negative at $x_1 = 1$ no matter what value x_2 takes. The same holds for f_2. Hence, there must be an interior solution. Setting the first-order conditions equal to zero, we obtain $(\frac{17}{47}, \frac{8}{47})$ as the solution. We have

$$\begin{aligned} f_1 &= 0 \\ (\frac{17}{47} - 0) f_1 &= (1 - \frac{17}{47}) f_1 = 0 \\ f_2 &= 0 \\ (\frac{8}{47} - 0) f_2 &= (1 - \frac{8}{47}) f_2 = 0 \end{aligned}$$

which satisfies the necessary conditions for a maximum.

(c)

$$\max y = 2x_1 + x_2 - 3x_1^2 - 4x_2^2 + x_1 x_2$$

$$\text{s.t. } 1 \leq x_1 \leq 2 \quad 1 \leq x_2 \leq 2$$

The function is identical to the one in part (b). However, the new constraints exclude the old optimum. Hence, we can guess that the lower bound of the feasible set is optimal, i.e.,

$$x_1^* = 1 \quad x_2^* = 1$$

$$f_1(1, 1) = -3 \leq 0 \quad (1 - 1)(-3) = 0$$

$$f_2(2, 2) = -5 \leq 0 \quad (1 - 1)(-6) = 0$$

$(1, 1)$ fulfills the necessary conditions for a maximum.

(d)

$$\max y = 2x_1 + x_2 - 3x_1^2 - 4x_2 + x_1 x_2$$

$$\text{s.t. } 0 \leq x_1 \leq 1 \quad 1 \leq x_2 \leq 2$$

$$f_1 = 2 - 6x_1 + x_2$$

$$f_2 = -3 + x_1$$

For every feasible value of x_1, f_2 is negative. Thus, $x_2^* = 1$. Inserting in $f_1 = 0$ yields $x_1^* = 0.5$. Since we have

$$f_1(0.5, 1) = 0 \quad (0.5 - 0) f_1 = (1 - 0.5) f_1 = 0$$

$$f_2(0.5, 1) = -2.5 \leq 0 \quad (1 - 1) f_2 = 0$$

$(0.5, 1)$ fulfills the necessary conditions for a maximum.

(e)

$$\max y = 2x_1^3 - 3x_1 x_2 + x_1^2 - 2x_2^2$$

$$\text{s.t. } 0 \leq x_1 \leq 1 \quad 0 \leq x_2 \leq 1$$

$$\begin{aligned} f_1 &= 6x_1^2 - 3x_2 + 2x_1 \\ f_2 &= -3x_1 - 4x_2 \end{aligned}$$

f_2 is always negative which implies $x_2^* = 0$. Substituting into f_1 we have that f_1 will be positive throughout the interval giving $x_1^* = 1$. The necessary conditions for a maximum are fulfilled since

$$f_1(1, 0) = 8 \geq 0 \quad (1 - 1) f_1 = 0$$

$$f_2(1, 0) = -3 \leq 0 \quad (0 - 0) f_2 = 0$$

(f)

$$\max y = 2x_1^3 - 3x_1 x_2 + x_1^2 - 2x_2^2$$

$$\text{s.t. } 0 \leq x_1 \leq 1 \quad 1 \leq x_2 \leq 2$$

By the same argument as in part (e) we have $x_2^* = 1$. Thus

$$f_1 = 6x_1^2 + 2x_1 - 3$$

Checking at the borders we find that at

$$x_1 = 0: \quad f_1 = -3 \Rightarrow (0, 1)$$

is a local maximum while at

$$x_1 = 1: \quad f_1 = 5 \Rightarrow (1, 1)$$

is a local maximum. Inserting, yields $f(0, 1) = -2$ and $f(1, 1) = -2$. Thus, both are also global maxima.

Finally note that $f_1 = 0$ at

$$x_1 = -\frac{1}{6} + \frac{\sqrt{76}}{12} \doteq 0.56$$

However, this cannot be a maximum since $f_{11} > 0$.

3. If unit costs are constant at c, the monopolist's problem is

$$\max \pi(q_1, q_2) = p_1(q_1) q_1 + p_2(q_1) q_2 - c(q_1 + q_2)$$

The first-order conditions are

$$\begin{aligned} \frac{\partial \pi}{\partial q_1} &= \frac{\partial p_1}{\partial q_1} q_1 + p_1 - c = 0 \\ \frac{\partial \pi}{\partial q_2} &= \frac{\partial p_2}{\partial q_2} q_2 + p_2 - c = 0 \end{aligned}$$

If a quota is imposed on q_1 (such that the second first-order condition is no longer fulfilled), the first first-order condition remains unchanged. Thus, the optimal solution q_1^* remains unchanged and so does $p_1(q_1)$.

This compares to the example in the text. There the effect of the quota was to alter marginal costs and Thus, change the optimal quantity in country 1. Here, however, marginal costs are independent of output. Hence, there is no way in which the quota in country 2 can affect optimal quantity and price in country 1.

Review Exercises (page 493)

1. (a) $y = 0.5x_1^2 + 2x_2^2$

$$f_1 = x_1 = 0 \quad f_2 = 4x_2 = 0$$

Thus, $(0, 0)$ is the only stationary value.

$$f_{11} = 1 \quad f_{12} = f_{21} = 0$$

$$f_{22} = 4 \quad \text{for any } x \in \mathbb{R}^n$$

$$F = \begin{vmatrix} 1 & 0 \\ 0 & 4 \end{vmatrix} = 4 > 0$$

for any $x \in \mathbb{R}^n$. Thus, the Hessian matrix is positive definite. $(0, 0)$ is a minimum.

(b) $y = x_1 + x_2 - x_1^2 - x_2^2 + x_1 x_2$

$$\begin{aligned} f_1 &= 1 - 2x_1 + x_2 = 0 \\ f_2 &= 1 - 2x_2 + x_1 = 0 \end{aligned}$$

so that $x_1^* = 1$ and $x_2^* = 1$. Thus, $(1, 1)$ is the only stationary value.

$$f_{11} = -2 \quad f_{12} = f_{21} = 1$$

$$f_{22} = -2 \quad \text{for any } x \in \mathbb{R}^n$$

$$F = \begin{vmatrix} -2 & 1 \\ 1 & -2 \end{vmatrix} = 3 > 0$$

for any $x \in \mathbb{R}^n$. Thus, the Hessian matrix is negative definite. $(1, 1)$ is a maximum.

(c) $y = 10x_1 + 2x_2 - 0.5x_1^2 - 2x_2^2 + 5x_1x_2$

$$f_1 = 10 - x_1 + 5x_2 = 0$$
$$f_2 = 2 - 4x_2 + 5x_1 = 0$$

or

$$\begin{bmatrix} -1 & 5 \\ 5 & -4 \end{bmatrix} \begin{bmatrix} x_1 \\ x_2 \end{bmatrix} = \begin{bmatrix} -10 \\ -2 \end{bmatrix}$$

Using Cramer's rule yields

$$x_1^* = -\frac{50}{21} \quad x_2^* = -\frac{52}{21}$$

The only stationary value is $(-\frac{50}{21}, -\frac{52}{21})$.

$$f_{11} = -1 \quad f_{12} = f_{21} = 5$$

$$f_{22} = -4$$

$$F = \begin{vmatrix} -1 & 5 \\ 5 & -4 \end{vmatrix} = -21 < 0$$

Thus, $(-\frac{50}{21}, -\frac{52}{21})$ is neither an extremum nor a saddle point.

(d) $y = 2x_1x_2 - x_1^3 - x_2^2$

$$f_1 = 2x_2 - 3x_1^2 = 0$$
$$f_2 = 2x_1 - 2x_2 = 0$$

Solving the second equation for x_2 yields $x_2 = x_1$. Substituting into the first we obtain

$$2x_1 - 3x_1^2 = x_1(2 - 3x_1) = 0$$

Thus, $(0, 0)$ and $(\frac{2}{3}, \frac{2}{3})$ are stationary values.

$$f_{11} = -6x_1 \quad f_{12} = f_{21} = 2$$

$$f_{22} = -2$$

At $(0, 0)$

$$F = \begin{vmatrix} 0 & 2 \\ 2 & -2 \end{vmatrix} = -4 < 0$$

Thus, $(0, 0)$ is neither an extremum nor a saddle point.
At $(\frac{2}{3}, \frac{2}{3})$

$$F = \begin{vmatrix} -4 & 2 \\ 2 & -2 \end{vmatrix} = 4$$

The Hessian matrix is negative definite. Thus, $(\frac{2}{3}, \frac{2}{3})$ yields a maximum.

(e) $y = x_1^3 + x_2^3 - 4x_1x_2$

$$f_1 = 3x_1^2 - 4x_2 = 0$$
$$f_2 = 3x_2^2 - 4x_1 = 0$$

Solving the second equation for x_1 yields

$$x_1 = \frac{3}{4}x_2^2 \Rightarrow x_1^2 = \frac{9}{16}x_2^4$$

Substituting into the first yields

$$\frac{27}{16}x_2^4 - 4x_2 = x_2(\frac{27}{16}x_2^3 - 4) = 0$$

Obviously $(0, 0)$ is a stationary point. A second stationary point is obtained by setting the expression in brackets equal to zero:

$$\frac{27}{16}x_2^3 = 4$$

or

$$\frac{27}{16}x_2^3 = 4$$

or

$$x_2^3 = \frac{64}{27} = (\frac{4}{3})^3$$

or

$$x_2 = \frac{4}{3}$$

Calculating the corresponding value of x_1 yields $(\frac{4}{3}, \frac{4}{3})$ as the second stationary point.

$$f_{11} = 6x_1 \quad f_{12} = f_{21} = -4$$

$$f_{22} = 6x_2$$

At $(0, 0)$

$$F = \begin{vmatrix} 0 & -4 \\ -4 & 0 \end{vmatrix} = -16 < 0$$

Thus, $(0, 0)$ is neither an extremum nor a saddle point.
At $(\frac{4}{3}, \frac{4}{3})$

$$F = \begin{vmatrix} 8 & -4 \\ -4 & 8 \end{vmatrix} = 48 > 0$$

Thus, the Hessian matrix is positive definite. $(\frac{4}{3}, \frac{4}{3})$ yields a minimum.

(f) $y = x_1x_2 + 2/x_1 + 4/x_2$

$$f_1 = x_2 - 2x_1^{-2} = 0$$
$$f_2 = x_1 - 4x_2^{-2} = 0$$

We can solve the first equation for x_2 which yields $x_2 = 2x_1^{-2}$. Rearranging the second equation to $x_1x_2^2 = 4$ and substituting we get

$$x_1(2x_2^{-2})^2 = x_1 4x_1^{-4} = 4x_1^{-3} = 4$$

so that $x_1^* = 1$ and $x_2^* = 2$. Thus, $(1, 2)$ is the only stationary value.

$$f_{11} = 4x_1^{-3} \quad f_{12} = f_{21} = 1$$

$$f_{22} = 8x_2^{-3}$$

At $(1, 2)$

$$F = \begin{vmatrix} 4 & 1 \\ 1 & 1 \end{vmatrix} = 3 > 0$$

The Hessian matrix is positive definite. Thus, $(1, 2)$ yields a minimum.

(g) $y = 2x_1^2 - 4x_2^2$

$$f_1 = 4x_1 = 0$$
$$f_2 = -8x_2 = 0$$

Obviously (0, 0) is a stationary point.

$$f_{11} = 4 \quad f_{12} = f_{21} = 0 \quad f_{22} = -8$$

$$\begin{vmatrix} 4 & 0 \\ 0 & -8 \end{vmatrix} = -32 < 0$$

Thus, (0, 0) is neither an extremum nor a saddle point.

3. (a) The upper limit to the amount the firm would bid for the franchise is just the profit the firm makes at the optimum. Thus, we must solve the firm's problem

$$\max_{p_C, p_H} \pi(p_C, p_H) = p_C D_C - u_C D_C + p_H D_H - u_H D_H$$

where $u_C = 0.5$, the constant unit cost of cola and $u_H = 0.1$, the constant unit cost of hot dogs. Using the demand functions we obtain

$$\begin{aligned} \pi(p_C, p_H) &= (20 - 4p_C - p_H)(p_C - 0.5) + \\ &\quad (15 - p_C - 5p_H)(p_H - 0.1) \\ &= -11.5 + 22.1p_C - 4p_C^2 - 2p_C p_H + \\ &\quad 16p_H - 5p_H^2 \end{aligned}$$

The first-order conditions are

$$\begin{aligned} \frac{\partial \pi}{\partial p_C} &= 22.1 - 8p_C - 2p_H = 0 \\ \frac{\partial \pi}{\partial p_H} &= 16 - 2p_C - 10p_H = 0 \end{aligned}$$

or

$$\begin{bmatrix} -8 & -2 \\ -2 & -10 \end{bmatrix} \begin{bmatrix} p_C \\ p_H \end{bmatrix} = \begin{bmatrix} -22.1 \\ -16 \end{bmatrix}$$

Using Cramer's rule we obtain

$$p_H^* = \frac{83.8}{76} = 1.10$$

$$p_C^* = \frac{189}{76} = 2.49$$

We check that this is indeed a maximum by using the second-order conditions:

$$\begin{aligned} \frac{\partial^2 \pi}{\partial p_C^2} &= -8 \\ \frac{\partial^2 \pi}{\partial p_C \partial p_H} &= -2 \\ \frac{\partial^2 \pi}{\partial p_H \partial p_C^2} &= -2 \\ \frac{\partial^2 \pi}{\partial p_H} &= -10 \end{aligned}$$

$$F = \begin{vmatrix} -8 & -2 \\ -2 & -10 \end{vmatrix} = 76$$

Thus, the Hessian matrix is negative definite and $p_H^* = 1.10$ and $p_C^* = 2.49$ are the profit-maximizing prices.

Profits at the optimum are $\pi(2.49, 1.10) = 24.8$. Thus, \$24.8 is the upper limit to the amount the firm would bid for the franchise.

(b) The demand for cola depends negatively on the price for hot dogs and vice versa. Cola and hot dogs are complements.

(c) If coke and hot dogs must be supplied by separate firms, one firm chooses only its price but is affected by the other firm's price. Thus, we must deal with the issue of cooperation or noncooperation by the firms.

Noncooperation:

If firms do not cooperate and each firm chooses its price given the other firm's price, we have the so called Bertrand competition with differentiated goods. The solution concept is the same as underlying the Cournot duopoly:

The cola firm maximizes profits

$$\begin{aligned} \max_{p_C} \pi_C &= (20 - 4p_C - p_H)(p_C - 0.5) \\ &= -10 + 22p_C - 4p_C^2 - p_H p_C + 0.5p_H \end{aligned}$$

given the price of a hot dog.

Similarly, the hot dog firm has the objective

$$\begin{aligned} \max \pi_H &= (15 - p_C - 5p_H)(p_H - 0.1) \\ &= -1.5 + 15.5p_H - 5p_H^2 - p_C p_H + 0.1p_C \end{aligned}$$

given the price of a can of cola.

The first-order conditions are

$$\begin{aligned} \frac{\partial \pi}{\partial p_C} &= 22 - 8p_C - p_H = 0 \\ \frac{\partial \pi}{\partial p_H} &= 15.5 - p_C - 10p_H = 0 \end{aligned}$$

or

$$\begin{bmatrix} -8 & -1 \\ -1 & -10 \end{bmatrix} \begin{bmatrix} p_C \\ p_H \end{bmatrix} = \begin{bmatrix} -22 \\ -15.5 \end{bmatrix}$$

Now the equilibrium prices are the solution of this pair of simultaneous equations. Using Cramer's rule we obtain

$$p_H^* = \frac{102}{79} = 1.29$$

$$p_C^* = \frac{204.5}{79} = 2.59$$

Substituting into the profit functions, the equilibrium profits and Thus, the largest bids by the firms are

$$\pi_H^* = 7.09 \qquad \pi_C^* = 17.45$$

Compared to the case where cola and hot dogs are supplied by one firm, the largest revenue by selling the franchise $\pi_H^* + \pi_C^* = 24.54$ is lower. Also consumers are worse off since prices are higher. The reason is that noncooperation leaves room for unexploited "gains of trade": If both firms simultaneously set a lower price and thus, stimulate the demand for the complementary good, each firm's profits would increase. Thus, contrary to intuition, the introduction of duopolistic competition would make everyone worse off in this case.

Cooperation:

If the firms cooperate, they would choose the same prices as the monopolist did in (a) to maximize their joint profits. The sum of the upper limits of the franchise bids would then be identical to the profits in (a). The upper limit for the franchise bid of each firm, however, does not have to be identical to the profits each firm makes at these prices according to the profit functions above. If we take as a benchmark the Bertrand equilibrium, we can only say that profits must be at least as high as in the Bertrand model. We cannot say, however, how the possible gains of trade of $24.8 - 17.45 - 7.09 = 0.26$ are divided among the firms. The sharing rule could include a profit-transfer from the hot dog firm to the cola firm or vice versa.

5. (a) Rewriting the demand functions in their inverse form

$$p_1 = 10 - q_1 \quad p_2 = 10 - 2q_2$$

the firm's objective is

$$\begin{aligned}\max_{q_1,q_2} \pi(q_1, q_2) &= p_1q_1 + p_2q_2 - C \\ &= (10 - q_1)q_1 + (10 - 2q_2)q_2 \\ &\quad -0.5(q_1 + q_2)^2\end{aligned}$$

The first-order conditions are

$$\begin{aligned}\frac{\partial\pi}{\partial q_1} &= 10 - 2q_1 - (q_1 + q_2) \\ &= 10 - 3q_1 - q_2 = 0 \\ \frac{\partial\pi}{\partial q_2} &= 10 - 4q_2 - (q_1 + q_2) \\ &= 10 - q_1 - 5q_2 = 0\end{aligned}$$

or

$$\begin{bmatrix} -3 & -1 \\ -1 & -5 \end{bmatrix}\begin{bmatrix} q_1 \\ q_2 \end{bmatrix} = \begin{bmatrix} -10 \\ -10 \end{bmatrix}$$

Using Cramer's rule we obtain

$$q_1^* = \frac{40}{14} = \frac{20}{7} = 2\frac{6}{7}$$

$$q_2^* = \frac{20}{14} = \frac{10}{7} = 1\frac{3}{7}$$

Note that the 2×2 matrix above is also the Hessian matrix. It is negative definite. Hence, we have indeed found a maximum.

The profit-maximizing prices are obtained by substituting q_1^* and q_2^* in the inverse demand functions:

$$p_1^* = 7\frac{1}{7} \quad p_2^* = 7\frac{1}{7}$$

(b) In market 1 the firm now faces the kinked demand function

$$p_1 = \begin{cases} b & \text{for } 0 \le q_1 \le 10 - b \\ 10 - q_1 & \text{for } q_1 \ge 10 - b \end{cases}$$

as represented in Figure 12.R.5. Hence, the firm's profit function is

$$\pi(q_1, q_2) = \begin{cases} bq_1 + (10 - 2q_2)q_2 & \\ -0.5(q_1 + q_2)^2 & \text{for } 0 \le q_1 \le 10 - b \\ (10 - q_1)q_1 + (10 - 2q_2)q_2 & \\ -0.5(q_1 + q_2)^2 & \text{for } q_1 \ge 10 - b \end{cases}$$

Analysis of the falling part of the demand function in market 1:

We start by examining the part of the demand function which is falling, i.e., the case where $q_1 \ge 10 - b$. Here the problem is to

$$\begin{aligned}\max_{q_1,q_2} \pi(q_1, q_2) &= (10 - q_1)q_1 + (10 - q_2)q_2 \\ &\quad -0.5(q_1 + q_2)^2\end{aligned}$$

subject to $q_1 \ge 10 - b$. The conditions for a maximum are

$$\frac{\partial\pi}{\partial q_1} = 10 - 3q_1 - q_2 \le 0$$

and

$$(q_1 - 10 + b_1)\frac{\partial\pi}{\partial q_1} = 0$$

Using the second first-order condition we obtain

$$\frac{\partial\pi}{\partial q_2} = 10 - 5q_2 - q_1 = 0$$

or $q_2 = 2 - 0.2q_1$. Substituting into the first derivative $\partial\pi/\partial q_1$ yields

$$\frac{\partial\pi}{\partial q_1} = 8 - 2.8q_1 \le 0$$

which implies $q_1 \ge 2\frac{6}{7}$. As in part (a) we have that $\partial\pi/\partial q_1 = 0$ when $q_1^* = 2\frac{6}{7}$ and $p_1^* = 7\frac{1}{7}$. Thus, as long as $b \ge 7\frac{1}{7}$ nothing has changed. However, for $b < 7\frac{1}{7}$ we must have $\partial\pi/\partial q_1 < 0$ because we have the constraint $q_1 \ge 10 - b > 2\frac{6}{7}$. Then the optimum must be at $q_1^* = 10 - b$. That is, the constraint is binding. The price is b.

Analysis of the horizontal part of the demand function in market 1:

Now we want to ask if the firm can do better by reducing its output when $b < 7\frac{1}{7}$. Therefore, we focus on the horizontal part of the demand function. Here the firm's problem is

$$\begin{aligned}\max_{q_1,q_2} \pi(q_1, q_2) &= bq_1 + (10 - 2q_2)q_2 \\ &\quad -0.5(q_1 + q_2)^2\end{aligned}$$

subject to $0 \le q_1 \le 10 - b$. The conditions for a maximum are

$$\begin{aligned}\frac{\partial\pi}{\partial q_1} &= b - q_1 - q_2 \le 0 \\ q_1\frac{\partial\pi}{\partial q_1} &= 0 \\ \frac{\partial\pi}{\partial q_1} &= b - q_1 - q_2 \ge 0 \\ (10 - b - q_1)\frac{\partial\pi}{\partial q_1} &= 0\end{aligned}$$

Again we use the second first-order condition

$$\frac{\partial\pi}{\partial q_2} = 10 - 5q_2 - q_1 = 0$$

which implies $q_2 = 2 - 0.2q_1$. Substituting into the first derivative we obtain

$$\frac{\partial\pi}{\partial q_1} = b - 2 - 0.8q_1$$

Here we can see that the solution crucially depends on b. We will now determine the values for b where $\partial\pi/\partial q_1$ is smaller, larger, or equal to zero.

(i) Range of b where $q_1^* = 0$:

If $b \leq 2$, we have $\partial\pi/\partial q_1 \leq 0$ and therefore $q_1^* = 0$.

(ii) Range of b where $0 < q_1 < 10 - b$:

We will have an interior solution if

$$\frac{\partial\pi}{\partial q_1} = b - 2 - 0.8q_1 = 0$$

or $q_1^* = 1.25b - 2.5$. Noting that for an interior solution we must have $q_1 < 10 - b$, we can determine the range of prices for which we have an interior solution:

$$10 - b > 1.25b - 2.5$$

or $b < \frac{50}{9} = 5\frac{5}{9}$.

(iii) Range of b where $q_1^* = 10 - b$:

Finally, we consider the case of a solution where the upper bound is binding. We then have

$$\frac{\partial\pi}{\partial q_1} = b - 2 - 0.8q_1 \geq 0$$

Substituting $10 - b$ for q_1 we obtain

$$b \geq \frac{50}{9} = 5\frac{5}{9}$$

Thus, we have that for $b < 7\frac{1}{7}$ the firm might do better by reducing its output in market 1. The table below summarizes the results. It also gives you the corresponding values of q_2^*, p_1^* and p_2^*. Note that for $b \geq 7\frac{1}{7}$ we have the same results as in the unconstrained case (a).

	$b \leq 2$	$2 \leq b \leq 5\frac{5}{9}$	$5\frac{5}{9} \leq b \leq 7\frac{1}{7}$
Home market:			
quantity	0	$\frac{5b-10}{4}$	$10 - b$
price	-	b	b
Foreign market:			
quantity	2	$\frac{10-b}{4}$	$b/5$
price	6	$5 + 0.5b$	$10 - 0.4b$

Congratulations if you have obtained this result!

Figure 12.R.5 illustrates the result. It shows you the quantities in the two markets as a functions of b.

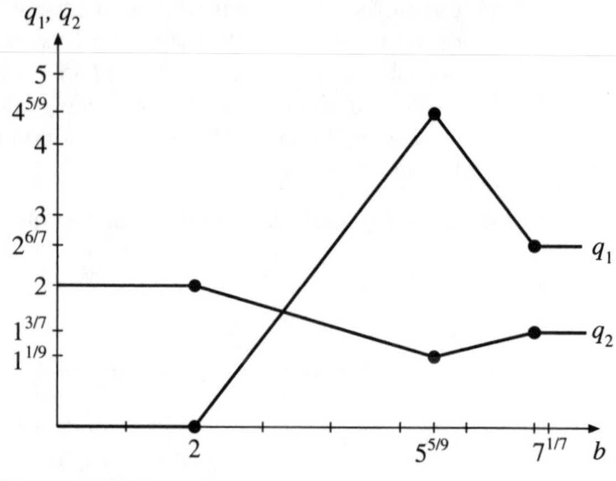

Figure 12.R.5

Interpretation of the results:

As expected the price regulation is useless when $b \geq 7\frac{1}{7}$. The firm chooses the same prices and quantities as in the unconstrained case.

If b is below $7\frac{1}{7}$ but close to $7\frac{1}{7}$, the monopolist does best if he satisfies demand in the home market. The quantity demanded is higher than in the unconstrained case. Thus marginal costs are higher. This implies that the optimal quantity sold abroad will shrink and $p_2$will go up.

When b is lower, it will no longer be optimal to satisfy home demand since marginal costs would be higher than price (verify that at $b = 5\frac{5}{9}$ marginal costs are $5\frac{5}{9}$ at the optimum). Thus for $b < 5\frac{5}{9}$ the optimal quantity will shrink. This means that marginal costs will be lower and therefore the optimal quantity in the foreign market will grow again and p_2 will fall.

Finally, if b becomes too small (< 2), the monopolist produces that output that it would choose if the foreign market were his only market, i.e., $q_2^* = 2$. Marginal costs are also 2 in this case and evidently he cannot gain anything by selling a unit in the home market.

CHAPTER 13

Section 13.1 (page 520)

1. (a)

$$\max y = 2x_1 + 3x_2 \quad \text{s.t. } 2x_1^2 + 5x_2^2 = 10$$

The Lagrangean function is

$$\mathcal{L} = 2x_1 + 3x_2 + \lambda(10 - 2x_1^2 - 5x_2^2)$$

and the first-order conditions are

$$\begin{aligned} 2 - 4\lambda x_1 &= 0 \\ 3 - 10\lambda x_2 &= 0 \\ 10 - 2x_1^2 - 5x_2^2 &= 0 \end{aligned}$$

Solving the first two of these equations to eliminate λ yields

$$x_1 = \frac{5}{3}x_2$$

and substituting into the third equation gives the solution

$$x_1^* = \frac{5}{3}\sqrt{\frac{18}{19}} \quad x_2^* = \sqrt{\frac{18}{19}}$$

(b)

$$\max y = x_1^{0.25}x_2^{0.75} \quad \text{s.t. } 2x_1^2 + 5x_2^2 = 10$$

The Lagrangean function is

$$\mathcal{L} = x_1^{0.25}x_2^{0.75} + \lambda(10 - 2x_1^2 - 5x_2^2)$$

The first-order conditions are

$$\begin{aligned} 0.25x_1^{-0.75}x_2^{0.75} - 4\lambda x_1 &= 0 \\ 0.75x_1^{0.25}x_2^{-0.25} - 10\lambda x_2 &= 0 \\ 10 - 2x_1^2 - 5x_2^2 &= 0 \end{aligned}$$

Solving the first two of these equations to eliminate λ yields

$$x_2^2 = 1.2x_1^2$$

Inserting into the third equation gives the solution

$$x_1^* = \frac{\sqrt{5}}{2} \quad x_2^* = \frac{\sqrt{6}}{2}$$

(c)

$$\min y = 2x_1 + 4x_2 \quad \text{s.t. } x_1^{0.25}x_2^{0.75} = 10$$

The Lagrangean function is

$$\mathcal{L} = 2x_1 + 4x_2 + \lambda(10 - x_1^{0.25}x_2^{0.75})$$

The first-order conditions are

$$\begin{aligned} 2 - 0.25\lambda x_1^{-0.75}x_2^{0.75} &= 0 \\ 4 - 0.75\lambda x_1^{0.25}x_2^{-0.25} &= 0 \\ 10 - x_1^{0.25}x_2^{0.75} &= 0 \end{aligned}$$

Solving the first two of these equations to eliminate λ yields

$$x_2 = 1.5x_1$$

Inserting into the third equation gives

$$x_1^* = 7.38 \quad x_2^* = 11.07$$

(d)

$$\max y = (x_1 + 2)(x_2 + 1) \quad \text{s.t. } x_1 + x_2 = 21$$

The Lagrangean function is

$$\mathcal{L} = (x_1 + 2)(x_2 + 1) + \lambda(21 - x_1 - x_2)$$

The first-order conditions are

$$\begin{aligned} (x_2 + 1) - \lambda &= 0 \\ (x_1 + 2) - \lambda &= 0 \\ 21 - x_1 + x_2 &= 0 \end{aligned}$$

Solving the first two of these equations to eliminate λ yields

$$x_2 = x_1 + 1$$

Inserting into the third equation we obtain

$$x_1^* = 10 \quad x_2^* = 11$$

3. In order to solve for output, we substitute into the constraint the input requirement functions:

$$l_i = y_i^{1/a_i}$$

Thus, the problem is

$$\max r_1y_1 + r_2y_2$$

s.t. $y_1^{1/a_1} + y_2^{1/a_2} = \bar{l}$

The Lagrange function is

$$\mathcal{L} = r_1y_1 + r_2y_2 + \lambda(\bar{l} - y_1^{1/a_1} - y_2^{1/a_2})$$

The first-order conditions are

$$\begin{aligned} r_1 - \lambda\frac{1}{a_1}y_1^{1/a_1-1} &= 0 \\ r_2 - \lambda\frac{1}{a_2}y_2^{1/a_2-1} &= 0 \\ y_1^{1/a_1} + y_2^{1/a_2} &= \bar{l} \end{aligned}$$

Solving the first two of these equations to eliminate λ yields

$$y_1^{1/a_1} = \left(\frac{r_1a_1}{r_2a_2}y_2^{1/a_2-1}\right)^{1/(1-a_1)}$$

Inserting into the third equation implicitly gives you y_2 and thus y_1:

$$y_2 = \left[\bar{l} - \left(\frac{r_1a_1}{r_2a_2}y_2^{1/a_2-1}\right)^{1/(1-a_1)}\right]^{a_2}$$

Figure 13.1.3 (a) shows you the level set of the net revenue function TT which is called the transformation curve. The maximum is reached at point A where the isoprofit line PP with the slope $-r_1/r_2$ is tangent to TT.

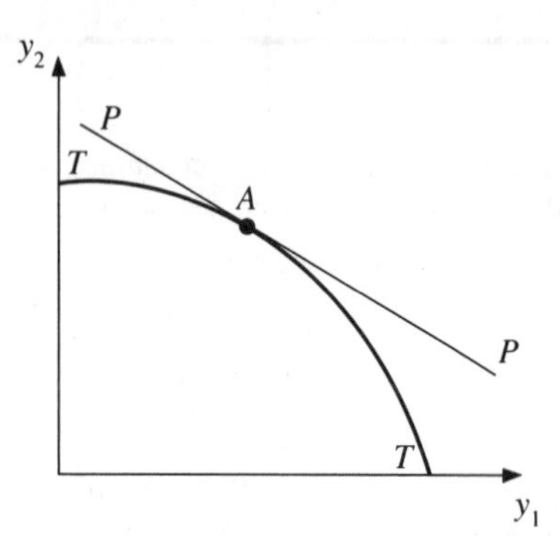

Figure 13.1.3 (a)

Figure 13.1.3 (b) illustrates the consequences of a rise of r_1/r_2. The isoprofit lines become steeper. The new optimum is at point B where the isoprofit line $P'P'$ is tangent to TT. The production of y_1 is higher while less is produced of y_2.

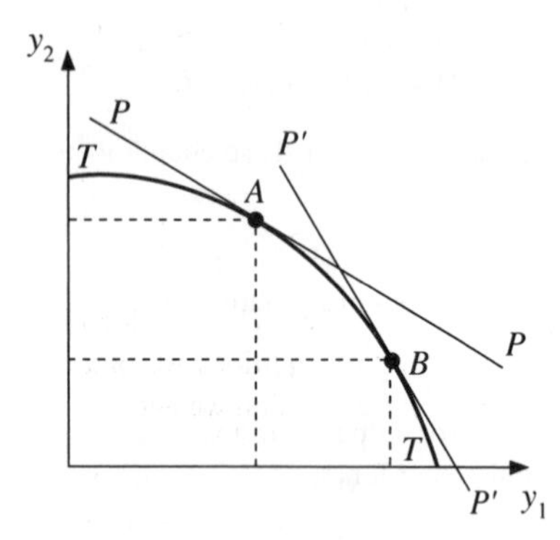

Figure 13.1.3 (b)

5. The problem

$$\max u = (x_1 - c_1)^{\alpha}(x_2 - c_2)^{1-\alpha}$$

s.t. $p_1x_1 + p_2x_2 = m$ can be reformulated by setting

$$\tilde{x}_1 = x_1 - c_1 \text{ and } \tilde{x}_2 = x_2 - c_2$$

yielding the equivalent problem

$$\max \tilde{x}_1^{\alpha}\tilde{x}_2^{1-\alpha} \text{ s.t. } p_1(\tilde{x}_1 + c_1) + p_2(\tilde{x}_2 + c_2) = m$$

Now note that the constraint is equivalent to

$$p_1\tilde{x}_1 + p_2\tilde{x}_2 = m - p_1c_1 - p_2c_2$$

Hence, we can use the results of the Cobb-Douglas utility function to obtain the demand functions for $\tilde{x}_1$ and $\tilde{x}_2$:

$$\begin{aligned} \tilde{x}_1 &= \alpha\frac{m - p_1c_1 - p_2c_2}{p_1} \\ \tilde{x}_2 &= (1-\alpha)\frac{m - p_1c_1 - p_2c_2}{p_2} \end{aligned}$$

Substituting for $\tilde{x}_1$ and $\tilde{x}_2$ we obtain the demand functions for x_1 and x_2:

$$\begin{aligned} x_1 &= \tilde{x}_1 + c_1 = \alpha\frac{m - p_2c_2}{p_1} + (1-\alpha)c_1 \\ x_2 &= \tilde{x}_2 + c_2 = (1-\alpha)\frac{m - p_1c_1}{p_2} + \alpha c_2 \end{aligned}$$

Properties and comparison to the Cobb-Douglas utility function

Own price elasticity:

Differentiating the first demand function with respect to p_1 yields

$$\frac{\partial x_1}{\partial p_1} = -\alpha\frac{m - p_2c_2}{p_1^2}$$

Thus

$$\begin{aligned} \epsilon_1 &= -\frac{p_1}{x_1}\frac{\partial x_1}{\partial p_1} \\ &= \frac{1}{1 + \frac{1-\alpha}{\alpha}\frac{p_1c_1}{m - p_2c_2}} < 1 \end{aligned}$$

and similarly for the second demand function. This compares to the Cobb-Douglas case where the own price elasticities are 1.

Expenditure shares of the goods:

The expenditure share of good 1 is

$$\begin{aligned} \frac{p_1x_1}{m} &= \alpha\frac{m - p_2c_2}{m} + (1-\alpha)\frac{p_1c_1}{m} \\ &= \alpha + \frac{(1-\alpha)p_1c_1 - \alpha p_2c_2}{m} \end{aligned}$$

and similarly for the second good. Hence, the expenditure shares vary with income. In the Cobb-Douglas case they are constant.

Expenditure and income elasticities:

Since the expenditure varies with income, the elasticities of expenditure with respect to income cannot be 1. Similarly, income elasticities of demand must be different from 1.

Demand and the other good's price:

As can be seen from the demand functions, demand for one good depends negatively on the other good's price. For example, as p_2 rises, demand for x_1 falls. The reason is that the rise in p_2 increases the amount of money needed to finance the subsistence level of x_2. This reduces disposable income and thus demand for good 1.

7.

$$\min C = rK + wL \text{ s.t. } \bar{y} = f(K, L)$$

The Lagrangean function is

$$\mathcal{L} = rK + wL + \lambda[\bar{y} - f(K, L)]$$

The first-order conditions are

$$\begin{aligned} r - \lambda\frac{\partial f}{\partial K} &= 0 \quad \text{(I)} \\ w - \lambda\frac{\partial f}{\partial L} &= 0 \quad \text{(II)} \\ \bar{y} - f(K, L) &= 0 \quad \text{(III)} \end{aligned}$$

Multiplying (I) by dK and (II) by dL and adding (I) and (II) yields

$$r\,dK + w\,dL = \lambda^*\left[\frac{\partial f}{\partial K}dK + \frac{\partial f}{\partial L}dL\right]$$

On the left-hand side of the equation we have the rise in costs caused by the additional inputs. On the right-hand side the term in square brackets is dy, i.e., the additional output achieved by increasing capital input by dK and labor input by dL. Thus, we have $dC = \lambda^* dy$. Dividing by dy we obtain the desired result

$$\frac{dC}{dy} = \lambda^*$$

However, this derivation is somewhat imprecise because we have not said anything about the way in which we choose dK and dL. Fortunately, the Envelope Theorem in Chapter 15 will tell you that this does not matter.

Alternatively, there is a less intuitive, but more precise way to show that λ^* equals marginal cost. We can argue as follows:

Solving the first-order conditions (I) to (III) yields

$$\lambda^* = \lambda^*(r, w, \bar{y})$$

$$K^* = K^*(r, w, \bar{y})$$

and

$$L^* = L^*(r, w, \bar{y})$$

Inserting these in the Lagrangean and differentiating with respect to $\bar{y}$ yields

$$\begin{aligned}\frac{d\mathcal{L}}{d\bar{y}} &= r\frac{\partial K^*}{\partial \bar{y}} + w\frac{\partial L^*}{\partial \bar{y}} \\ &\quad + \frac{\partial \lambda^*}{\partial \bar{y}}[\bar{y} - f(K, L)] + \lambda^* \\ &\quad - \lambda^*\left(\frac{\partial f}{\partial K}\frac{\partial K}{\partial \bar{y}} + \frac{\partial f}{\partial L}\frac{\partial L}{\partial \bar{y}}\right) \\ &= \lambda^*\end{aligned}$$

Observe that the optimized Lagrangean has the same value as the cost function for all $\bar{y}$. Hence, we have

$$\frac{d\mathcal{L}}{d\bar{y}} = \frac{dC}{d\bar{y}} = \lambda^* \text{ for all } \bar{y}$$

Section 13.2 (page 526)

1. (a) The Hessian is

$$H = \begin{bmatrix} -4\lambda & 0 & -4 \\ 0 & -10\lambda & -10 \\ -4 & -10 & 0 \end{bmatrix}$$

From the first-order condition we have

$$\lambda^* = \frac{3}{10x_2^*}$$

Inserting $x_2^* = \sqrt{\frac{18}{19}}$ yields $\lambda^* = 0.31$. Thus, at the optimum we have

$$H^* = \begin{bmatrix} -1.23 & 0 & -4x_1 \\ 0 & -3.1 & -10x_2 \\ -4x_1 & -10x_2 & 0 \end{bmatrix}$$

and so $|H^*| = 247.0 > 0$. Therefore, we have a true maximum.

(b) The Hessian is

$$H = \begin{bmatrix} -\frac{3}{16}x_1^{-1.75}x_2^{0.75} - 4\lambda & \frac{3}{16}x_1^{-0.75}x_2^{-0.25} & -4x_1 \\ \frac{3}{16}x_1^{-0.75}x_2^{-0.25} & -\frac{3}{16}x_1^{0.25}x_2^{-1.25} - 10\lambda & -10x_2 \\ -4x_1 & -10x_2 & 0 \end{bmatrix}$$

At the optimum we have $x_1^* = \sqrt{5}/2$ and $x_2^* = \sqrt{6}2$. From the first-order conditions we obtain $\lambda^* = 0.06$. Thus, at the optimum we get

$$H^* = \begin{bmatrix} -0.420 & 0.164 & -4.472 \\ 0.164 & -0.75 & -12.247 \\ -4.472 & -12.247 & 0 \end{bmatrix}$$

yielding $|H^*| = 959.58 > 0$. Therefore, we have a true maximum.

(c) The Hessian is

$$H = \begin{bmatrix} \frac{3}{16}\lambda x_1^{-1.75}x_2^{0.75} & -\frac{3}{16}\lambda x_1^{-0.75}x_2^{-0.25} & -\frac{1}{4}x_1^{-0.75}x_2^{0.75} \\ -\frac{3}{16}\lambda x_1^{-0.75}x_2^{-0.25} & \frac{3}{16}\lambda x_1^{0.25}x_2^{-1.25} & -\frac{3}{4}x_1^{0.25}x_2^{-0.25} \\ -\frac{1}{4}x_1^{-0.75}x_2^{0.75} & -\frac{3}{4}x_1^{0.25}x_2^{-0.25} & 0 \end{bmatrix}$$

At the optimum we have $x_1^* = 7.38$ and $x_2^* = 11.07$. From the first-order conditions we obtain $\lambda^* = 5.90$. Thus, at the optimal solution we get

$$H^* = \begin{bmatrix} 0.203 & 0.135 & -0.339 \\ 0.135 & 0.090 & -0.678 \\ -0.339 & -0.678 & 0 \end{bmatrix}$$

yielding $|H^*| = -0.042 < 0$ and so we have a true minimum.

(d) The Hessian is independent of the values of x_1, x_2, and λ. Thus, we obtain

$$H^* = \begin{bmatrix} 0 & 1 & -1 \\ 1 & 0 & -1 \\ -1 & -1 & 0 \end{bmatrix}$$

and so $|H^*|2 > 0$. Hence, the optimal solution is a maximum.

3. The Hessian matrix is

$$H = \begin{bmatrix} r_1 a_1(a_1 - 1)l_1^{a_1-2} & 0 & -1 \\ 0 & r_2(a_2 - 1)a_2 l_2^{a_2-2} & -1 \\ -1 & -1 & 0 \end{bmatrix}$$

We have

$$|H| = -r_2 a_2(a_2 - 1)l_2^{a_2-2} - r_1 a_1(a_1 - 1)l_1^{a_1-2} > 0$$

since $(a_2 - 1) < 0$ and $(a_1 - 1) < 0$. In particular we have $|H^*| > 0$. Thus, the solutions l_1^* and l_2^* yield a true constrained maximum.

Section 13.3 (page 531)

1. In answering this question we take only the case in which $x_1 \geq 0, x_2 \geq 0$.

(a) The objective function in the problem is linear and therefore quasiconcave. The condition for the constraint function $f(x_1, x_2)$ to be strictly quasiconvex is

$$f_{11}f_2^2 - 2f_{12}f_1f_2 + f_{22}f_1^2 > 0$$

For the function given here we have

$$f_1 = 4x_1;\ f_2 = 10x_2;\ f_{11} = 4;\ f_{22} = 10;\ f_{12} = f_{21} = 0$$

Thus, we have

$$4(10x_2)^2 + 10(4x_1)^2 > 0$$

and the constraint function is strictly quasiconvex. Applying Theorem 13.7 we know that the solution we obtained earlier is a global as well as a local maximum.

(b) The objective function here is a Cobb-Douglas function which we know to be strictly quasiconcave. The constraint function is linear and therefore quasiconvex. Thus, the conditions of Theorem 13.7 are satisfied.

(c) The counterpart of Theorem 13.7 for a minimum is that if the objective function is quasiconvex and the constraint function quasiconcave, then every local minimum is a global minimum. Here the objective function is linear and therefore quasiconvex, the constraint function is Cobb-Douglas and therefore strictly quasiconcave and so the condition of the theorem is satisfied.

(d) The constraint function is linear and therefore quasiconvex. For the objective function we have

$$f_1 = x_2 + 1;\ f_2 = x_1 + 2;\ f_{12} = f_{21} = 1;\ f_{11} = f_{22} = 0$$

For the function $f(x_1, x_2)$ to be strictly quasiconcave we require

$$f_{22}f_1^2 - 2f_{12}f_1f_2 + f_{11}f_2^2 < 0$$

Thus, we have

$$-2(x_1 + 2)(x_2 + 1) < 0$$

and so the objective function is strictly quasiconcave.

3. An example of a case in which the f-function is not quasiconcave is shown in Figure 13.3.3. Here both g and f are quasiconvex functions.

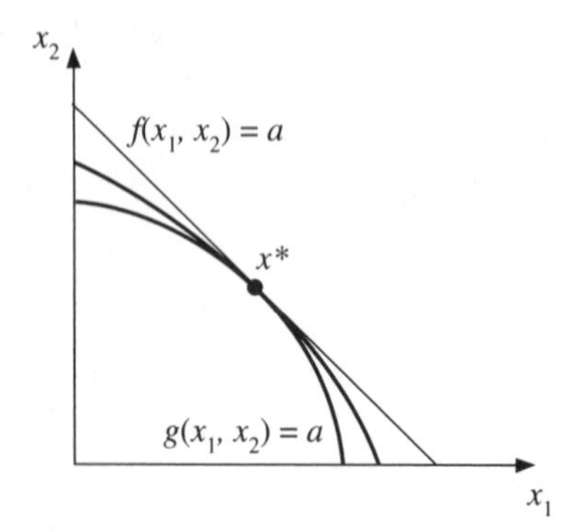

Figure 13.3.3

Review Exercises (page 532)

1. With the production functions

$$\begin{aligned} y_1 &= f(l_1) \\ y_2 &= g(l_2 \end{aligned}$$

and the constraint $\bar{l} = l_1 + l_2$ the Lagrange function is

$$\mathcal{L} = p_1 f(l_1) + p_2 g(l_2) + \lambda(\bar{l} - l_1 - l_2)$$

The first-order conditions are

$$\begin{aligned} p_1 f'(l_1) - \lambda^* &= 0 \quad \text{(I)} \\ p_2 g'(l_2) - \lambda^* &= 0 \quad \text{(II)} \\ \bar{l} - l_1 - l_2 &= 0 \quad \text{(III)} \end{aligned}$$

From (I) and (II) we obtain

$$p_1 f'(l_1) = p_2 g'(l_2)$$

That is, the marginal value product of the input must be the same in the production of the two goods; otherwise, the firm could increase its profits by shifting the use of the input to the production of the good in which the marginal value product of the input is higher.

The increase in profits resulting from a small increase in the amount of input:

The Lagrange multiplier at the optimum λ^* equals the change in the value of the objective function if the constraint is loosened by one unit (e.g., see Section 13.1, Exercise 7). Thus, in this problem λ^* equals the increase in profits resulting from a small increase in the amount of input available. From the first-order conditions (I) and (II) we have

$$\lambda^* = p_1 f'(l_1) = p_2 g'(l_2)$$

Thus, the increase in profit that would result from a small increase in the amount of fixed input the firm has available equals the marginal value product.

3. With A being the amount invested in the productive investment, $Y(A)$ the profit of the investment in one year's time, B the money put in the bank at the rate of interest r, and M the fixed sum of money available to the enterpreneur, his problem is

$$\max_{A,B} Y(A) + (1+r)B \text{ s.t. } A + B = M$$

The Lagrange function is

$$\mathcal{L} = Y(A) + (1+r)B + \lambda(M - A - B)$$

The first-order conditions are

$$\begin{aligned} Y'(A^*) - \lambda &= 0 \quad \text{(I)} \\ 1 + r - \lambda &= 0 \quad \text{(II)} \\ M - A - B &= 0 \quad \text{(III)} \end{aligned}$$

From (I) and (II) we obtain

$$Y'(A^*) = 1 + r$$

That is, the marginal product of the last dollar invested in the productive investment equals $(1 + r)$. Note that this condition is independent of M. M only determines whether the enterpreneur will borrow or lend money: If $A^* > M$, he will borrow money ($B < 0$), if $A^* < M$, he will lend money.

CHAPTER 14

Section 14.1 (page 542)

1. In the simple Keynesian model of income determination equilibrium national income is

$$Y^* = \frac{I}{1-c}$$

Thus, with $c = 0.8$, we obtain for $I_0 = 1000$ an equilibrium national income of $Y_0^* = 5000$. For $I_1 = 1200$, the equilibrium level is $Y_1^* = 6000$. The multiplier is

$$\frac{dY^*}{dI} = \frac{1}{1-c} = \frac{1}{0.2} = 5$$

Hence, the increase in investment by 200 leads to a rise in national income of 1000. Figure 14.1.1 illustrates these results.

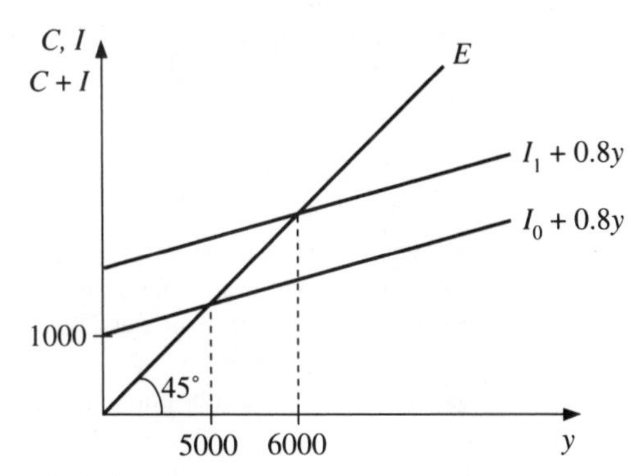

Figure 14.1.1

3. The monopolist's problem is to

$$\max_q \pi = (p-t)q - C$$

Substituting from the functions given in the exercise, we have

$$\max_q \pi = (100 - q - t)q - 0.5q^2$$

or

$$\max_q \pi = (100-t)q - 1.5q^2$$

The first-order condition is

$$\frac{d\pi}{dq} = 100 - t - 3q = 0$$

which gives the solution

$$q^* = \frac{100-t}{3}$$

Thus, for $t = \$1$ the profit-maximizing output is $q_1 = 33$.

At $t = \$2$ $q_2 = 32\frac{2}{3}$ maximizes profits. Figure 14.1.3 (a) illustrates this result in terms of profits as a function of output. Figure 14.1.3 (b) demonstrates the result in terms of the marginal-profit functions $\frac{d\pi}{dq} = 100 - t - 3q$. Finally, Figure 14.1.3 (c) restates the result in terms of the usual diagram with marginal revenue and cost curves. Marginal costs are

$$\frac{dC}{dq} = \text{MC} = q$$

For marginal revenue we obtain

$$\frac{dR}{dq} = \text{MR} = 100 - t - 2q$$

The condition for an optimum is MC = MR. It is equivalent to the condition $\frac{d\pi}{dq} = 0$ above since $\frac{d\pi}{dq} = \text{MR} - \text{MC}$. In Figure 14.1.3 (c) the tax rise results in a downward shift of the marginal revenue curve by $dt = 1$. The marginal cost curve does not shift since marginal costs do not depend on t. At the new point of intersection the optimal quantity is lower.

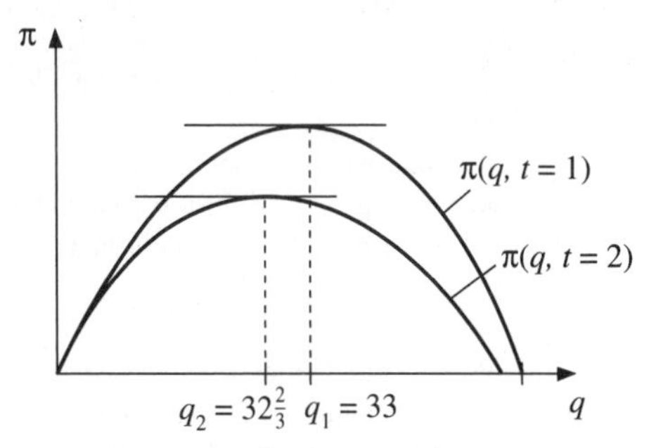

Figure 14.1.3 (a)

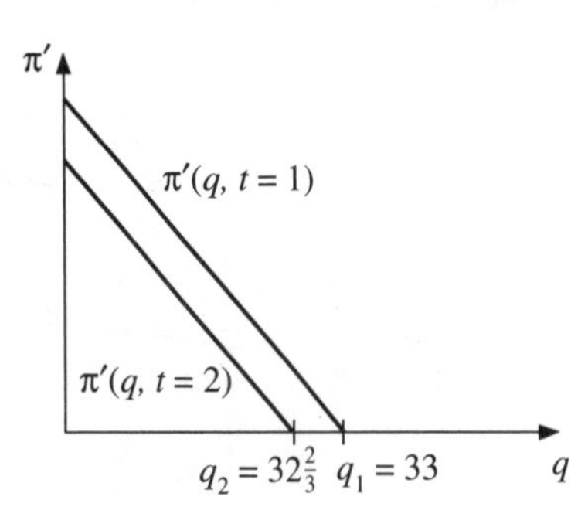

Figure 14.1.3 (b)

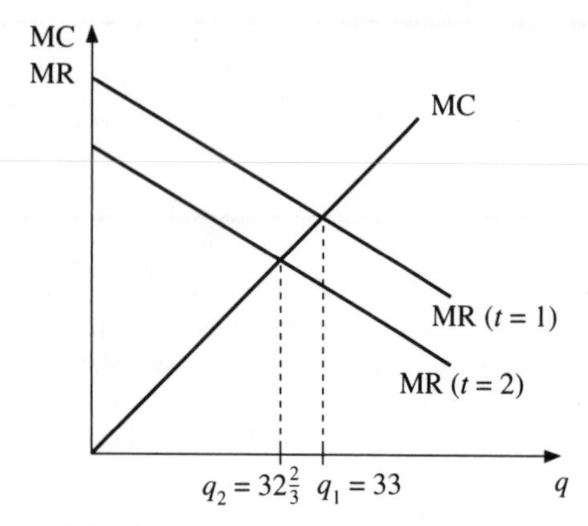

Figure 14.1.3 (c)

5. (a) Equilibrium national income is

$$Y^* = \frac{I}{1-c} = I(1-c)^{-1}$$

Using the Chain Rule we have

$$\frac{dY^*}{dc} = I(-1)(1-c)^{-2}(-1) = \frac{I}{(1-c)^2} > 0$$

A fall in c lowers Y^*. This is illustrated in Figure 14.1.5. The initial aggregate demand line is IC. The fall in c leads to the new line IC', thus yielding a lower equilibrium income Y.

Note that while the savings rate $s \equiv 1 - c$ rises, aggregate savings in equilibrium do not change. To see this rewrite the equilibrium condition as

$$(1-c)Y^* = sY^* = I$$

Thus, equilibrium output is where savings are sufficient to finance investment. Since planned investment remains the same, aggregate savings sY^*, will not change. However, with a higher savings rate, necessary savings are reached at a lower level of output.

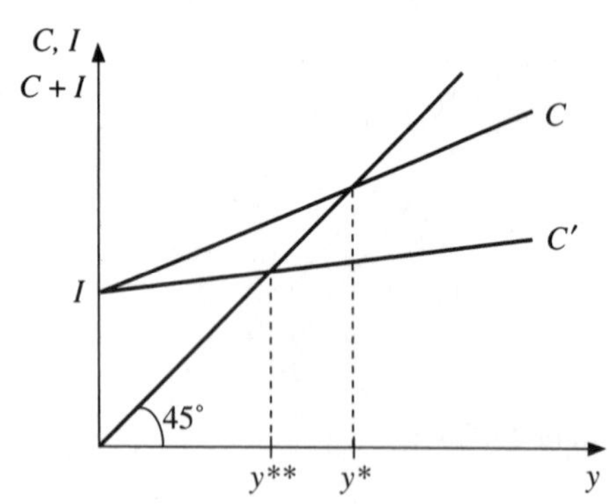

Figure 14.1.5

(b) In the linear market model, the demand and supply functions are

$$\begin{aligned} D &= a - bp + cy \qquad a, b, c > 0 \\ S &= \alpha + \beta p \qquad \alpha, \beta > 0 \quad \alpha < a + cy \end{aligned}$$

The equilibrium price is

$$p^* = \frac{a - \alpha + cy}{b + \beta}$$

The effects of a change in b:
Using the Chain Rule we obtain

$$\frac{dp^*}{db} = -\frac{a - \alpha + cy}{(b+\beta)^2} = -\frac{p^*}{b+\beta} < 0$$

Since a change in b only shifts the demand but not the supply curve, the change in b represents a movement along the supply curve. We can determine the effect on equilibrium output by calculating the effect of the price change through b on supply:

$$\frac{dq^*}{db} = \frac{\partial S}{\partial p}\frac{\partial p*}{\partial b} = \beta\left(-\frac{p^*}{b+\beta}\right) = -\frac{\beta p*}{b+\beta} < 0$$

A rise in b implies that consumers become more sensitive to price and causes demand to shrink. As a consequence price will go down. At a lower price producers will supply less. Hence, quantity will shrink, too.

The effects of a change in β:
Using the Chain Rule we have

$$\frac{dp^*}{d\beta} = -\frac{a - \alpha + cy}{(b+\beta)^2} = -\frac{p^*}{b+\beta} < 0$$

Note that here it is the supply curve which shifts while the demand curve remains in its place. Hence we move along the demand curve and find the effect on output by determining the effect of the price change through β on demand:

$$\frac{dq^*}{d\beta} = \frac{\partial D}{\partial p}\frac{\partial p^*}{\partial \beta} = -b\left(-\frac{p^*}{b+\beta}\right) = \frac{bp^*}{b+\beta} > 0$$

When β rises, supply increases at any given price. Since consumers are not willing to buy more at the former equilibrium price, price must sink and quantity grows.

The effects of a change in c:
The effect on the equilibrium price is

$$\frac{dp^*}{dc} = \frac{y}{b+\beta} > 0$$

Because the change in c only affects the demand curve we move along the supply curve and obtain the effect on equilibrium quantity by

$$\frac{dq^*}{dc} = \frac{\partial S}{\partial p}\frac{\partial p^*}{\partial c} = \beta\frac{y}{b+\beta} = \frac{\beta y}{b+\beta} > 0$$

When c rises, consumers would like to buy more at a given price. Then price must rise to make producers willing to supply the larger quantity.

(c) In the monopoly model with tax, the inverse demand and the total cost functions are

$$\begin{aligned} p &= a - bq \qquad a, b > 0 \\ C &= cq^2 \qquad c > 0 \end{aligned}$$

Equilibrium output is

$$q^* = \frac{a - t}{2(b+c)}$$

The effects of a change in a:
The effect on equilibrium quantity is

$$\frac{dq^*}{da} = \frac{1}{2(b+c)} > 0$$

The effect on equilibrium price is obtained by

$$\frac{dp^*}{da} = \frac{\partial p}{\partial a} - b\frac{dq^*}{da}$$
$$= 1 - \frac{b}{2(b+c)} = \frac{b+2c}{2(b+c)} > 0$$

Increasing demand causes output and price to grow.

The effects of a change in c:

Using the Chain Rule we obtain

$$\frac{dq^*}{dc} = \frac{a-t}{2}\left[-\frac{1}{(b+c)^2}\right]$$
$$= \frac{a-t}{2(b+c)}\left[-\frac{1}{(b+c)}\right]$$
$$= -\frac{q^*}{b+c} < 0$$

The effect on the equilibrium price is

$$\frac{dp^*}{dc} = \frac{\partial p}{\partial q}\frac{\partial q*}{dc}$$
$$= -b\left(-\frac{q^*}{b+c}\right) = \frac{bq^*}{b+c} > 0$$

Increasing marginal costs causes the new intersection of the marginal revenue and the marginal cost curve to be to the left of the old intersection. Hence, output decreases. Since less output can be sold at a higher price, the equilibrium price increases.

7. *Interpretation of the coefficients of r:*

The demand for a country's exports is given by

$$X = a - br + cY_R \qquad a, b, c > 0$$

Exports depend negatively on the rate of exchange since the price foreigners must pay in *their* currency for the country's good is (domestic price) × (rate of exchange). If r rises, i.e., the country's currency appreciates, exports become more expensive in the foreign country and demand decreases. Since exports are defined as (domestic price of exports) × (quantity of exports demanded), export demand must fall.

The country's demand for imports is given by

$$Q = \alpha + \beta r + \gamma Y_D \qquad \alpha, \beta, \gamma > 0$$

The domestic price for foreign goods is (foreign price) / (rate of exchange). A rise in r lowers the domestic price. Hence, more import goods will be demanded. The effect on import demand which is defined as (domestic price of imports) × (quantity of imports) is thus unclear. Since the sign of the coefficient of r is positive, it is assumed that the quantity effect of an increase in r is stronger.

(Observe that this definition of r is just one possible definition of the "rate of exchange". Here it states the quantity of foreign currency which can be bought for one dollar where dollars are the domestic currency. Sometimes the reciprocal definition is used for the rate of exchange. Then it gives you the amount of dollars which must be paid for one unit of foreign currency. If this definition were used, the signs of the coefficients would have to be reversed.)

The exchange rate that achieves equilibrium of the balance of trade:

Setting $X = Q$ and solving for r yields

$$\bar{r} = \frac{a - \alpha + cY_R - \gamma Y_D}{b+\beta}$$

The comparative-static effect of a change in the income of the rest of the world Y_R is

$$\frac{d\bar{r}}{dY_R} = \frac{c}{b+\beta} > 0$$

This is illustrated in Figure 14.1.7 (a) which shows the export demand X and import demand Q as a function of r. At $\bar{r}$ the balance of trade is initially in equilibrium. An increase in Y_R leads to a rise in export demand at any given rate of exchange. Thus, the export demand curve shifts to the right to X'. For balance of trade equilibrium to be restored it is necessary that r rises to reduce export demand and promote import demand. At $\bar{r}'$ is the new equilibrium.

The comparative-static effect of a change in the domestic income Y_D is

$$\frac{d\bar{r}}{dY_D} = -\frac{\gamma}{b+\beta} < 0$$

This is shown in Figure 14.1.7 (b). The import demand curve shifts to the right to Q' because at any given r, import demand is higher due to the rise in Y_D. Balance of trade equilibrium requires r to sink to repress import demand and to increase export demand. At $\bar{r}''$ equilibrium is restored.

The level of domestic income that achieves equilibrium of the balance of trade:

Assuming that r and Y_R are exogenous, the national domestic income consistent with balance of trade equilibrium is obtained by setting $X = Q$ and solving for Y_D. This yields

$$\bar{Y}_D = \frac{a - \alpha - (b+\beta)r + cY_R}{\gamma}$$

The comparative-static effect of a change in r is

$$\frac{d\bar{Y}_D}{dr} = -\frac{\beta + b}{\gamma} < 0$$

This is illustrated by Figure 14.1.7 (c) which shows export and import demand as a function of Y_D. The X-curve is vertical since export demand does not depend on Y_D. The initial equilibrium is at $\bar{Y}_D$. An increase in r shifts the export demand curve to X' because at any given Y_D less exports are demanded. Likewise, import demand shifts to Q' since at any given Y_D import demand has risen. Thus, a revaluation of the domestic currency causes a deficit in the balance of trade. To restore equilibrium Y_D must sink to repress imports. The new equilibrium is at $\bar{Y}'_D$. The comparative-static effect of a change in Y_R is

$$\frac{d\bar{Y}_D}{dY_R} = \frac{c}{\gamma} > 0$$

This is shown in Figure 14.1.7 (d). An increase in Y_R increases export demand at given Y_D and r. Thus the export demand curve shifts to X' and the trade balance is in surplus at the initial level of domestic income $\bar{Y}_D$. To reestablish equilibrium, imports must rise. This is accomplished by a rise in Y_D. The new equilibrium is at $\bar{Y}''_D$.

Level of domestic income and rate of exchange endogenous and equilibrium of the balance of trade:

If exchange rate and the country's income are treated as endogenous there is a continuum of (Y_D, r)-pairs described by

$$r = \frac{a - \alpha + cY_R - \gamma Y_D}{b+\beta}$$

which are compatible with equilibrium of the trade balance. A unique solution cannot be determined.

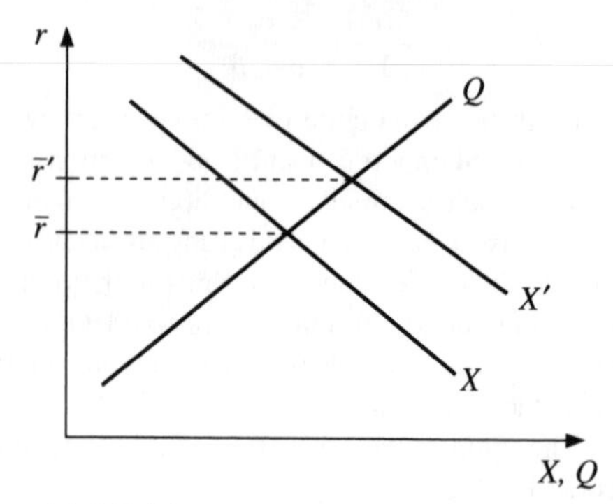

Figure 14.1.7 (a)

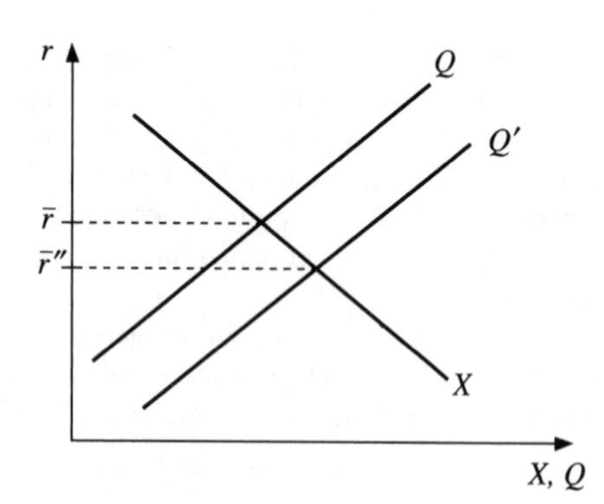

Figure 14.1.7 (b)

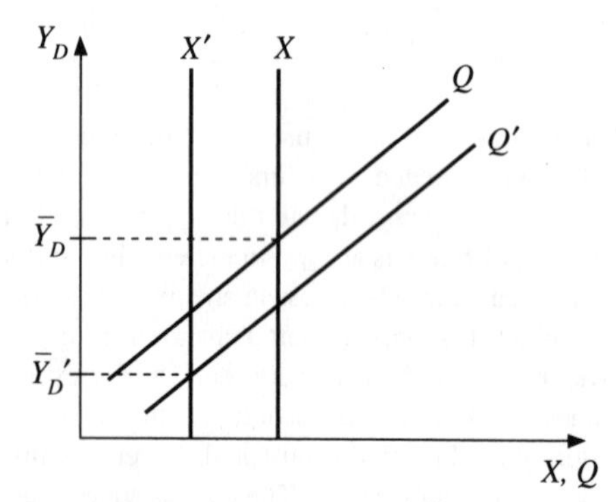

Figure 14.1.7 (c)

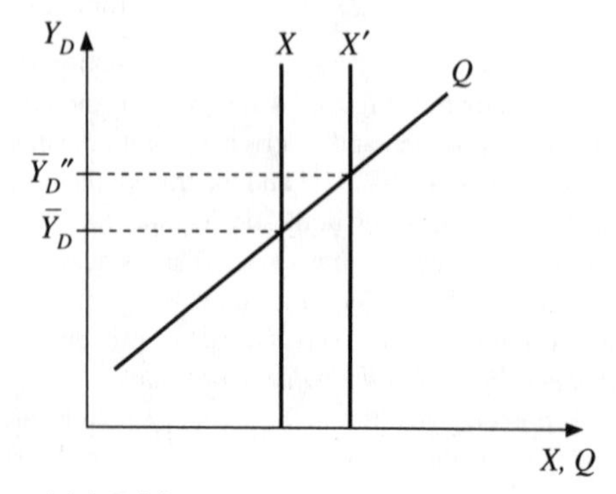

Figure 14.1.7 (d)

Section 14.2 (page 556)

1. In the general IS-LM model, the equilibrium values of Y^* and R^* are determined by

$$\begin{aligned} Y^* - \bar{E} - E(Y^*, R^*) &= 0 \\ L(Y^*, R^*) - \bar{M} &= 0 \end{aligned}$$

Applying the general method of comparative-statics we have

$$\begin{bmatrix} (1-E_Y) & -E_R \\ L_Y & L_R \end{bmatrix} \begin{bmatrix} \partial Y^*/\partial \bar{M} \\ \partial R^*/\partial \bar{M} \end{bmatrix} = \begin{bmatrix} 0 \\ 1 \end{bmatrix}$$

The determinant $|D|$ is

$$|D| = (1-E_Y)L_R + L_Y E_R < 0$$

The comparative-static effects are

$$\begin{aligned} \frac{\partial Y^*}{\partial \bar{M}} &= \begin{vmatrix} 0 & -E_R \\ 1 & L_R \end{vmatrix} / |D| \\ &= \frac{E_R}{(1-E_Y)L_R + L_Y E_R} > 0 \\ \frac{\partial R^*}{\partial \bar{M}} &= \begin{vmatrix} (1-E_Y) & 0 \\ L_Y & 1 \end{vmatrix} / |D| \\ &= \frac{1-E_Y}{(1-E_Y)L_R + L_Y E_R} < 0 \end{aligned}$$

3. (a) The consumer's problem is

$$\max x_1 x_2 \quad \text{s.t. } p_1 x_1 + p_2 x_2 = m$$

Applying the Lagrangean method yields the first-order conditions

$$\begin{aligned} x_2^* - \lambda^* p_1 &= 0 \\ x_1^* - \lambda^* p_2 &= 0 \\ m - p_1 x_1^* - p_2 x_2^* &= 0 \end{aligned}$$

We obtain the effect of changes in prices and income on the demands by applying the general method of comparative statics. We restrict ourselves to the demand for good 1 because the utility function is symmetric. Thus, we have

$$\begin{aligned} \frac{\partial x_1^*}{\partial p_1} &= \begin{vmatrix} \lambda^* & 1 & -p_1 \\ 0 & 0 & -p_2 \\ x_1^* & -p_2 & 0 \end{vmatrix} / |D| \\ &= -\frac{\lambda^* p_2^2}{|D|} - x_1^* \frac{p_2}{|D|} \end{aligned}$$

where

$$|D| = \begin{vmatrix} 0 & 1 & -p_1 \\ 1 & 0 & -p_2 \\ -p_1 & -p_2 & 0 \end{vmatrix} = 2p_1 p_2 > 0$$

The income effect

$$\left(-x_1^* \frac{p_2}{|D|}\right)$$

is negative. Therefore, $\partial x_1^*/\partial p_1 < 0$. Due to the symmetry of the utility function, the same applies to $\partial x_2^*/\partial p_2$.

(b) The consumer's problem is

$$\max \sqrt{x_1} + x_2 \quad \text{s.t. } p_1 x_1 + p_2 x_2 = m$$

Applying the Lagrangean method gives the first-order conditions

$$\begin{aligned} 0.5(x_1^*)^{-0.5} - \lambda^* p_1 &= 0 \\ 1 - \lambda^* p_2 &= 0 \\ m - p_1 x_1^* - p_2 x_2^* &= 0 \end{aligned}$$

Applying the standard method we obtain the effects of changes in prices and income on the demand of good 1:

$$\begin{aligned} \frac{\partial x_1^*}{\partial p_1} &= \begin{vmatrix} \lambda^* & 0 & -p_1 \\ 0 & 0 & -p_2 \\ x_1^* & -p_2 & 0 \end{vmatrix} /|D| \\ &= -\frac{\lambda p_2^2}{|D|} < 0 \\ \frac{\partial x_1^*}{\partial m} &= \begin{vmatrix} 0 & 0 & -p_1 \\ 0 & 0 & -p_2 \\ 1 & -p_2 & 0 \end{vmatrix} /|D| = \frac{0}{|D|} = 0 \end{aligned}$$

where

$$|D| = \begin{vmatrix} -0.25(x_1^*)^{-1.5} & 0 & -p_1 \\ 0 & 0 & -p_2 \\ -p_1 & -p_2 & 0 \end{vmatrix} = 0.25 p_2^2 x_1^{-1.5}$$

Since the income effect is zero, we must have $\partial x_1^*/\partial p_1 < 0$. Similarly, we obtain for good 2

$$\begin{aligned} \frac{\partial x_2^*}{\partial p_2} &= \begin{vmatrix} -0.25(x_1^*)^{-1.5} & 0 & -p_1 \\ 0 & \lambda^* & -p_2 \\ -p_1 & x_2^* & 0 \end{vmatrix} /|D| \\ &= -\frac{\lambda^* p_1^2}{|D|} - x_2^* \frac{0.25 x_1^{-1.5} p_2}{|D|} \end{aligned}$$

The income effect $\left(-x_2^* \frac{0.25 x_1^{-1.5} p_2}{|D|}\right)$ is negative which implies $\partial x_2^*/\partial p_2 < 0$.

5. Consider variations of p_1 and m only. The first-order conditions are

$$\begin{aligned} \mathcal{L}_1 &= u_1 - \lambda^* p_1 = 0 \\ \mathcal{L}_2 &= u_2 - \lambda^* p_2 = 0 \\ \mathcal{L}_3 &= u_3 - \lambda^* p_3 = 0 \\ \mathcal{L}_\lambda &= m - p_1 x_1^* - p_2 x_2^* - p_3 x_3^* = 0 \end{aligned}$$

These characterize a unique global maximum. To get the comparative static effects of a change in p_1, we use the following equation system

$$\begin{bmatrix} u_{11} & u_{12} & u_{13} & -p_1 \\ u_{21} & u_{22} & u_{23} & -p_2 \\ u_{31} & u_{32} & u_{33} & -p_3 \\ -p_1 & -p_2 & -p_3 & 0 \end{bmatrix} \begin{bmatrix} \partial x_1^*/\partial p_1 \\ \partial x_2^*/\partial p_1 \\ \partial x_3^*/\partial p_1 \\ \partial \lambda^*/\partial p_1 \end{bmatrix} = \begin{bmatrix} \lambda^* \\ 0 \\ 0 \\ x_1^* \end{bmatrix}$$

Note that a unique global maximum is a unique local maximum. So we have from the second-order conditions for a local maximum that the determinant $|D|$ of the coefficient matrix on the left-hand side is smaller than zero

$$|D| = \begin{vmatrix} u_{11} & u_{12} & u_{13} & -p_1 \\ u_{21} & u_{22} & u_{23} & -p_2 \\ u_{31} & u_{32} & u_{33} & -p_3 \\ -p_1 & -p_2 & -p_3 & 0 \end{vmatrix} < 0$$

Applying Cramer's rule we get

$$\frac{\partial x_1^*}{\partial p_1} = \frac{\lambda^* \begin{vmatrix} u_{22} & u_{23} & -p_2 \\ u_{32} & u_{33} & -p_3 \\ -p_2 & -p_3 & 0 \end{vmatrix}}{|D|} - x_1^* \frac{\begin{vmatrix} u_{12} & u_{13} & -p_1 \\ u_{22} & u_{23} & -p_2 \\ u_{32} & u_{33} & -p_3 \end{vmatrix}}{|D|}$$

From the second-order conditions for a local maximum we have

$$\begin{vmatrix} u_{22} & u_{23} & -p_2 \\ u_{32} & u_{33} & -p_3 \\ -p_2 & -p_3 & 0 \end{vmatrix} > 0$$

so the first term is invariably negative.

Now let's turn to the comparative-static effects of changes in m

$$\begin{bmatrix} u_{11} & u_{12} & u_{13} & -p_1 \\ u_{21} & u_{22} & u_{23} & -p_2 \\ u_{31} & u_{32} & u_{33} & -p_3 \\ -p_1 & -p_2 & -p_3 & 0 \end{bmatrix} \begin{bmatrix} \partial x_1^*/\partial p_1 \\ \partial x_2^*/\partial p_1 \\ \partial x_3^*/\partial p_1 \\ \partial \lambda^*/\partial p_1 \end{bmatrix} = \begin{bmatrix} 0 \\ 0 \\ 0 \\ -1 \end{bmatrix}$$

Applying Cramer's rule yields

$$\frac{\partial x_1^*}{\partial m} = \frac{\begin{vmatrix} u_{12} & u_{13} & -p_1 \\ u_{22} & u_{23} & -p_2 \\ u_{32} & u_{33} & -p_3 \end{vmatrix}}{|D|}$$

the sign being indefinite. So we have the Slutsky equation

$$\frac{\partial x_1^*}{\partial p_1} = \underbrace{\lambda^* \frac{\begin{vmatrix} u_{22} & u_{23} & -p_2 \\ u_{32} & u_{33} & -p_3 \\ -p_2 & -p_3 & 0 \end{vmatrix}}{|D|}}_{\text{substitution effect } < 0} \underbrace{-x_1^* \frac{\partial x_1^*}{\partial m}}_{\text{sign indefinite}}$$

7. The equilibrium conditions are

$$\begin{aligned} Y_1^* + Q_1^* - C_1^* - I_1^0 - Q_2^* &= 0 \quad \text{for country 1} \\ Y_2^* + Q_2^* - C_2^* - I_2^0 - Q_1^* &= 0 \quad \text{for country 2} \end{aligned}$$

Inserting the consumption and import functions we obtain

$$\begin{aligned} 0.5Y_1^* - 0.5Y_2^* - I_1^0 &= 0 \\ 0.8Y_2^* - 0.3Y_1^* - I_2^0 &= 0 \end{aligned}$$

Applying the general method of comparative-statics we get for a change in I_1^0.

$$\begin{bmatrix} 0.5 & -0.5 \\ -0.3 & 0.8 \end{bmatrix} \begin{bmatrix} \partial Y_1^*/\partial I_1^0 \\ \partial Y_2^*/\partial I_1^0 \end{bmatrix} = \begin{bmatrix} 1 \\ 0 \end{bmatrix}$$

The determinant is $|D| = 0.25$. The comparative-statics effects are

$$\frac{\partial Y_1^*}{\partial I_1^0} = \begin{vmatrix} 1 & -0.5 \\ 0 & 0.8 \end{vmatrix} /|D|$$

$$= \frac{0.8}{0.25} = 3.2$$

$$\frac{\partial Y_2^*}{\partial I_1^0} = \begin{vmatrix} 0.5 & 1 \\ -0.3 & 0 \end{vmatrix} /|D|$$

$$= \frac{0.3}{0.25} = 1.2$$

Similarly, we obtain for a change in I_2^0.

$$\begin{bmatrix} 0.5 & -0.5 \\ -0.3 & 0.8 \end{bmatrix} \begin{bmatrix} \partial Y_1^*/\partial I_2^0 \\ \partial Y_2^*/\partial I_2^0 \end{bmatrix} = \begin{bmatrix} 0 \\ 1 \end{bmatrix}$$

The comparative-static effects are

$$\frac{\partial Y_1^*}{\partial I_2^0} = \begin{vmatrix} 0 & -0.5 \\ 1 & 0.8 \end{vmatrix} /|D|$$

$$= \frac{0.5}{0.25} = 2$$

$$\frac{\partial Y_2^*}{\partial I_2^0} = \begin{vmatrix} 0.5 & 0 \\ -0.3 & 1 \end{vmatrix} /|D|$$

$$= \frac{0.5}{0.25} = 2$$

Thus, a rise of exogenous investment in any country raises equilibrium income levels in both countries. The reason is that the increase in domestic income caused by the increase of investment leads to an increase in the demand for imports and thus to higher demand in the other country.

The effect on income in both countries depends inter alia on the marginal propensity to import $dQ_i/dY_i, i = 1, 2$, in the country where exogenous investment is increased. The lower dQ_i/dY_i, the stronger the increase in domestic income and the weaker the rise of foreign income.

Section 14.3 (page 567)

1. The central planner's problem is

$$\max Y = x_1 + 2x_2$$

s.t.

$$x_1 = 100L_1^{0.5}$$
$$x_2 = 50L_2^{0.5}$$
$$L_1 + L_2 = 1000$$

The Lagrange function is

$$\mathcal{L} = 100L_1^{0.5} + 100L_2^{0.5} + \lambda(1000 - L_1 - L_2)$$

The first-order conditions are

$$50L_1^{-0.5} - \lambda = 0$$
$$50L_2^{-0.5} - \lambda = 0$$
$$1000 - L_1 - L_2 = 0$$

Obviously, we must have $L_1^* = L_2^* = 500$. Solving for λ yields $\lambda^* = 2.236$ which is the shadow wage rate.

We obtain the optimal output by substituting the optimal amount of labor used in the production functions. This yields $x_1 = 2236$ and $x_2 = 1118$.

We will now confirm this result by using the value function which has been derived in the text. With the given exogenous parameters we have

$$V(p_1, p_2, L^0) = 100p_1 \left(\left[1 + \left(\frac{2p_1}{p_2} \right)^{-2} \right]^{-1} L^0 \right)^{0.5} +$$

$$50p_2 \left[\left(1 - \left[1 + \left(\frac{2p_1}{p_2} \right)^{-2} \right]^{-1} \right) L^0 \right]^{0.5}$$

Applying the Envelope Theorem, the optimal output of good 1 is given by

$$\frac{\partial V}{\partial p_1} = 100 \left(\left[1 + \left(\frac{2p_1}{p_2} \right)^{-2} \right]^{-1} L^0 \right)^{0.5}$$

$$= 100(500)^{0.5} = 2236$$

Similarly, we obtain for good 2

$$\frac{\partial V}{\partial p_2} = 50 \left[\left(1 - \left[1 + \left(\frac{2p_1}{p_2} \right)^{-2} \right]^{-1} \right) L^0 \right]^{0.5}$$

$$= 50(500)^{0.5} = 1118$$

Finally, we calculate the shadow wage rate. We have

$$\frac{\partial V}{\partial L^0} = \left[50p_1 \left(\left[1 + \left(\frac{2p_1}{p_2} \right)^{-2} \right]^{-1} \right)^{0.5} \right.$$
$$\left. +25p_2 \left(1 - \left[1 + \left(\frac{2p_1}{p_2} \right)^{-2} \right]^{-1} \right)^{0.5} \right] (L^0)^{-0.5}$$

$$= \left[50(0.5)^{0.5} + 50(0.5)^{0.5} \right] (1000)^{-0.5}$$

$$= (70.71)(0.0316) = 2.236$$

3. The competitive firm maximizes profits

$$\pi = px - wL - rK$$

subject to its production function

$$x = L^{0.5}K^{0.3}$$

The Lagrange function is

$$\mathcal{L} = px - wL - rK - \lambda(x - L^{0.5}K^{0.3})$$

The first-order conditions are

$$p - \lambda = 0$$
$$\lambda(0.5L^{-0.5}K^{0.3}) - w = 0$$
$$\lambda(0.3L^{0.5}K^{-0.7}) - r = 0$$
$$x - L^{0.5}K^{0.3} = 0$$

Substituting p for λ in the second and third conditions and noting that we have a linear system in logarithms, we obtain

$$\begin{bmatrix} -0.5 & 0.3 & 0 \\ 0.5 & -0.7 & 0 \\ -0.5 & -0.3 & 1 \end{bmatrix} \begin{bmatrix} \hat{L} \\ \hat{K} \\ \hat{x} \end{bmatrix} = \begin{bmatrix} \hat{w} - \hat{p} - \log 0.5 \\ \hat{r} - \hat{p} - \log 0.3 \\ 0 \end{bmatrix}$$

where a "hat" over a variable denotes its log. Denoting the determinant by $|D|$ we have $|D| = 0.2$. Solving the system yields

$$\hat{L} = \begin{vmatrix} \hat{w} - \hat{p} - \log 0.5 & 0.3 & 0 \\ \hat{r} - \hat{p} - \log 0.3 & -0.7 & 0 \\ 0 & -0.3 & 1 \end{vmatrix} / |D|$$
$$= (-0.7\hat{w} + \hat{p} - 0.3\hat{r} + 0.7 \log 0.5 + 0.3 \log 0.3)/0.2$$
$$= -3.5\hat{w} + 5\hat{p} - 1.5\hat{r} + 3.5 \log 0.5 + 1.5 \log 0.3$$

$$\hat{K} = \begin{vmatrix} -0.5 & \hat{w} - \hat{p} - \log 0.5 & 0 \\ 0.5 & \hat{r} - \hat{p} - \log 0.3 & 0 \\ -0.5 & 0 & 1 \end{vmatrix} / |D|$$
$$= (-0.5\hat{w} + \hat{p} - 0.5\hat{r} + 0.5 \log 0.5 + 0.5 \log 0.3)/0.2$$
$$= -2.5\hat{w} + 5\hat{p} - 2.5\hat{r} + 2.5 \log 0.5 + 2.5 \log 0.3$$

$$\hat{x} = \begin{vmatrix} -0.5 & 0.3 & \hat{w} - \hat{p} - \log 0.5 \\ 0.5 & -0.7 & \hat{r} - \hat{p} - \log 0.3 \\ -0.5 & -0.3 & 0 \end{vmatrix} / |D|$$
$$= (-0.5\hat{w} + 0.8\hat{p} - 0.3\hat{r} + 0.5 \log 0.5 + 0.3 \log 0.3)/0.2$$
$$= -2.5\hat{w} + 4\hat{p} - 1.5\hat{r} + 2.5 \log 0.5 + 1.5 \log 0.3$$

Taking antilogs gives the input demand and output supply functions

$$L = 0.0145 w^{-3.5} r^{-1.5} p^5$$
$$K = 0.0087 w^{-2.5} r^{-2.5} p^5$$
$$x = 0.0290 w^{-2.5} r^{-1.5} p^4$$

To obtain the profit function we substitute into the expression $\pi = px - wL - rK$ to get

$$\pi = 0.0058 w^{-2.5} r^{-1.5} p^5$$

5. In Exercise 5 of Section 13.1, we derived the demand functions

$$x_1 = \alpha \frac{m - p_2 c_2}{p_1} + (1 - \alpha) c_1$$
$$x_2 = (1 - \alpha) \frac{m - p_1 c_1}{p_2} + \alpha c_2$$

Substituting these into the utility function yields the indirect utility function

$$V(\mathbf{p}, m) = \left(\alpha \frac{m - p_1 c_1 - p_2 c_2}{p_1}\right)^{\alpha} \times \left[(1 - \alpha) \frac{m - p_1 c_1 - p_2 c_2}{p_2}\right]^{1-\alpha}$$
$$= \alpha^{\alpha} (1 - \alpha)^{1-\alpha} p_1^{-\alpha} p_2^{\alpha - 1} (m - p_1 c_1 - p_2 c_2)$$

Observe that $V(\mathbf{p}, m)$ is not defined for $m < p_1 c_1 + p_2 c_2$. Rewriting the equation above as

$$m = V(\mathbf{p}, m) \alpha^{-\alpha} (1 - \alpha)^{\alpha - 1} p_1^{\alpha} p_2^{1-\alpha} + p_1 c_1 + p_2 c_2$$

and substituting u for $V(\mathbf{p}, m)$ and $E(\mathbf{p}, u)$ for m, we obtain the expenditure function

$$E(\mathbf{p}, u) = \alpha^{-\alpha} (1 - \alpha)^{\alpha - 1} p_1^{\alpha} p_2^{1-\alpha} u + p_1 c_1 + p_2 c_2$$

7. From the optimality of the long-run total cost curve, we know that

$$C(y_a) = \min c(y_a, K) \equiv c(y_a, K_a) \quad \text{(I)}$$

(a) The first-order condition of the minimization problem in (I) is

$$\frac{\partial c(y_a, K_a)}{\partial K} = 0 \quad \text{(II)}$$

Now differentiate (I) at y_a with respect to y

$$\frac{dC}{dy} = \frac{\partial c}{\partial y} + \frac{\partial c}{\partial K} \frac{\partial K}{\partial y}$$

From (II) we know that the second term on the right-hand side is zero which proves

$$\frac{dC}{dy} = \frac{\partial c}{\partial y}$$

i.e., long-run marginal cost equals short-run marginal cost at y_a. Figure 14.3.7 illustrates.

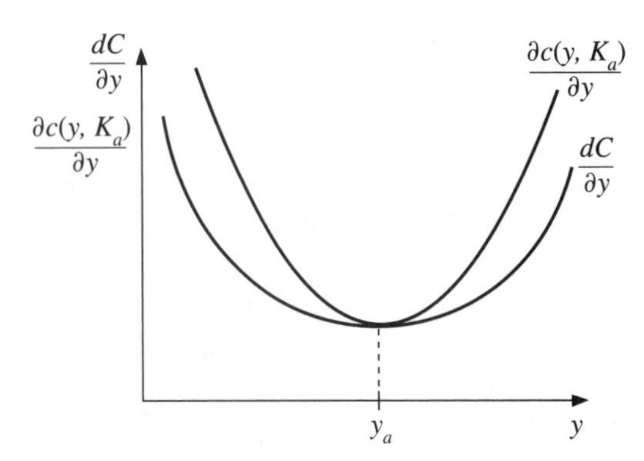

Figure 14.3.7

(b) The equality of long-run average cost and short-run average cost can be obtained simply by dividing equation (I) by y_a. This yields

$$\frac{C(y_a)}{y_a} = \frac{c(y_a, K_a)}{y_a}$$

Review Exercises (page 569)

1. (a)

$$\max y = x_1^{0.25} x_2^{0.75} \quad \text{s.t. } 2x_1 + 4x_2 = \alpha$$

The Lagrange function is

$$\mathcal{L} = x_1^{0.25} x_2^{0.75} + \lambda(\alpha - 2x_1 - 4x_2)$$

The first-order conditions are

$$0.25 x_1^{-0.75} x_2^{0.75} - 2\lambda = 0$$
$$0.75 x_1^{0.25} x_2^{-0.25} - 4\lambda = 0$$
$$\alpha - 2x_1 - 4x_2 = 0$$

Solving the second condition for λ and substituting into the first yields

$$x_2 = 1.5 x_1$$

Inserting into the third condition we obtain

$$x_1 = \frac{\alpha}{8}$$

and

$$x_2 = \frac{3\alpha}{16}$$

Inserting into the objective functions yields

$$V(\alpha) = \left(\frac{\alpha}{8}\right)^{0.25} \left(\frac{3\alpha}{16}\right)^{0.75} = 0.17\alpha$$

The comparative-static effect of a change in the α is

$$\frac{\partial V}{\partial \alpha} = 0.17$$

(b)

$$\max y = 2x_1 + 3x_2 \quad \text{s.t. } \alpha_1 x_1^2 + 5x_2^2 = \alpha_2$$

The Lagrange function is

$$\mathcal{L} = 2x_1 + 3x_2 + \lambda(\alpha_2 - \alpha_1 x_1^2 - 5x_2^2)$$

The first-order conditions are

$$\begin{aligned} 2 - 2\lambda\alpha_1 x_1 &= 0 \\ 3 - 10\lambda x_2 &= 0 \\ \alpha_2 - \alpha_1 x_1^2 - 5x_2^2 &= 0 \end{aligned}$$

Solving for λ in the second condition and substituting into the first yields

$$x_1 = \frac{10}{3\alpha_1} x_2$$

Inserting into the third conditions we obtain the two solutions

$$x_2 = \pm\sqrt{\frac{9\alpha_1\alpha_2}{100 + 45\alpha_1}}$$

Short inspection shows that the positive root is a candidate for a maximum while the negative root will give us a minimum. Hence

$$x_1 = \frac{10}{3\alpha_1}\sqrt{\frac{9\alpha_1\alpha_2}{100 + 45\alpha_1}}$$

and

$$x_2 = \sqrt{\frac{9\alpha_1\alpha_2}{100 + 45\alpha_1}}$$

Substituting into the objective function yields

$$V(\alpha_1, \alpha_2) = \left(\frac{20}{3\alpha_1} + 3\right)\sqrt{\frac{9\alpha_1\alpha_2}{100 + 45\alpha_1}}$$

We obtain the comparative-static effects of changes in α_1 and α_2 by rewriting the value function as

$$V(\alpha_1, \alpha_2) = \left(20\alpha_1^{-0.5}\alpha_2^{0.5} + 9\alpha_1^{0.5}\alpha_2^{0.5}\right)(100+45\alpha_1)^{-0.5}$$

and differentiating with respect to α_1 and α_2

$$\begin{aligned} \frac{\partial V}{\partial \alpha_1} &= \left(-10\alpha_1^{-1.5}\alpha_2^{0.5} + 4.5\alpha_1^{-0.5}\alpha_2^{0.5}\right) \times \\ &\quad (100 + 45\alpha_1)^{-0.5} \\ &\quad - \left(20\alpha_1^{-0.5}\alpha_2^{0.5} + 9\alpha_1^{0.5}\alpha_2^{0.5}\right) \times \\ &\quad 0.5(45)(100 + 45\alpha_1)^{-1.5} \\ \frac{\partial V}{\partial \alpha_2} &= 0.5\alpha_2^{-0.5}\left(20\alpha_1^{-0.5} + 9\alpha_1^{0.5}\right) \times \\ &\quad (100 + 45\alpha_1)^{-0.5} \end{aligned}$$

(c)

$$\max y = x_1^{0.25} x_2^{0.75} \quad \text{s.t. } 2x_1^2 + 5x_2^2 = \alpha$$

The Lagrange function is

$$\mathcal{L} = x_1^{0.25} x_2^{0.75} + \lambda(\alpha - 2x_1^2 - 5x_2^2)$$

The first-order conditions are

$$\begin{aligned} 0.25x_1^{-0.75}x_2^{0.75} - 4\lambda x_1 &= 0 \\ 0.75x_1^{0.25}x_2^{-0.25} - 10\lambda x_2 &= 0 \\ \alpha - 2x_1^2 - 5x_2^2 &= 0 \end{aligned}$$

Solving for λ in the first condition and substituting into the second yields

$$x_2 = \pm\sqrt{\frac{6}{5}x_1^2}$$

Note that the objective function is only defined for the positive root. Inserting the positive root into the third condition we obtain

$$x_1 = \sqrt{\frac{\alpha}{8}}$$

and

$$x_2 = \sqrt{\frac{3}{20}\alpha}$$

Inserting into the objective function yields

$$V(\alpha) = 0.38\sqrt{\alpha}$$

The comparative-static effect of a change in α is

$$\frac{\partial V}{\partial \alpha} = \frac{0.19}{\sqrt{\alpha}}$$

(d)

$$\min y = 2x_1 + 4x_2 \quad \text{s.t. } x_1^{0.25}x_2^{0.75} = \alpha$$

The Lagrange function is

$$\mathcal{L} = 2x_1 + 4x_2 + \lambda(\alpha - x_1^{0.25}x_2^{0.75})$$

The first-order conditions are

$$\begin{aligned} 2 - 0.25\lambda x_1^{-0.75}x_2^{0.75} &= 0 \\ 4 - 0.75\lambda x_1^{0.25}x_2^{-0.25} &= 0 \\ \alpha - x_1^{0.25}x_2^{0.75} &= 0 \end{aligned}$$

Eliminating λ in the first two conditions we obtain

$$x_2 = 1.5x_1$$

Substituting into the third condition yields

$$x_1 = (1.5)^{-0.75}\alpha$$

and

$$x_2 = (1.5)^{0.25}\alpha$$

Inserting into the objective function we get

$$V(\alpha) = 2(1.5)^{-0.75}\alpha + 4(1.5)^{0.25}\alpha = 5.91\alpha$$

The comparative-static effect of a change in α is

$$\frac{\partial V}{\partial \alpha} = 5.91$$

(e)

$$\max y(x_1 + 2)(x_2 + 1) \quad \text{s.t. } x_1 + x_2 = \alpha$$

The Lagrange function is

$$\mathcal{L} = (x_1 + 2)(x_2 + 1) + \lambda(\alpha - x_1 - x_2)$$

The first-order conditions are

$$\begin{aligned} x_2 + 1 - \lambda &= 0 \\ x_1 + 2 - \lambda &= 0 \\ \alpha - x_1 - x_2 &= 0 \end{aligned}$$

Solving for λ in the first condition and substituting into the second we obtain

$$x_2 = x_1 + 1$$

Inserting in the third equation yields

$$x_1 = \frac{\alpha - 1}{2}$$

and

$$x_2 = \frac{\alpha + 1}{2}$$

Substituting into the objective function we obtain

$$V(\alpha) = \left(\frac{\alpha + 3}{2}\right)^2$$

The comparative-static effect of a change in α is

$$\frac{\partial V}{\partial \alpha} = \frac{\alpha + 3}{2}$$

(f)

$$\min y = \alpha_1 x_1 + x_2 \quad \text{s.t. } (x_1 + 2)(x_2 + 1) = \alpha_2$$

The Lagrange function is

$$\mathcal{L} = \alpha_1 x_1 + x_2 + \lambda[\alpha_2 - (x_1 + 2)(x_2 + 1)]$$

The first-order conditions are

$$\begin{aligned} \alpha_1 - \lambda(x_2 + 1) &= 0 \\ 1 - \lambda(x_1 + 2) &= 0 \\ \alpha_2 - (x_1 + 2)(x_2 + 1) &= 0 \end{aligned}$$

Eliminating λ in the first two conditions and solving for x_2 yields

$$x_2 = \alpha_1(x_1 + 2) - 1$$

Inserting in the third condition we obtain the solutions

$$x_1 = \pm\sqrt{\alpha_2/\alpha_1} - 2$$

and

$$x_2 = \pm\sqrt{\alpha_1 \alpha_2} - 1$$

Obviously, the negative roots give us the minimum. Thus, we obtain the value function

$$V(\alpha_1, \alpha_2) = (-2)\left(\sqrt{\alpha_1 \alpha_2} + \alpha_1\right) - 1$$

The comparative-static effects are

$$\begin{aligned} \frac{\partial V}{\partial \alpha_1} &= -\alpha_1^{-0.5}\alpha_2^{0.5} - 2 \\ \frac{\partial V}{\partial \alpha_2} &= -\alpha_1^{0.5}\alpha_2^{-0.5} \end{aligned}$$

Note that in this case the Envelope Theorem implies $\partial V/\partial \alpha_1 = x_1$.

3. (a) The maximization problem is

$$\max_{x_{11}, x_{12}, x_{21}, x_{22}} u(x_{11}, x_{12}) + \beta u(x_{21}, x_{22})$$

subject to

$$\begin{aligned} p_1 x_{11} + p_2 x_{12} &= m_1 \\ p_1 x_{21} + p_2 x_{22} &= m_2 \end{aligned}$$

Solving the problem yields the optimal values of x_{11}, x_{12}, x_{21} and x_{22} as functions of the exogenous variables p_1, p_2, m_1 and m_2. The impossibility of shifting income across periods and the additivity of the utility function imply that

$$\frac{\partial x_{ti}}{\partial m_s} = 0, \quad t \neq s, \quad \text{all } i \text{ (goods)}$$

So we have the indirect utility function v

$$\begin{aligned} &v(p_1, p_2, m_1, m_2) = \\ &u[x_{11}^*(p_1, p_2, m_1), x_{12}^*(p_1, p_2, m_1)] \\ &+\beta u[x_{21}^*(p_1, p_2, m_2), x_{22}^*(p_1, p_2, m_2)] \end{aligned}$$

The Lagrange multipliers λ_1 and λ_2 inform us about the increase in utility that can be obtained by a marginal increase in m_1 and m_2. They are equal if the between-period marginal rates of substitution are

$$\left|\frac{dx_{2i}}{dx_{1i}}\right|_{du=0,\beta\in(0,1)} = \frac{1}{\beta}\left|\frac{dx_{2i}}{dx_{1i}}\right|_{du=0,\beta\in(0,1)} \qquad i = 1, 2$$

(b) Introducing the capital market allows exchange of income across periods as a result of which the two budget constraints of the former problem are lumped together in one:

$$[m_2 - p_1 x_{21} - p_2 x_{22}] + (1+r)[m_1 - p_1 x_{11} - p_2 x_{12}] = 0$$

Introduction of the capital market allows the equalization of the marginal utility of income across periods, this being a necessary (and here also sufficient) condition for a utility maximum. The consumer is therefore generally better off than without the capital market. He retains his utility level if the marginal utilities of income were by coincidence equalized without capital market. He cannot be worse off.

Define $m \equiv m_2 + (1 + r)m_1$. The indirect utility function v is

$$\begin{aligned} v(p_1, p_2, m) = \; & u[x_{11}^*(p_1, p_2, m), x_{12}^*(p_1, p_2, m)] \\ & +\beta u[x_{21}^*(p_1, p_2, m), x_{22}^*(p_1, p_2, m)] \end{aligned}$$

5. Consider the dual problem: "Minimize $\bar{m}_2$ subject to the constraint that utility u can be achieved":

$$\min \bar{m}_2 = m_2 + (1 + r)(m_1 - \bar{m}_1)$$

s.t.

$$v(m_1) + \beta v(m_2) \geq u$$

Solving this problem gives demand functions

$$m_1 = H_1(r, u)$$

and

$$m_2 = H_2(r, u)$$

which are called "Hicksian" or compensated demand functions. Inserting into the objective function we obtain the expenditure function

$$m_2^* = H_2(r, u) + (1 + r)(H_1(r, u) - \bar{m}_1)$$

By the Envelope Theorem we get Shephard's Lemma:

$$\frac{\partial m_2^*}{\partial r} = H_1(r, u) - \bar{m}_1$$

From the problem

$$\max u = v(m_1) + \beta v(m_2)$$

s.t.

$$m_1 + \frac{m_2}{1+r} = \bar{m}_1 + \frac{\bar{m}_2}{1+r}$$

we can derive the so-called Marshallian or uncompensated demand functions

$$m_1 = m_1(r, \bar{m}_1, \bar{m}_2)$$

If the consumer is just given enough wealth to reach utility level u and the interest rate is equal, Hicksian must equal Marshallian demand. Thus, for demand for consumption in period 1 we have

$$H_1(r, u) = m_1[r, \bar{m}_1, m_2^*(r, u)]$$

Differentiating with respect to r yields

$$\frac{\partial H_1}{\partial r} = \frac{\partial m_1}{\partial r} + \frac{\partial m_1}{\partial \bar{m}_2}\frac{\partial m_2^*}{\partial r}$$

Insert Shephard's Lemma and rearrange to get the Slutsky equation for the effect of a change in the interest rate on the choice of consumption in period 1:

$$\frac{\partial m_1}{\partial r} = \frac{\partial H_1}{\partial r} - \frac{\partial m_1}{\partial \bar{m}_2}(H_1 - \bar{m}_1)$$

Analogously, we obtain for consumption in period 2:

$$\frac{\partial m_2}{\partial r} = \frac{\partial H_2}{\partial r} - \frac{\partial m_2}{\partial \bar{m}_2}(H_1 - \bar{m}_1)$$

The left-hand side of these Slutsky equations is the total effect of a change in the interest rate. It can be split into a substitution and an income effect—the two terms on the right-hand side.

First, we interpret the Slutsky equation for the effect of a change in the interest rate on choice of consumption in period 1. Here the substitution effect $\partial H_1/\partial r$ is unambigiously negative. The interest rate is a "price" of consumption in period 1. If one consumes \$1 more today, one will forgo \$$(1+r)$ of consumption tomorrow. An increase in the price will reduce the "compensated" Hicksian demand (draw a picture of an indifference curve and a wealth constraint and observe how the tangential point shifts as r is increased).

The income effect

$$\left[-\frac{\partial m_1}{\partial \bar{m}_2}(H_1 - \bar{m}_1)\right]$$

is undetermined. $\partial m_1/\partial \bar{m}_2$ can be safely assumed to be positive: If one is richer, one will probably consume more in each period. Note the difference to the usual Slutsky equation where you cannot say a priori that more income implies more consumption of the good. Why then does the income effect not have to be negative? The reason is that we do not know whether the increasing interest rate makes the individual richer or poorer: A lender ($H_1 - \bar{m}_1 < 0$) will become richer while a borrower ($H_1 - \bar{m}_1 > 0$) becomes poorer. Thus, for a borrower, both effects work in the same direction; a rise in the interest rate will reduce optimal borrowing. For a lender, however, the effects work in opposite directions. It is unclear which effect dominates.

In the Slutsky equation for the effect of a change in the interest rate on choice of consumption in period 2, the substitution effect $\partial H_2/\partial r$ is positive: If the price of consumption in period 2 $1/(1+r)$ goes down, compensated demand for consumption in period 2 will rise. Again, the income effect

$$\left[-\frac{\partial m_2}{\partial \bar{m}_2}(H_1 - \bar{m}_1)\right]$$

is not determined. As above, we can assume that $\partial m_2/\partial \bar{m}_2$ is positive. Thus, the income effect will be positive if the individual is a lender ($H_1 - \bar{m}_1 < 0$) while it will be negative if she is a borrower ($H_1 - \bar{m}_1 > 0$). Hence, both effects work in the same direction for a lender. If the interest rises she will consume more tomorrow. For a borrower, the effects work in opposite directions and we cannot say which effect dominates.

CHAPTER 15

Section 15.1 (page 577)

1. $\max x_1 + x_2$ s.t. $-(x_1^2 + x_2) \geq 0$ and $x_1, x_2 \geq 0$

The Lagrange function is

$$\mathcal{L} = x_1 + x_2 + \lambda\left[-(x_1^2 + x_2^2)\right]$$

and the Kuhn-Tucker conditions are

$$\frac{\partial \mathcal{L}}{\partial x_1} = 1 - \lambda^* 2x_1 \leq 0 \quad x_1^* \geq 0 \quad x_1^*(1 - \lambda^* 2x_1) = 0$$

$$\frac{\partial \mathcal{L}}{\partial x_2} = 1 - \lambda^* 2x_2 \leq 0 \quad x_2^* \geq 0 x_2^*(1 - \lambda^* 2x_2) = 0$$

$$\frac{\partial \mathcal{L}}{\partial \lambda} = -(x_1^{*2} + x_2^{*2}) \geq 0 \quad \lambda^* \geq 0$$

$$\lambda^*\left[-(x_1^{*2} + x_2^{*2})\right] = 0$$

Only $x_1 = x_2 = 0$ solves $\partial\mathcal{L}/\partial\lambda \geq 0$. However, this implies that $\partial\mathcal{L}/\partial x_1 = \partial\mathcal{L}/\partial x_2 = 1$. Therefore, the K-T conditions are not fulfilled and we do not obtain a solution to the problem. That is, the first two conditions yield $1 \leq 0$, which is nonsense. The reason is that Slater's condition is not satisfied because there exists no point (x_1, x_2) such that $-(x_1^2 + 2x_2^2) > 0$.

3. $\max u = (x_1 + a)x_2^b$ s.t. $p_1x_1 + p_2x_2 \leq m$ and $x_1, x_2 \geq 0$

The Lagrange function is

$$\mathcal{L} = (x_1 + a)x_2^b + \lambda(m - p_1x_1 - p_2x_2)$$

and the Kuhn-Tucker conditions are

$$\frac{\partial \mathcal{L}}{\partial x_1} = (x_2^*)^b - \lambda^* p_1 \leq 0 \quad x_1^* \geq 0$$

$$x_1^*\left[(x_2^*)^b - \lambda^* p_1)\right] = 0$$

$$\frac{\partial \mathcal{L}}{\partial x_2} = b(x_1^* + a)(x_2^*)^{b-1} - \lambda^* p_2 \leq 0 \quad x_2^* \geq 0$$

$$x_2^*\left[b(x_1^* + a)(x_2^*)^{b-1}\right] = 0$$

$$\frac{\partial \mathcal{L}}{\partial \lambda} = m - p_1x_1^* - p_2x_2^* \geq 0 \quad \lambda^* \geq 0$$

$$\lambda^*(m - p_1x_1 - p_2x_2) = 0$$

First, we show that $\lambda^* > 0$. Assume that $\lambda^* = 0$. Then from the first condition we have $x_2^* = 0$ which implies that $u = 0$. However, since $(x_1 + a)$ is strictly positive, we can find an $x_2 > 0$ fulfilling the budget constraint such that $u > 0$. Thus, we must have $\lambda^* > 0$ which implies that

$$m - p_1x_1^* - p_2x_2^* = 0 \quad \text{(I)}$$

Now, we check the three possible cases implied by $x_1^* \geq 0$ and $x_2^* \geq 0$:

(i) $x_1^* > 0, x_2^* > 0$

Then we have

$$(x_2^*)^b - \lambda^* p_1 = 0$$

$$b(x_1^* + a)(x_2^*)^{b-1} - \lambda^* p_2 = 0$$

Solving for x_1^* and x_2^* using (I) yields

$$x_1^* = \frac{m - abp_1}{(1+b)p_1}$$

and

$$x_2^* = \frac{(m + ap_1)b}{(1+b)p_2}$$

Thus, we see that case (i) is only possible if $m > abp_1$.

(ii) $x_1^* > 0, x_2^* = 0$.

In this case we have

$$(x_2^*)^b - \lambda^* p_1 = 0$$

Inserting for $x_2^* = 0$, leads to the contradiction $\lambda^* p_1 = 0$. Hence, case (ii) cannot be an optimal solution.

(iii) $x_1^* = 0, x_2^* > 0$.

Then we have

$$b(x_1^* + a)(x_2^*)^{b-1} - \lambda^* p_2 = 0$$

Inserting $x_1^* = 0$ does not lead to a contradiction as in (ii). Using (i) and the result of (i) we have $x_1^* = 0$ and $x_2^* = m/p_1$ for $m \leq abp_1$.

Economic Interpretation:

The utility function implies that the consumer will always consume good 2 (case (ii)). If she does not, her utility would be zero which cannot be an optimal solution if she has positive income. Thus, good 2 is an essential good for the consumer. This is not the case for good 1. It will be consumed or not depending on its price:

If its price is relatively low (case (i)), she will consume a positive amount of good 1. As you can check, we then have

$$u_1/u_2 = p_1/p_2$$

i.e., the marginal rate of substitution between the goods is equal to their price-ratio.

If good 1's price is relatively high (case (iii)), the consumer will not consume good 1. We can derive

$$u_1/u_2 \leq p_1/p_2$$

or

$$u_1/p_1 \leq u_2/p_2$$

This says that even if the consumer spends all her money on good 2, the last dollar spent on good 2 yields a marginal utility at least as high as if she spent it on good 1.

5. (a) Substituting for x_1 and x_2 the problem is

$$\max V = 40L_1 - 2L_1^2 + 40L_2 - 3.75L_2^2$$

s.t. $L_1 + L_2 \leq 12$. The Lagrange function is

$$\mathcal{L} = 40L_1 - 2L_1^2 + 40L_2 - 3.75L_2^2 + \lambda(12 - L_1 - L_2)$$

The Kuhn-Tucker conditions are

$$\frac{\partial \mathcal{L}}{\partial L_1} = 40 - 4L_1^* - \lambda^* \leq 0 \quad L_1^* \geq 0$$
$$L_1^*(40 - 4L_1^* - \lambda^*) = 0$$
$$\frac{\partial \mathcal{L}}{\partial L_2} = 40 - 7.5L_2^* - \lambda^* \leq 0 \quad L_2^* \geq 0$$
$$L_2^*(40 - 7.5L_2^* - \lambda^*) = 0$$
$$\frac{\partial \mathcal{L}}{\partial \lambda} = 12 - L_1^* - L_2^* \geq 0 \quad \lambda^* \geq 0$$
$$\lambda^*(12 - L_1^* - L_2^*) = 0$$

First, we show that $\lambda^* > 0$. Assuming that $\lambda^* = 0$ we have from the first two conditions

$$40 - 4L_1^* \leq 0 \quad \Leftrightarrow \quad L_1^* \geq 10$$
$$40 - 7.5L_2^* \leq 0 \quad \Leftrightarrow \quad L_2^* \geq 5\frac{1}{3}$$

which implies

$$L_1^* + L_2^* \geq 15\frac{1}{3}$$

This violates the third condition. Thus, we have $\lambda^* > 0$ and

$$12 - L_1^* - L_2^* = 0 \qquad (I)$$

Now note that neither $L_1^* = 0$ nor $L_2^* = 0$ can be part of the optimal solution. We then would have $L_2^* = 12$ or $L_1^* = 12$. As can be seen by inspection the K-T conditions would not be fulfilled. Thus the optimal solution is characterized by $L_1^* > 0$ and $L_2^* > 0$. This implies

$$40 - 4L_1^* - \lambda^* = 0$$
$$40 - 7.5L_2^* - \lambda^* = 0$$

Using (I) the optimal solution is $L_1^* = 7.83$, $L_2^* = 4.17$, $\lambda^* = 8.7$.

(b) The Lagrange function is

$$\mathcal{L} = 40L_1 - 2L_1^2 + 40L_2 - 3.75L_2^2 + \lambda(20 - L_1 - L_2)$$

and the Kuhn-Tucker conditions are

$$\frac{\partial \mathcal{L}}{\partial L_1} = 40 - 4L_1^* - \lambda^* \leq 0 \quad L_1^* \geq 0$$
$$L_1^*(40 - 4L_1^* - \lambda^*) = 0$$
$$\frac{\partial \mathcal{L}}{\partial L_2} = 40 - 7.5L_2^* - \lambda^* \leq 0 \quad L_2^* \geq 0$$
$$L_2^*(40 - 7.5L_2^* - \lambda^*) = 0$$
$$\frac{\partial \mathcal{L}}{\partial \lambda} = 20 - L_1^* - L_2^* \geq 0 \quad \lambda^* \geq 0$$
$$\lambda^*(20 - L_1^* - L_2^* = 0$$

In this problem $\lambda^* = 0$ is possible as can be seen by comparing to problem (a). Also by the same argument as above we must have $L_1^* > 0$ and $L_2^* > 0$. We will find the solution by comparing the case $\lambda^* > 0$ and $\lambda^* = 0$.

(i) $\lambda^* > 0$:

In this case the optimal solution would be characterized by

$$40 - 4L_1^* - \lambda^* = 0$$
$$40 - 7.5L_2^* - \lambda^* = 0$$
$$20 - L_1^* - L_2^* = 0$$

yielding $L_1^* = 13.04$, $L_2^* = 6.96$, $\lambda^* = -12.2$.

Note that the shadow price of labor is negative, i.e., the value of total output can be increased by reducing the amount of labor used in the economy. Thus, case (i) cannot be the optimal solution and we must have $\lambda^* = 0$.

(ii) $\lambda^* = 0$:

In this case, the optimal solution is described by

$$40 - 4L_1^* - \lambda^* = 0$$
$$40 - 7.5L_2^* - \lambda^* = 0$$
$$\lambda^* = 0$$

which yields $L_1^* = 10$, $L_2^* = 5\frac{1}{3}$, $\lambda^* = 0$.

(c) The production functions x_1 and x_2 have a unique maximum at $L_1 = 10$ and $L_2 = 5\frac{1}{3}$. Using more labor decreases output. Thus, the maximum amount of labor that can be fruitfully employed is $15\frac{1}{3}$. In exercise (a) the available labor is smaller than this maximum amount, so the constraint is binding. In exercise (b) the available labor is larger than this maximum amount. Thus, the constraint is not binding and part of the available labor remains idle.

This is shown in Figures 15.1.5 (a) and 15.1.5 (b). The length of the x-axis corresponds to the amount of labor available. In Figure 15.1.5 (a) the length is 12 units while in Figure 15.1.5 (b) it is 20 units. From the left we measure $\partial V/\partial L_1 = 40 - 4L_1$, the marginal contribution of labor in the production of good 1 to the value of total output. From the right we measure $\partial V/\partial L_2 = 40 - 7.5L_2$. Note that it can never be optimal to have $\partial V/\partial L_i < 0, i = 1, 2$ because the value of total output would rise if L_i were reduced.

In problem (a) we can have full employment without $\partial V/\partial L_i < 0, i = 1, 2$, as Figure 15.1.5 (a) demonstrates. The optimal solution is characterized by the intersection of the two curves at point A. Otherwise V could be increased by shifting labor between the two sectors.

In problem (b), however, it is not possible to employ all labor in both sectors such that $\partial V/\partial L_i \geq 0, i = 1, 2$, as is shown in Figure 15.1.5 (b). It is optimal to employ labor in both sectors until $\partial V/\partial L_i = 0$, as indicated by points B and C, and to leave $\overline{BC}$ units of labor unemployed.

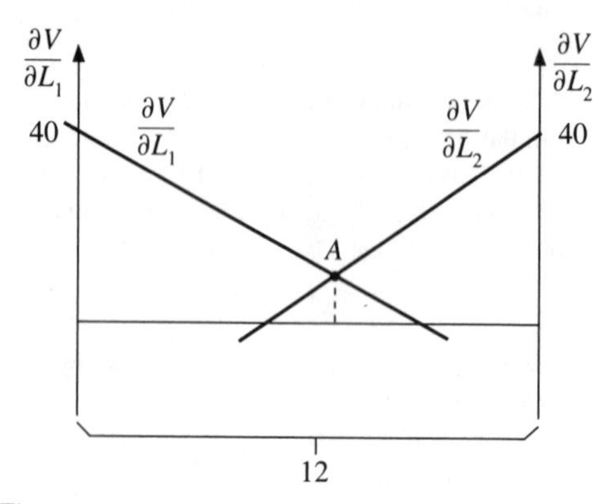

Figure 15.1.5 (a)

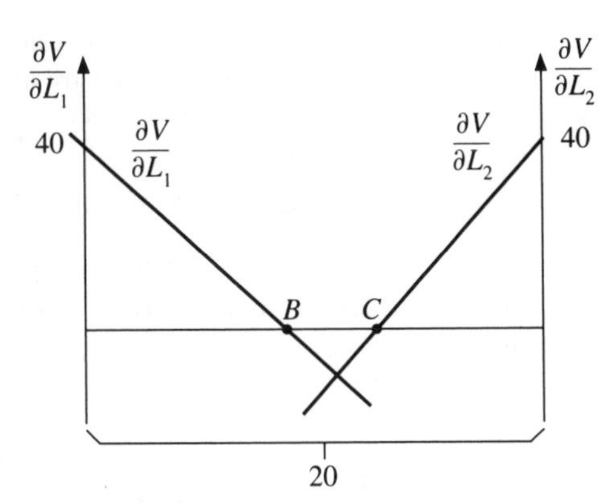

Figure 15.1.5 (b)

Section 15.2 (page 585)

1. The consumer's problem is

$$\max x_1 x_2 \quad \text{s.t.} \quad 2x_1 + 3x_2 \le 100$$
$$x_1 + 4x_2 \le 80$$

The Lagrange function is

$$\mathcal{L} = x_1 x_2 + \lambda_1(1000 - 2x_1 - 3x_2) + \lambda_2(80 - x_1 - 4x_2)$$

and the Kuhn-Tucker conditions are

$$\frac{\partial \mathcal{L}}{\partial x_1} = x_2^* - 2\lambda_1^* - \lambda_2^* \le 0 \quad x_1^* \ge 0$$
$$x_1^*(x_2^* - 2\lambda_1^* - \lambda_2^*) = 0$$
$$\frac{\partial \mathcal{L}}{\partial x_2} = x_1^* - 3\lambda_1^* - 4\lambda_2^* \le 0 \quad x_2^* \ge 0$$
$$x_2^*(x_1^* - 3\lambda_1^* - 4\lambda_2^*) = 0$$
$$\frac{\partial \mathcal{L}}{\partial \lambda_1} = 100 - 2x_1^* - 3x_2^* \ge 0 \quad \lambda_1^* \ge 0$$
$$\lambda_1^*(100 - 2x_1^* - 3x_2^*) = 0$$
$$\frac{\partial \mathcal{L}}{\partial \lambda_2} = 80 - x_1^* - 4x_2^* \ge 0$$
$$\lambda_2^* \ge 0$$
$$\lambda_2^*(80 - x_1^* - 4x_2^*) = 0$$

We assume that neither of the x-values is zero. Thus, the conditions $\partial\mathcal{L}/\partial x_1 \le 0$ and $\partial\mathcal{L}/\partial x_2 \le 0$ become strict equalities. Because the utility function implies $u_i > 0, i = 1, 2$, and both x-values are positive, it is not possible that both λ-values are zero. At least one constraint must bind. Thus, we must consider three cases:

(i) $\lambda_1^* > 0, \lambda_2^* = 0$:
This implies

$$\begin{aligned} x_1^* - 2\lambda_1^* - \lambda_2^* &= 0 \\ x_1^* - 3\lambda_1^* - 4\lambda_2^* &= 0 \\ 100 - 2x_1^* - 3x_2^* &= 0 \\ \lambda_2^* &= 0 \end{aligned}$$

Solving for x_1^*, x_2^*, and λ_1^* we obtain $x_1^* = 25$, $x_2^* = 16\frac{2}{3}$, $\lambda_1^* = 8\frac{1}{3}$. However, we have

$$\frac{\partial \mathcal{L}}{\partial \lambda_2} = -11\frac{2}{3} < 0$$

which contradicts the fourth K-T condition. Thus, case (i) does not yield the optimal solution.

(ii) $\lambda_1^* = 0, \lambda_2^* > 0$:
We have

$$\begin{aligned} x_2^* - 2\lambda_1^* - \lambda_2^* &= 0 \\ x_1^* - 3\lambda_1^* - 4\lambda_2^* &= 0 \\ \lambda_1^* &= 0 \\ 80 - x_1^* - 4x_2^* &= 0 \end{aligned}$$

Solving for x_1^*, x_2^* and λ_2^* we get $x_1^* = 40, x_2^* = 10, \lambda_2^* = 10$. But we have $\partial\mathcal{L}/\partial\lambda_1 = -10 < 0$ which contradicts the third K-T condition. Hence, case (ii) does not yield the optimal solution either.

(iii) $\lambda_1^* > 0, \lambda_2^* > 0$:
Then all constraints are binding. The solution is characterized by

$$\frac{\partial \mathcal{L}}{\partial x_1} = \frac{\partial \mathcal{L}}{\partial x_2} = \frac{\partial \mathcal{L}}{\partial \lambda_1} = \frac{\partial \mathcal{L}}{\partial \lambda_2} = 0$$

Cramer's rule yields $x_1^* = 32$, $x_2^* = 12$, $\lambda_1^* = 2.3$, $\lambda_2^* = 7.4$ which is the optimal solution.

3. The planner's problem is

$$\max x_1 x_2 \quad \text{s.t.} \quad y_1 = 10 l_1^{0.5}$$
$$y_2 = 5 l_2^{0.5}$$
$$l_1 + l_2 \le 1000$$
$$10(x_1 - y_1) + 8(x_2 - y_2) \le 0$$
$$x_1, x_2 \ge 0$$

Substituting the production functions (first two constraints) into the balance-of-trade constraint (fourth constraint), the Lagrange functions is

$$\begin{aligned} \mathcal{L} = {} & x_1 x_2 + \lambda_1(1000 - l_1 - l_2) \\ & + \lambda_2(100 l_1^{0.5} + 40 l_2^{0.5} - 10x_1 - 8x_2) \end{aligned}$$

The Kuhn-Tucker conditions are

$$\frac{\partial \mathcal{L}}{\partial x_1} = x_2^* - 10\lambda_2^* \le 0$$
$$x_1^* \ge 0 \quad x_1^*(x_2^* - 10\lambda_2^*) = 0$$
$$\frac{\partial \mathcal{L}}{\partial x_2} = x_1^* - 8\lambda_2^* \le 0$$
$$x_2^* \ge 0 \quad x_2^*(x_1^* - 8\lambda_2^*) = 0$$
$$\frac{\partial \mathcal{L}}{\partial l_1} = -\lambda_1^* + 50\lambda_2^*(l_1^*)^{-0.5} \le 0$$
$$l_1^* \ge 0 \quad l_2^*[-\lambda_1^* + 50\lambda_2^*(l_1^*)^{-0.5}] = 0$$
$$\frac{\partial \mathcal{L}}{\partial l_2} = -\lambda_1^* + 20\lambda_2^*(l_2^*)^{-0.5} \le 0$$
$$l_2^* \ge 0 \quad l_2^*[-\lambda_1^* + 20\lambda_2^*(l_2^*)^{-0.5}] = 0$$
$$\frac{\partial \mathcal{L}}{\partial \lambda_1} = 1000 - l_1^* - l_2^* \ge 0$$
$$\lambda_1^* \ge 0 \quad \lambda_1^*(1000 - l_1^* - l_2^*) = 0$$
$$\frac{\partial \mathcal{L}}{\partial \lambda_2} = 100(l_1^*)^{0.5} + 40(l_2^*)^{0.5} - 10x_1^* - 8x_2^* \ge 0$$
$$\lambda_2^* \ge 0$$
$$\lambda_2^*[100(l_1^*)^{0.5} + 40(l_2^*)^{0.5} - 10x_1^* - 8x_2^*] = 0$$

Note that we must have $x_1^* > 0$ and $x_2^* > 0$ since $x_1 = 0$ or $x_2 = 0$ implies $u = 0$. This cannot be an optimum because it is always possible to choose $l_1 > 0$ and $l_2 > 0$ to achieve a positive consumption level for both goods. Thus, we have

$$\frac{\partial \mathcal{L}}{\partial x_1} = x_2^* - 10\lambda_2^* = 0$$
$$\frac{\partial \mathcal{L}}{\partial x_2} = x_1^* - 8\lambda_1^* = 0$$

implying that $\lambda_2^* > 0$. At the optimum there will be no trade surplus. Since the economy is consuming both goods and the balance of trade is zero, at least one good is produced, i.e., $l_1^* > 0$ and/or $l_2^* > 0$. Because of

$$\frac{\partial \mathcal{L}}{\partial l_1} = -\lambda_1^* + 50\lambda_2^*(l_1^*)^{-0.5} \leq 0$$
$$\frac{\partial \mathcal{L}}{\partial l_2} = -\lambda_1^* + 20\lambda_2^*(l_2^*)^{-0.5} \leq 0$$

we must have $\lambda_1^* > 0$.

Now we will proceed on the assumption that both goods are produced, i.e., $l_1^* > 0$ and $l_2^* > 0$. The optimal solution is then characterized by all derivatives of the Lagrange function being equal to zero in the K-T conditions. Solving for the optimal values yields

$$y_1^* = 293.6 \qquad y_2^* = 58.7$$
$$l_1^* = 862.1 \qquad l_2^* = 137.9$$
$$x_1^* = 212.9 \qquad x_2^* = 170.3$$

shadow wage rate $\lambda_1^* = 36.25$.

The result can be summarized as follows. The economy exports part of its production of good 1 to finance imports of good 2. All labor is employed. The trade balance is in balance.

To see the intuition behind the solution, note that $\partial \mathcal{L}/\partial l_1 = \partial \mathcal{L}/\partial l_2 = 0$ implies

$$50(l_1^*)^{-0.5} = 20(l_2^*)^{-0.5}$$

or

$$10\left[5(l_1^*)^{-0.5}\right] = 8\left[2.5(l_2^*)^{-0.5}\right]$$

or

$$p_1\frac{dy_1}{dl_1} = p_2\frac{dy_2}{dl_2}$$

i.e., the marginal value product of labor in the production of the two goods is equated. This means that the planner maximizes the total value of production or income of the economy. With that maximized income goods are bought on the domestic and the world markets.

Review Exercises (page 587)

1. The worker's problem is

$$\max u = x_1x_2$$

s.t.

$$x_1 = m + w(T - x_2)$$
$$x_2 \geq T - H$$
$$x_2 \leq T$$

The first constraint arises from the assumption of nonsatiation, which implies that the budget constraint will bind. The second constraint says that the worker will have at least as much leisure as implied by the maximum hours she can work, and the third says that the worker cannot have more leisure time than total time available.

The Lagrange function is

$$\mathcal{L} = x_1x_2 + \lambda(m - x_1 + wT - wx_2) + \mu_1(x_2 + H - T) + \mu_2(T - x_2)$$

We assume that at the optimum we have $x_1^* > 0$ and $x_2^* > 0$. Thus the Kuhn-Tucker conditions are

$$\frac{\partial \mathcal{L}}{\partial x_1} = x_2^* - \lambda^* = 0$$
$$\frac{\partial \mathcal{L}}{\partial x_2} = x_1^* - \lambda^* + \mu_1^* - \mu_2^* = 0$$
$$\frac{\partial \mathcal{L}}{\partial \lambda} = m - x_1^* + wT - wx_2^* = 0$$
$$\frac{\partial \mathcal{L}}{\partial \mu_1} = x_2^* + H - T \geq 0$$
$$\mu_1^* \geq 0 \quad \mu_1^*(x_2^* + H - T) = 0$$
$$\frac{\partial \mathcal{L}}{\partial \mu_2} = T - x_2^* \geq 0$$
$$\mu_1^* \geq 0 \quad \mu_2^*(T - x_2^*) = 0$$

Now we need to consider the three cases implied by $\mu_1^* \geq 0$ and $\mu_2^* \geq 0$.

(i) $\mu_1^* = \mu_2^* = 0$, i.e., the time constraints on x_2 do not bind. Then we obtain from the first two conditions

$$x_1^* = wx_2^*$$

Inserting into the third conditions yields

$$x_1^* = \frac{m + wT}{2}$$

and

$$x_2^* = \frac{m + wT}{2w}$$

Noting that we must have $x_2^* \leq T$ and $x_2^* \geq T - H$, this yields the optimal solution if $m/T \leq w \leq m/(T - 2H)$.

(ii) $\mu_1^* = 0$, $\mu_2^* > 0$, i.e., the worker will use all her available time for leisure. Then we have $x = m$ and $x_2 = T$. From (i) we know that this will only be the case if $w < m/T$.

(iii) $\mu_1^* > 0$, $\mu_2^* = 0$, i.e., the worker will work the maximum number of hours. Thus, we have $x_1^* = m + wH$ and $x_2^* = T - H$ which will be the optimal solution if $w > m/(T - 2H)$.

Thus, we can summarize the result

$$x_1^* = \begin{cases} m & \text{if } w < m/T \\ (m + wT)/2 & \text{if } m/T \leq w \leq m/(T - 2H) \\ m + wH & \text{if } w > m/(T - 2H) \end{cases}$$

and

$$x_2^* = \begin{cases} T & \text{if } w < m/T \\ (m + wT)/2w & \text{if } m/T \leq w \leq m/(T - 2H) \\ T - H & \text{if } w > m/(T - 2H) \end{cases}$$

If the wage rate is relatively low, the worker will, therefore, not work at all. For an intermediate wage rate, she will work but less than the maximum number of hours possible. If the wage rate is relatively high, she will work the maximum hours she can.

Figures 15.R.1 (a) to 15.R.1 (c) illustrate the three cases. The kinked budget line reflects the fact that the individual has nonwage income and can only work H hours. The slope of the budget line in its "normal" part is w.

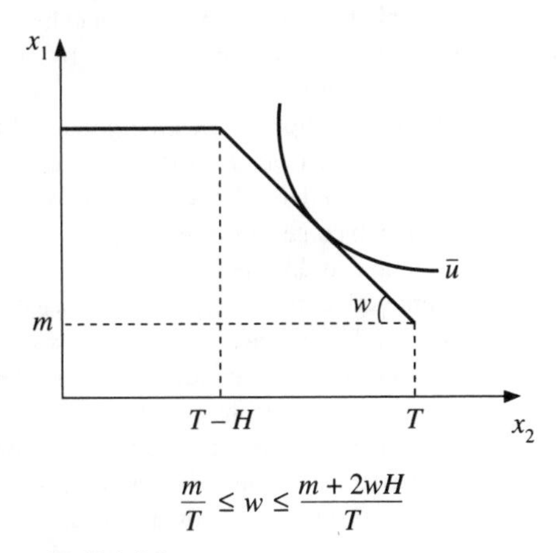

Figure 15.R.1 (a)

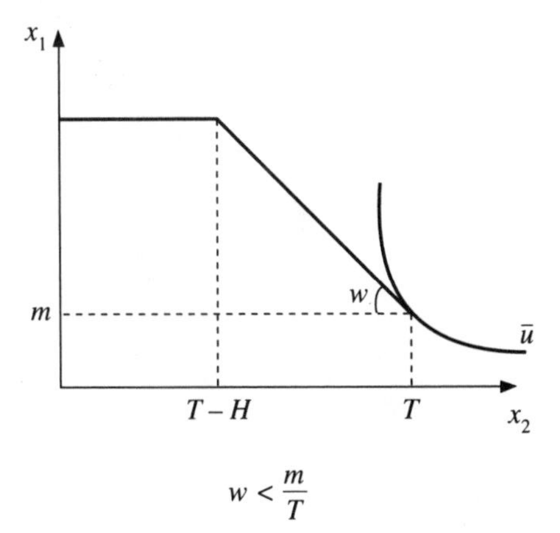

Figure 15.R.1 (b)

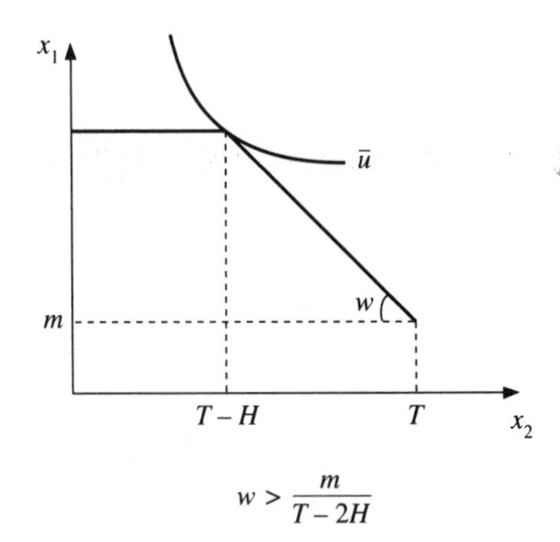

Figure 15.R.1 (c)

3. Figure 15.R.3 shows you that the kinked budget line is the interesting feature of this problem. Our approach is to suppose that the investor is a borrower. Then her wealth constraint is

$$x_2 - 1000 \geq 1.15(x_1 - 1000)$$

That is, the income in the second period she does not consume must suffice to pay back the loan. Clearly, this constraint will bind at the optimum. Also we must have $x_1 \geq 1000$ since we suppose she is a borrower. Thus, the Lagrange function is

$$\begin{aligned} \mathcal{L} &= \log x_1 + 0.8 \log x_2 \\ &\quad + \lambda(x_2 - 1.15x_1 - 150) + \mu(x_1 - 1000) \end{aligned}$$

The first-order and Kuhn-Tucker conditions are

$$\begin{aligned} \frac{\partial \mathcal{L}}{\partial x_1} &= \frac{1}{x_1^*} - 1.15\lambda^* + \mu^* = 0 \\ \frac{\partial \mathcal{L}}{\partial x_2} &= \frac{0.8}{x_2^*} - \lambda^* = 0 \\ \frac{\partial \mathcal{L}}{\partial \lambda} &= x_2^* - 1.15x_1^* - 150 = 0 \\ \frac{\partial \mathcal{L}}{\partial \mu} &= x_1^* - 1000 \geq 0 \quad \mu^* \geq 0 \\ & \mu^*(x_1^* - 1000) \end{aligned}$$

Now we check the two cases $\mu^* > 0$ and $\mu^* = 0$

(i) $\mu^* > 0$: Then we have $x_1^* = 1000$ and $x_2^* = 1000$. Substituting into the second condition yields $\lambda^* = 0.0008$. Inserting into the first condition we obtain $\mu^* = -0.00008$ which contradicts $\mu^* > 0$. Thus $\mu^* = 0$.

(ii) $\mu^* = 0$: From the first two conditions we get $x_2^* = 0.92x_1^*$. Inserting into the third condition yields $x_1^* = 1039$ and $x_2^* = 955$. The investor will borrow approximately \$39. This is the optimal solution. Note that we do not need to check the assumption that the investor is a lender. In the above formulation the investor is allowed to lend for $r_B = 0.15$ but does not choose to do so. This implies that at the even worse interest rate of $r_L = 0.05$ it cannot be optimal for her to lend money.

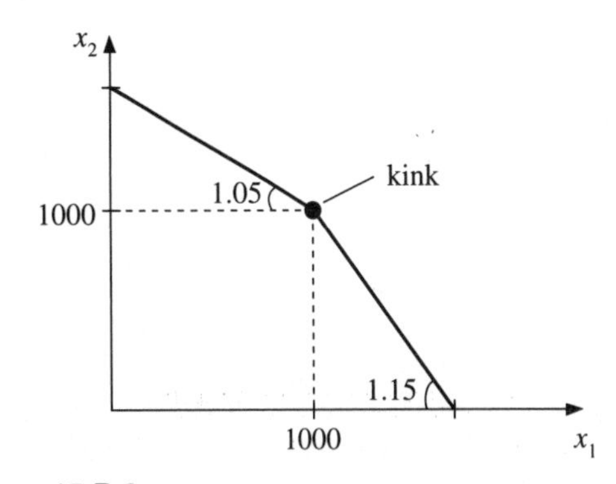

Figure 15.R.3

5. (a) $p_1 = 10$ and $p_2 = 1$:

The consumer's problem is

$$\max u = x_1 x_2$$

s.t.

$$\begin{aligned} 10x_1 + x_2 &= 80 \quad \text{(income constraint)} \\ 60x_1 + 20x_2 &\geq 1000 \text{(subsistence calorie constraint)} \end{aligned}$$

The Lagrange function is

$$\begin{aligned} \mathcal{L} &= x_1 x_2 + \lambda(80 - 10x_1 - x_2) \\ &\quad + \mu(60x_1 + 20x_2 - 1000) \end{aligned}$$

The Kuhn-Tucker conditions are

$$\frac{\partial \mathcal{L}}{\partial x_1} = x_2^* - 10\lambda^* + 60\mu^* = 0$$

$$\frac{\partial \mathcal{L}}{\partial x_2} = x_1^* - \lambda^* + 20\mu^* = 0$$

$$\frac{\partial \mathcal{L}}{\partial \lambda} = 80 - 10x_1^* - x_2^* = 0$$

$$\frac{\partial \mathcal{L}}{\partial \mu} = 60x_1^* + 20x_2^* - 1000 \geq 0 \quad \mu^* \geq 0$$

$$\mu^*(60x_1^* + 20x_2^* - 1000) = 0$$

We check the two cases $\mu^* > 0$ and $\mu^* = 0$.

(i) $\mu^* > 0$: From the last two conditions we obtain $x_1^* = 4.29$ and $x_2^* = 37.14$. Substituting into the first two equations yields $\mu^* = -0.04$. Thus, the assumption $\mu^* > 0$ leads to a contradiction and we must have $\mu^* = 0$.

(ii) $\mu^* = 0$: Using the first three conditions we obtain $x_1^* = 4$ and $x_2^* = 40$, which is the optimal solution.

(b) $p_1 = 10$ and $p_2 = 1.6$:

If the price of potatoes p_2 rises to 1.6, the only point satisfying both constraints is $x_1^* = 0$ and $x_2^* = 50$, which trivially is the constrained maximum of the utility function.

Thus, an increase in the price of potatoes leads to increased demand for potatoes, i.e., potatoes are a Giffen good. The reason is that an increase in the price of potatoes which yield more calories per dollar than meat reduces available real income such that the subsistence calorie requirement can be met only by spending all income on potatoes.

This explanation of a Giffen good is different to the one usually encountered in economics textbooks. There it is assumed that the consumer maximizes utility subject to an income constraint only. A Giffen good is then explained as an inferior good for which the positive income effect is stronger than the negative substitution effect. This requires very special assumptions on the utility function. Here, however, we explain Giffen goods by means of a second constraint. This is perfectly plausible since in reality there are additional constraints like a subsistence calorie constraint or time constraint. It also shows that Giffen goods are consistent with quite ordinary utility functions like the one used in this example.

CHAPTER 16

Section 16.1 (page 595)

1. In each case, you should check your work by differentiating your answer and making sure you get the integrand. We do this for part (a) only.

(a)

$$\int (x^4 + 2x^3 + 4x + 10)dx$$

$$= \int x^4 dx + \int 2x^3 dx + \int 4x dx + \int 10 dx$$

$$= x^5/5 + x^4/2 + 2x^2 + 10x + C$$

As a check, we differentiate the answer to get

$$\frac{d[x^5/5 + x^4/2 + 2x^2 + 10x + C]}{dx}$$

$$= 5x^4/5 + 4x^3/2 + 4x + 10$$
$$= (x^4 + 2x^3 + 4x + 10)$$

which is the integrand.

(b)

$$\int x^{2/3}dx = \frac{1}{(2/3+1)}x^{(2/3+1)} + C = \frac{3}{5}x^{5/3} + C$$

(c)

$$\int 10e^x dx = 10\int e^x dx = 10e^x + C$$

(d)

$$\int 6xe^{x^2}dx = 3\int 2xe^{x^2}dx = 3e^{x^2} + C$$

(e)

$$\int \frac{3x^2+2}{(x^3+2x)}dx = \ln(x^3 + 2x) + C$$

3. (a)

$$F(x) = \int 2dx = 2x + C$$

The condition $F(0) = 0$ implies that $2(0) + C = 0$ or $C = 0$. Therefore, making the substitution that $C = 0$ gives us $F(x) = 2x$.

(b)

$$F(x) = \int 6xdx = 3x^2 + C$$

The condition $F(0) = 5$ implies that $3(0)^2 + C = 5$ or $C = 5$. Therefore, making the substitution that $C = 5$ gives us $F(x) = 3x^2 + 5$.

(c)

$$F(x) = \int (5x^3 + 2x + 6)dx = (5/4)x^4 + x^2 + 6x + C$$

The condition $F(0) = 0$ implies that

$$(5/4)(0)^4 + (0)^2 + 6(0) + C = 0$$

or $C = 0$. Therefore, making the substitution that $C = 0$ gives us $F(x) = (5/4)x^4 + x^2 + 6x$.

(d)

$$F(x) = \int 2xdx = x^2 + C$$

The condition $F(3) = 10$ implies that $(3)^2 + C = 10$ or $C = 1$. Therefore, making the substitution that $C = 1$ gives us $F(x) = x^2 + 1$.

5. The marginal product-function is the derivative of the production function and so the production function is the antiderivative or integral of the marginal product-function. Therefore

$$Q(L) = \int 10L^{1/2}dL = (20/3)L^{3/2} + C$$

Since $Q = 0$ if $L = 0$, we have $Q(0) = (20/3)0^{3/2} + C = 0$ which implies that $C = 0$. Therefore,

$$Q(L) = (20/3)L^{3/2}$$

7. To prove the result

$$\int [f(x) \pm g(x)]dx = \int f(x)dx \pm \int g(x)dx$$

first rewrite the equation as

$$\int [f(x) \pm g(x)]dx = \int f(x)dx \pm \int g(x)dx = F(x) \pm G(x)$$

where $F(x)$ and $G(x)$ are the antiderivatives of $f(x)$ and $g(x)$, respectively. Thus, we have that $d[F(x)]/dx = f(x)$, $d[G(x)]/dx = g(x)$, and so $d[F(x) \pm G(x)]/dx = f(x) \pm g(x)$, which is the integrand of the expression and so this proves the result.

Section 16.2 (page 605)

1. (a) The area under the curve $f(x) = 3x+1$ over the interval [0,1] is 2.5, as seen in Figure 16.2.1 (a).

(b)

$$A_1 + A_2 + A_3 = 1/3 + 2/3 + 1 = 2$$

and so $S_{\min} = 2$.

$$B_1 + B_2 + B_3 = 2/3 + 1 + 4/3 = 3$$

and so $S_{\max} = 3$.

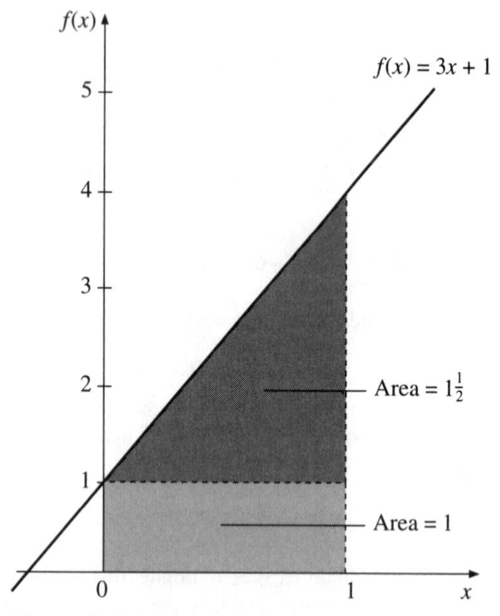

Figure 16.2.1 (a)

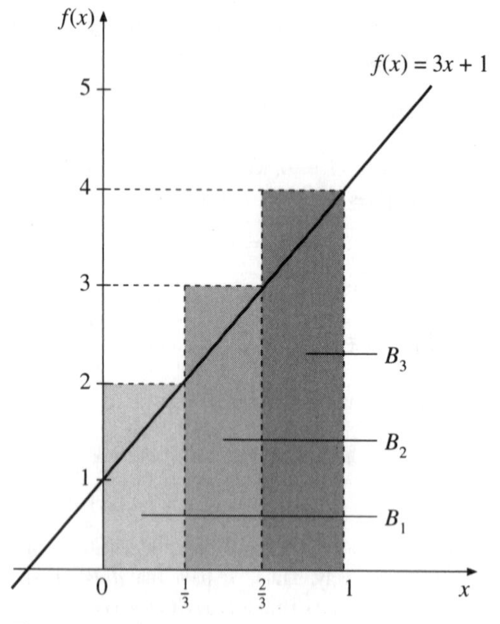

Figure 16.2.1 (b)ii

(c)

$$A_1 + A_2 + A_3 + A_4 + A_5$$

$$= 0.2 + 0.32 + 0.44 + 0.56 + 0.68 = 2.2$$

and so $S_{\min} = 2.2$.

$$B_1 + B_2 + B_3 + B_4 + B_5$$

$$= 0.32 + 0.44 + 0.56 + 0.68 + 0.80 = 2.8$$

and so $S_{\max} = 2.8$.

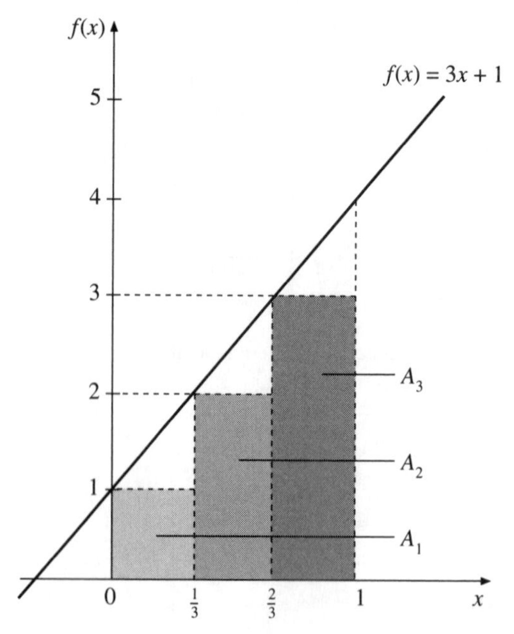

Figure 16.2.1 (b)i

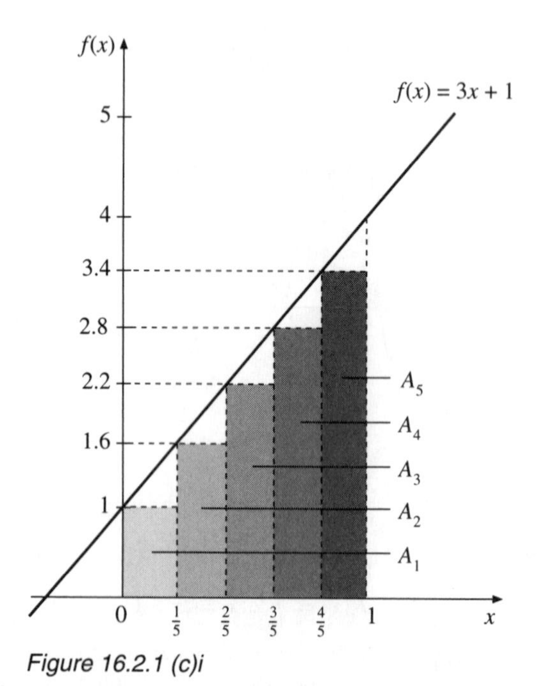

Figure 16.2.1 (c)i

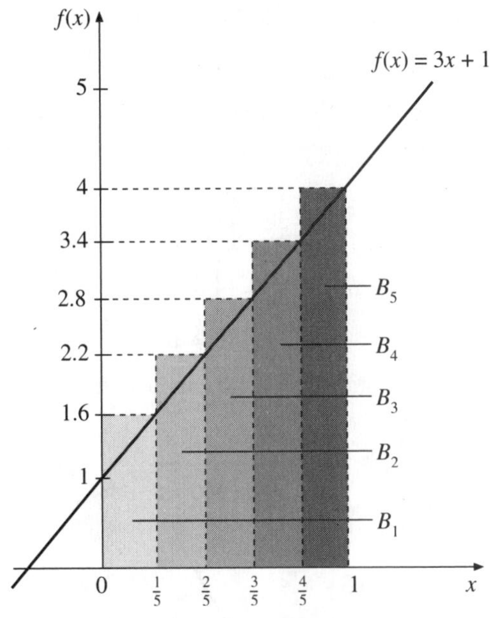

Figure 16.2.1 (c)ii

(d) The width of each interval in this partition is $\Delta = 1/n$. Thus, the lower and upper sums, respectively, are

$$\begin{aligned} S_{\min} &= \sum_{i=1}^{n}[3(i-1)/n+1](1/n) \\ &= \left[(3/n)\sum_{i=1}^{n}(i-1)+\sum_{i=1}^{n}1\right](1/n) \\ &= [(3/n)(n/2)(n-1)+n](1/n) \\ &= [(3/2)(n-1)+n](1/n) \\ &= (2.5n-1)(1/n) = 2.5-1.5/n \end{aligned}$$

$$\begin{aligned} S_{\max} &= \sum_{i=1}^{n}[3i/n+1](1/n) \\ &= \left[(3/n)\sum_{i=1}^{n}i+\sum_{i=1}^{n}1\right](1/n) \\ &= [(3/n)((n+1)/2)n+n](1/n) \\ &= [(3/2)(n+1)+n](1/n) \\ &= (2.5n+1.5)(1/n) = 2.5+1.5/n \end{aligned}$$

It follows from the above expressions that

$$\lim_{n\to\infty} S_{\min} = \lim_{n\to\infty} S_{\max} = 2.5$$

the area under the curve as determined in part (a). As $n \to \infty$, the length of the largest subinterval of this partition goes to zero. We see that both the upper (Riemann) sum and lower (Riemann) sum approach the value 2.5 and so any intermediate sum would also approach 2.5. Thus, the function is integrable and $\int_0^1(3x+1)dx$ represents the area under the curve $f(x) = 3x+1$ between the points $x = 0$ and $x = 1$.

3. To find the definite integral we simply find the difference in the value of the indefinite integral, ignoring the constant term, at the two limits of integration, i.e., the value at the upper limit minus the value at the lower limit. Thus, using the results of question 1 of exercises 16.1 we get

(a)

$$\begin{aligned} &\int_0^1 (x^4+2x^3+4x+10)dx \\ &= [x^5/5+x^4/2+2x^2+10x]_0^1 \\ &= [1/5+1/2+2+10]-[0] = 12.7 \end{aligned}$$

(b)

$$\begin{aligned} \int_0^8 x^{2/3}dx &= \left[\frac{3}{5}x^{5/3}\right]_0^8 \\ &= \left[\frac{3}{5}8^{5/3}\right]-\left[\frac{3}{5}0^{5/3}\right] \\ &= 19.2-0 = 19.2 \end{aligned}$$

(c)

$$\begin{aligned} \int_{-1}^0 10e^x dx &= \left[10e^x\right]_{-1}^0 \\ &= 10e^0-10e^{-1} \\ &\doteq 10-3.679 = 6.321 \end{aligned}$$

(d)

$$\begin{aligned} \int_1^2 6xe^{x^2}dx &= \left[3e^{x^2}\right]_1^2 \\ &= 3e^4-3e \\ &\doteq 163.8-8.155 = 155.645 \end{aligned}$$

(e)

$$\begin{aligned} \int_1^2 \frac{3x^2+2}{(x^3+2x)}dx &= [\ln(x^3+2x)]_1^2 \\ &= \ln(12)-\ln(3) \\ &\doteq 2.485-1.099 = 1.386 \end{aligned}$$

5. $t = 0$ represents the current point in time. The level of capital stock in 5 years' time, $K(5)$, equals the current level of capital stock, $K(0) = K^0$, plus accumulated net investment over the next 5 years, $\int_0^5 10t^{1/2}dt$, where $I(t) = 10t^{1/2}$ is the rate of net investment. Thus

$$\begin{aligned} K(5) &= K^0 + \int_0^5 10t^{1/2}dt \\ &= K^0 + \left[\frac{20}{3}t^{3/2}\right]_0^5 \\ &= K^0 + \left[\frac{20}{3}5^{3/2}-\frac{20}{3}0^{3/2}\right] \\ &= K^0 + \frac{20}{3}5^{3/2} \\ &\doteq K^0 + (20/3)1.18 \doteq K^0 + 74.53 \end{aligned}$$

7. From the definition of the derivative we have

$$F'(x) = \lim_{\Delta x\to 0}\frac{F(x+\Delta x)-F(x)}{\Delta x}$$

where

$$F(x) = \int_a^x f(t)dt$$

and

$$F(x+\Delta x) = \int_a^{x+\Delta x} f(t)dt$$

Property 1 from Section 16.3, states that

$$\int_a^c f(t)dt = \int_a^b f(t)dt + \int_b^c f(t)dt$$

where $a < b < c$. Thus, we get

$$F'(x) = \lim_{\Delta x \to 0} \frac{\int_a^{x+\Delta x} f(t)dt - \int_a^x f(t)dt}{\Delta x}$$

which, since

$$\int_a^{x+\Delta x} f(t)dt = \int_a^x f(t)dt + \int_x^{x+\Delta x} f(t)dt$$

gives us

$$F'(x) = \lim_{\Delta x \to 0} \frac{\int_x^{x+\Delta x} f(t)dt}{\Delta x}$$

For $f(t)$ continuous on the interval $t \in [x, x+\Delta x]$ it follows that, with $\Delta x \to 0$, $f(t) \doteq f(x)$ on this interval and so $\int_x^{x+\Delta x} f(t)dt \doteq f(x)\cdot\Delta x$, as demonstrated in Figure 16.2.7, which gives us

$$\begin{aligned} F'(x) &= \lim_{\Delta x \to 0} \frac{\int_x^{x+\Delta x} f(t)dt}{\Delta x} \\ &= \lim_{\Delta x \to 0} \frac{f(x)\Delta x}{\Delta x} = f(x) \end{aligned}$$

which proves the result.

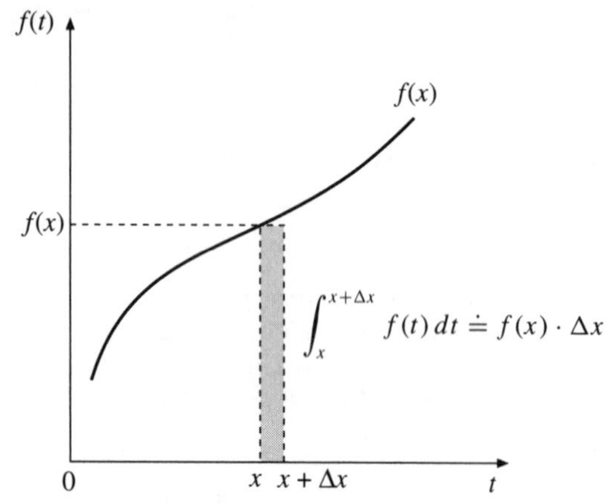

Figure 16.2.7

Section 16.3 (page 616)

1. (a) At price $p_0 = 7$, $\text{MC}(q_0) = p_0$ implies $q_0 = 2$ and so

$$\begin{aligned} \text{PS}(p_0) &= p_0q_0 - \int_0^{q_0} \text{MC}(q)dq \\ &= (7)(2) - \int_0^2 (q^2+3)dq \\ &= 14 - \left[\frac{q^3}{3} + 3q\right]_0^2 \\ &= 14 - \left(\left[\frac{2^3}{3} + 3(2)\right] - \left[\frac{0^3}{3} + 3(0)\right]\right) \\ &= 14 - 8\frac{2}{3} = 5\frac{1}{3} \end{aligned}$$

This corresponds to the area indicated in Figure 16.3.1 (a).

(b) At price $\hat{p} = 12$, $\text{MC}(\hat{q}) = \hat{p}$ implies $\hat{q} = 3$ and so

$$\begin{aligned} \text{PS}(\hat{p}) &= \hat{p}\hat{q} - \int_0^{\hat{q}} \text{MC}(q)dq \\ &= (12)(3) - \int_0^3 (q^2+3)dq \\ &= 36 - \left[\frac{q^3}{3} + 3q\right]_0^3 \\ &= 36 - \left(\left[\frac{3^3}{3} + 3(3)\right] - \left[\frac{0^3}{3} + 3(0)\right]\right) \\ &= 36 - 18 = 18 \end{aligned}$$

This corresponds to the area indicated in Figure 16.3.1 (b).

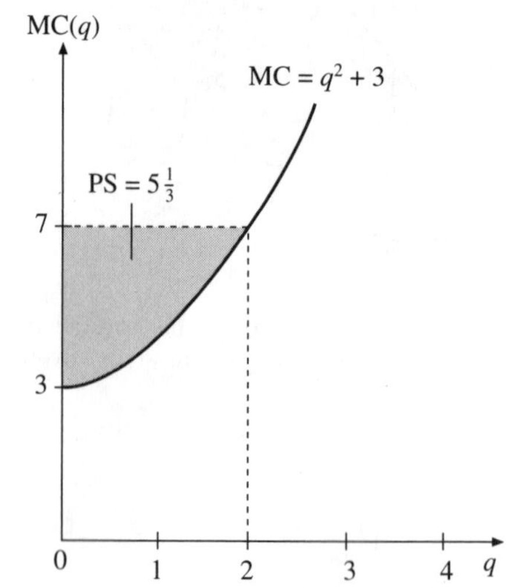

Figure 16.3.1 (a)

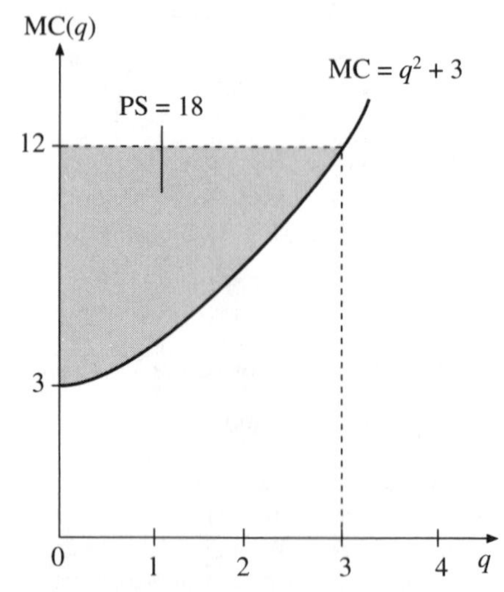

Figure 16.3.1 (b)

(c) Thus, the change in producer surplus is

$$\Delta\,\text{PS} = 18 - 5\frac{1}{3} = 12\frac{2}{3}$$

3. (a) At $p_0 = 1$, $q_0 = 8$. The inverse demand function is $p = 25 - 5q + q^2/4$ and so

$$\begin{aligned} &\mathrm{CS}(p_0 = 1) \\ &= \int_0^{q_0} D^{-1}(q)dq - p_0 q_0 \\ &= \int_0^{8} (25 - 5q + q^2/4)dq - (1)(8) \\ &= [25q - (5/2)q^2 + (1/12)q^3]_0^8 - 8 \\ &= [25(8) - (5/2)8^2 + (1/12)8^3] - 8 \\ &= \left[200 - 160 + 42\frac{2}{3}\right] - [0] - 8 \\ &= 82\frac{2}{3} - 8 = 74\frac{2}{3} \end{aligned}$$

This corresponds to the area in Figure 16.3.3 (a).

(b) At $\hat{p} = 4$, $\hat{q} = 6$.

$$\begin{aligned} \mathrm{CS}(\hat{p} = 1) &= \int_0^{\hat{q}} D^{-1}(q)dq - \hat{p}\hat{q} \\ &= \int_0^{6} (25 - 5q + q^2/4)dq - (4)(6) \\ &= [25q - (5/2)q^2 + (1/12)q^3]_0^6 - 24 \\ &= [25(6) - (5/2)6^2 + (1/12)6^3] - [0] - 24 \\ &= [150 - 90 + 18] - 24 = 78 - 24 = 54 \end{aligned}$$

This corresponds to the area in Figure 16.3.3 (b).

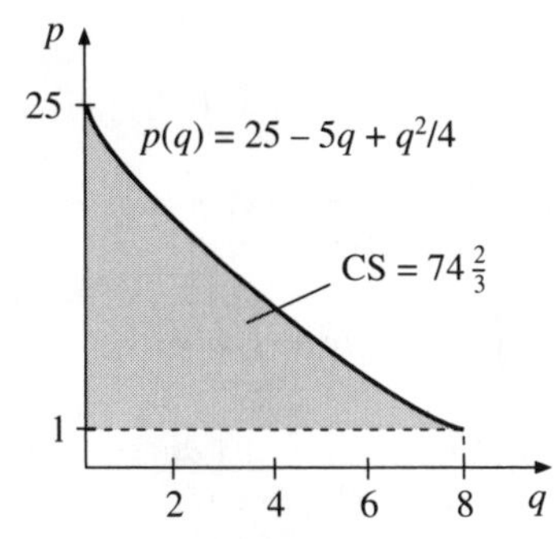

Figure 16.3.3 (a)

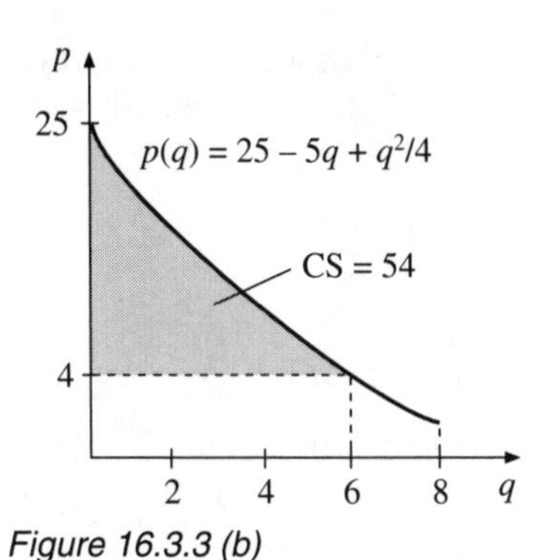

Figure 16.3.3 (b)

(c) Thus, the change in consumer surplus is

$$\Delta\,\mathrm{CS} = 54 - 74\frac{2}{3} = -20\frac{2}{3}$$

Section 16.4 (page 624)

1. Using the formula for consumer surplus that makes use of the demand function, rather than the inverse demand function, and noting that there is no finite valued choke price, we have that

$$\begin{aligned} \mathrm{CS} &= \lim_{\hat{p}\to\infty} \int_{p_0}^{\hat{p}} q(p)dp \\ &= \lim_{\hat{p}\to\infty} \int_{2}^{\hat{p}} 30p^{-2}dp \\ &= \lim_{\hat{p}\to\infty} 30[-p^{-1}]_2^{p} \\ &= 30 \lim_{\hat{p}\to\infty} [-1/\hat{p} + 1/2] \\ &= 30\left[\lim_{\hat{p}\to\infty} (-1/\hat{p}) + 1/2\right] = 30(0 + 1/2) = 15 \end{aligned}$$

3. Using the formula for consumer surplus that makes use of the demand function, rather than the inverse demand function, we have that

(a)

$$\begin{aligned} \Delta\,\mathrm{CS} &= \lim_{\hat{p}\to 0} \int_{\hat{p}}^{1} q(p)dp \\ &= \lim_{\hat{p}\to 0} \int_{\hat{p}}^{1} 10p^{-1/3}dp \\ &= \lim_{\hat{p}\to 0} [15p^{2/3}]_{\hat{p}}^{1} \\ &= \lim_{\hat{p}\to 0} [(15)1^{2/3} - 15\hat{p}^{2/3}] \\ &= 15 - 0 = 15 \end{aligned}$$

(b)

$$\begin{aligned} \Delta\mathrm{CS} &= \lim_{\hat{p}\to 0} \int_{\hat{p}}^{1} q(p)dp \\ &= \lim_{\hat{p}\to 0} \int_{\hat{p}}^{1} 25p^{-2}dp \\ &= \lim_{\hat{p}\to 0} [-25p^{-1}]_{p}^{1} \\ &= \lim_{\hat{p}\to 0} [-25(1/1) + 25(1/\hat{p})] \\ &= -25 + \infty = +\infty \end{aligned}$$

since $\lim_{\hat{p}\to 0} 1/\hat{p} = +\infty$. For the demand function in part (b) the ΔCS cannot be computed (i.e., it is not a finite value).

5. For payments spread evenly throughout the year, we can use the formula of equation (16.8) developed in this section to find the present value of this stream of payments, with $b = \$500$ and $r = 0.02$ to give

$$\mathrm{PVB} = b/r = \frac{\$500}{0.02} = \$25,000$$

Section 16.5 (page 628)

1. The substitution rule is used to evaluate integrals when the integral can be written in the form $\int f(u)\frac{du}{dx}dx = F(u)$, where $F(u)$ is the integral of $f(u)$. Thus, if it is easy to find the integral of $f(u)$, then it is easy to find the integral of $f(u)\frac{du}{dx}$.

(a) In this case, letting $u = (x^3 + 5x)$ and noting that $\frac{du}{dx} = (3x^2 + 5)$ allows us to write the integral

$$\int (x^3 + 5x)^{10}(3x^2 + 5)dx$$

as

$$F(u) = \int f(u)\frac{du}{dx}dx = \int u^{10}\frac{du}{dx}dx$$

where

$$F(u) = \int f(u)du = \int u^{10}du = (1/11)u^{11} + C$$

Substituting back for $u = (x^3 + 5x)$ gives us

$$\int (x^3 + 5x)^{10}(3x^2 + 5)dx = (1/11)(x^3 + 5x)^{11} + C$$

(b) In this case, letting $u = (e^{x^2} + 4x)$ and noting that $\frac{du}{dx} = (2xe^{x^2} + 4)$ allows us to write the integral

$$\int (e^{x^2} + 4x)(2xe^{x^2} + 4)dx$$

as

$$F(u) = \int f(u)\frac{du}{dx}dx = \int u\frac{du}{dx}dx$$

where

$$F(u) = \int f(u)du = \int udu = (1/2)u^2 + C$$

Substituting back for $u = (e^{x^2} + 4x)$ gives us

$$\int (e^{x^2} + 4x)(2xe^{x^2} + 4)dx = (1/2)(e^{x^2} + 4x)^2 + C$$

(c) In this case, letting $u = (e^x + 3x^2)$ and noting that $\frac{du}{dx} = (e^x + 6x)$ allows us to write the integral

$$\int 2(e^x + 3x^2)(e^x + 6x)dx = 2\int (e^x + 3x^2)(e^x + 6x)dx$$

as

$$F(u) = 2\int f(u)\frac{du}{dx}dx = 2\int u\frac{du}{dx}dx$$

where

$$F(u) = \int f(u)du = \int udu = (1/2)u^2 + C$$

Substituting back for $u = (e^x + 3x^2)$ and noting the factor 2 gives us

$$\int 2(e^x + 3x^2)(e^x + 6x)dx = (e^x + 3x^2)^2 + C$$

(d) In this case, letting $u = (x^2 + 2)$ and noting that $\frac{du}{dx} = (2x)$ allows us to write the integral

$$\int \frac{2x}{(x^2 + 2)^{10}}dx$$

as

$$F(u) = \int f(u)\frac{du}{dx}dx = \int u^{-10}\frac{du}{dx}dx$$

where

$$F(u) = \int f(u)du = \int u^{-10}du = -(1/9)u^{-9} + C$$

Substituting back for $u = (x^2 + 2)$ gives us

$$\int \frac{2x}{(x^2 + 2)^{10}}dx = -(1/9)(x^2 + 2)^{-9} + C$$

(e) In this case, letting $u = (x^3 + 4x)$ and noting that $\frac{du}{dx} = (3x^2 + 4)$ allows us to write the integral

$$\int \frac{6x^2 + 8}{(x^3 + 4x)^2}dx = 2\int \frac{3x^2 + 4}{(x^3 + 4x)^2}dx$$

as

$$2F(u) = 2\int f(u)\frac{du}{dx}dx = \int u^{-2}\frac{du}{dx}dx$$

where

$$F(u) = \int f(u)du = \int u^{-2}du = -u^{-1} + C$$

Substituting back for $u = (x^3 + 4x)$ gives us

$$\begin{aligned}\int \frac{6x^2 + 8}{(x^3 + 4x)^2}dx &= 2\int \frac{3x^2 + 4}{(x^3 + 4x)^2}dx \\ &= -2(x^3 + 4x)^{-1} + C\end{aligned}$$

3. Suppose the integrand of an integral can be broken into two functions which are multiplied by each other. Suppose further that if one of the functions was replaced by its derivative while the other was replaced by its antiderivative, then the resulting integral formed by these "new" functions could easily be determined. If this is so, then one should use integration by parts. Symbolically, we make use of the following relationship:

$$\int g(x)f'(x)dx = f(x)g(x) - \int f(x)g'(x)dx$$

or, in differential form ($u = f(x) \Leftrightarrow du = f'(x)dx$ and $v = g(x) \Leftrightarrow dv = g'(x)dx$)

$$\int udv = uv - \int vdu$$

(a) *Stage 1:* Letting $u = x^2$ and $dv = e^x dx$, implying $du = 2xdx$ and $v = e^x$, gives us

$$\int x^2e^x dx = x^2e^x - 2\int e^x xdx$$

Stage 2: We can now use integration by parts again to find $\int e^x xdx$. Letting $u = x$ and $dv = e^x dx$, implying $du = dx$ and $v = e^x$, gives us

$$\int e^x xdx = \int xe^x dx = xe^x - \int e^x dx$$

and so

$$\int e^x xdx = xe^x - e^x$$

Making this substitution gives us

$$\begin{aligned}\int x^2e^x dx &= x^2e^x - 2\int e^x xdx \\ &= x^2e^x - 2(xe^x - e^x) + C\end{aligned}$$

(b) Note that

$$\int \frac{x^3}{\sqrt{1 + x^2}}dx$$

can be written as $\int x^3(1+x^2)^{-1/2}dx$. Letting $u = x^2$ and $dv = x(1+x^2)^{-1/2}dx$, implying $du = 2xdx$ and $v = (1+x^2)^{1/2}$, gives us

$$\int x^3(1+x^2)^{-1/2}dx$$

$$= x^2(1+x^2)^{1/2} - \int 2x(1+x^2)^{1/2}dx$$

$$= x^2(1+x^2)^{1/2} - \frac{2}{3}(1+x^2)^{3/2} + C$$

(c) Letting $u = \ln x$ and $dv = xdx$, implying $du = \frac{dx}{x}$ and $v = \frac{x^2}{2}$, gives us

$$\begin{aligned}\int x\ln x dx &= \left(\frac{x^2}{2}\right)\ln x - \int\left(\frac{x^2}{2}\right)\frac{dx}{x}\\ &= \left(\frac{x^2}{2}\right)\ln x - (1/2)\int xdx\\ &= \left(\frac{x^2}{2}\right)\ln x - \frac{x^2}{4} + C\end{aligned}$$

Review Exercises (page 630)

1. (a) $\int x^2dx = x^3/3 + C$

(b)

$$\int (2x^3 + 5x^2 + x + 5)dx$$

$$= x^4/2 + (5/3)x^3 + x^2/2 + 5x + C$$

(c)

$$\begin{aligned}\int \sum_{i=0}^{n} a_i x^i dx &= \int (a_0 + a_1x + a_2x^2 + \ldots + a_nx^n)dx\\ &= a_0x + a_1(x^2/2) + a_2(x^3/3)\\ &\quad + \ldots + a_n(x^{n+1}/n+1)\\ &= \sum_{i=0}^{n} a_i\left[\frac{x^{i+1}}{i+1}\right]\end{aligned}$$

3. Since MP(L), the marginal-product function for L, is the derivative of TP(L), the production function (or total product of labor function), it follows that TP(L) is the antiderivative or integral of MP(L). Hence, we have:

$$\text{TP}(L) = \int \text{MP}(L)dL = \int 5L^{1/3}dL = (15/4)L^{4/3} + C$$

Since TP$(0) = 0$ it follows that $C = 0$ and so we can write TP$(L) = (15/4)L^{4/3}$

5. (a) At price $p_0 = 7$, MC$(q_0) = p_0$ implies $q_0 = 1$ and so

$$\begin{aligned}\text{PS}(p_0) &= p_0q_0 - \int_0^{q_0} \text{MC}(q)dq\\ &= (7)(1) - \int_0^1 (q^{3/2} + 6)dq\\ &= 7 - [(2/5)q^{5/2} + 6q]_0^1 = 7 - (6\frac{2}{5} - 0) = 3/5\end{aligned}$$

This corresponds to the area indicated in Figure 16.R.5 (a).

(b) At price $\hat{p} = 70$, MC$(\hat{q}) = \hat{p}$ implies $\hat{q} = 16$ and so

$$\begin{aligned}\text{PS}(\hat{p}) &= \hat{p}\hat{q} - \int_0^{\hat{q}} \text{MC}(q)dq\\ &= (70)(16) - \int_0^{16} (q^{3/2} + 6)dq\\ &= 1120 - [(2/5)q^{5/2} + 6q]_0^{16}\\ &= 1120 - ([(2/5)16^{5/2} + 6(16)] - [0])\\ &= 1120 - [409.6 + 96] = 614.4\end{aligned}$$

This corresponds to the area indicated in Figure 16.R.5 (b).

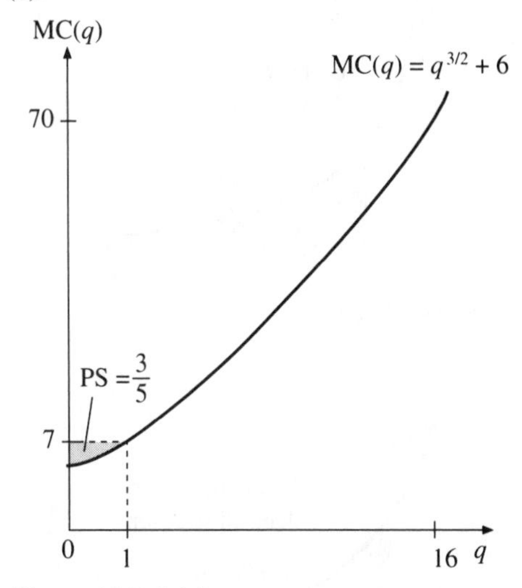

Figure 16.R.5 (a)

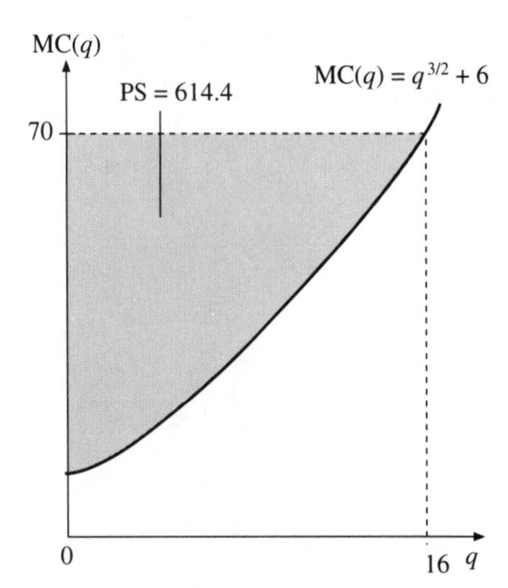

Figure 16.R.5 (b)

(c) Thus, the change in producer surplus is $\Delta\text{PS} = 614.4 - 0.6 = 613.8$

7. Using the formula for consumer surplus that makes use of the demand function, rather than the inverse demand function, and noting that there is no finite valued choke price, we have that

$$\text{CS} = \lim_{\hat{p}\to\infty} \int_{p_0}^{\hat{p}} q(p)dp$$

$$= \lim_{\hat{p}\to\infty} \int_1^{\hat{p}} 12p^{-3}dp$$

$$= \lim_{\hat{p}\to\infty} 12[-(1/2)p^{-2}]_1^{\hat{p}}$$

$$= \lim_{\hat{p}\to\infty} -6[1/p^2]_1^{\hat{p}}$$

$$= -6 \lim_{\hat{p}\to\infty} [1/\hat{p}^2 - 1/1] = 6$$

This result is illustrated in Figure 16.R.7 using both the demand function (Figure 16.R.7 (a)) and the inverse demand function (Figure 16.R.7 (b)).

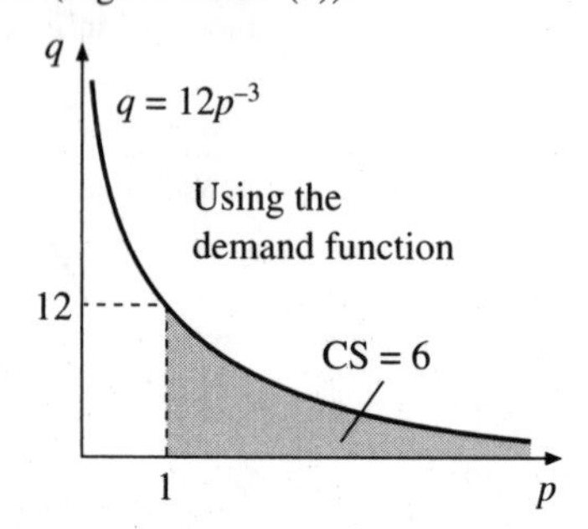

Figure 16.R.7 (a)

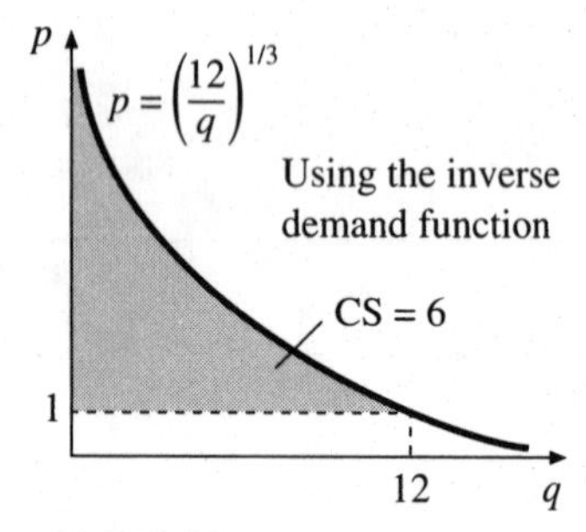

Figure 16.R.7 (b)

9. The substitution rule is used to evaluate integrals when the integral can be written in the form

$$\int f(u)\frac{du}{dx}dx = F(u)$$

where $F(u)$ is the integral of $f(u)$. In this case, letting $u = (x^3 + 4x^2 + 3)$ and noting that $\frac{du}{dx} = (3x^2 + 8x)$ allows us to write the integral

$$\int (x^3 + 4x^2 + 3)^4(3x^2 + 8x)dx$$

as

$$F(u) = \int f(u)\frac{du}{dx}dx = \int u^4\frac{du}{dx}dx$$

where

$$F(u) = \int f(u)du = \int u^4 du = (1/5)u^5 + C$$

and so making the substitution for $u = (x^3 + 4x^2 + 3)$ we get the result that

$$\int (x^3 + 4x^2 + 3)^4(3x^2 + 8x)dx = (1/5)(x^3 + 4x^2 + 3)^5 + C$$

CHAPTER 18

Section 18.1 (page 654)

1. (i) (a) $y_t = C2^t + 10(1 - 2^t)$
   (b) $y_t = (y_0 - 10)2^t + 10$
   (c) $\bar{y} = 10$. Does not converge.
   (ii) (a) $y_t = C$
   (b) $y_t = y_0$
   (c) $\bar{y}$ does not exist.
   (iii) (a) $y_t = C(0.5)^t + 2(1 - 0.5^t)$
   (b) $y_t = (y_0 - 2)(0.5)^t + 2$
   (c) $\bar{y} = 2$. Does converge.

3. $y_{t+1} = y_t(1 + r/n)$ where $r/n = 0.005$. Therefore,
$$y_{120} = 1000(1.005)^{120} = \$1819.40$$

5. $K_{t+1} = K_t(1 - \delta) + I$. The solution is
$$K_t = (K_0 - I/\delta)(1 - \delta)^t + I/\delta$$

7. $Y_{t+1} = C_{t+1} + I = A + BY_t + I$. The solution is
$$Y_t = \left(Y_0 - \frac{A + I}{1 - B}\right) B^t + \frac{A + I}{1 - B}$$
Since $B > 0$ is given in the model, we add the restriction $B < 1$ to ensure convergence. The short-run (one-period) impact is $\partial Y_t / \partial I = 1$; the long-run (steady-state) impact is $\partial \bar{Y} / \partial I = 1/(1 - B)$.

Section 18.2 (page 659)

1.
$$\begin{aligned} y_t &= \prod_{i=0}^{t-1} \frac{\alpha}{i+1} y_0 \\ &= \left(\frac{\alpha}{1}\frac{\alpha}{2}\frac{\alpha}{3}\frac{\alpha}{4}\cdots\frac{\alpha}{t}\right) y_0 \\ &= \frac{\alpha^t}{t!} y_0 \end{aligned}$$

3.
$$\begin{aligned} y_t &= \prod_{i=0}^{t-1}(i+1)y_0 + b\sum_{k=0}^{t-1}\prod_{i=k}^{t-1}\frac{i+1}{k+1} \\ &= t!y_0 + b\left[\frac{1}{1}(1\,2\,3\ldots t) + \frac{1}{2}(2\,3\,4\ldots t) + \cdots + \frac{1}{t}(t)\right] \\ &= t!y_0 + bt!\left[1 + \frac{1}{2!} + \frac{1}{3!} + \cdots + \frac{1}{t!}\right] \\ &= t!y_0 + bt!\sum_{k=0}^{t-1}\frac{1}{(k+1)!} \end{aligned}$$

5.
$$\begin{aligned} y_t &= \prod_{i=0}^{t-1}\alpha^i y_0 + b\sum_{k=0}^{t-1}\frac{\prod_{i=k}^{t-1}\alpha^i}{\alpha^k} \\ &= y_0\alpha^{t(t-1)/2} + b\sum_{k=0}^{t-1}\frac{\alpha^{[t(t-1)-k(k-1)]/2}}{\alpha^k} \end{aligned}$$

7.
$$y_t = 100\prod_{i=1}^{t-1}(1 + r_i) + 100\sum_{k=0}^{t-1}\prod_{i=k}^{t-1}\frac{1 + r_i}{1 + r_k}$$

Review Exercises (page 660)

1. (i) (a) $y_t = C(0.8)^t + 5(1 - 0.8^t)$
   (b) $y_t = (y_0 - 5)(0.8)^t + 5$
   (c) $\bar{y} = 5$. Does converge.
   (ii) (a) $y_t = C + 10t$
   (b) $y_t = y_0 + 10t$
   (c) $\bar{y}$ does not exist.
   (iii) (a) $y_t = C(-0.1)^t + 9[1 - (-0.1)^t]$
   (b) $y_t = (y_0 - 9)(-0.1)^t + 9$
   (c) $\bar{y} = 9$. Does converge.

3. (i) (a) $y_t = 2(-1)^t + [1 - (-1)^t]$
   (b) $\{y_t\} = \{0, 2, 0, 2, 0\}$
   (c) $y_5 = 0$.
   (ii) (a) $y_t = 3^t + 0.5(1 - 3^t)$
   (b) $\{y_t\} = \{2, 5, 14, 41, 122\}$
   (c) $y_5 = 122$.
   (iii) (a) $y_t = 50(0.5)^t + 100(1 - 0.5^t)$
   (b) $\{y_t\} = \{75, 87.5, 93.75, 96.88, 98.44\}$, $\bar{y} = 100$
   (c) $y_5 = 98.44$.
   (iv) (a) $y_t = 2(-2/3)^t + 0.2[1 - (-2/3)^t]$
   (b) $\{y_t\} = \{-1, 1, -0.33, 0.56, -0.04\}$, $\bar{y} = 0.2$
   (c) $y_5 = -0.04$.
   (v) (a) $y_t = 2 + \sum_{k=0}^{t-1}(-1)^k$
   (b) $\{y_t\} = \{3, 2, 3, 2, 3\}$
   (c) $y_5 = 3$.
   (vi) (a) $y_t = 2(-1)^t + \sum_{k=0}^{t-1}(-1)^{t-1}$
   (b) $\{y_t\} = \{-1, 0, 1, -2, 3\}$
   (c) $y_5 = 3$
   (vii) (a)
$$\begin{aligned} y_t &= 2\prod_{i=0}^{t-1}(-1)^i + \sum_{k=0}^{t-1}\frac{\prod_{i=k}^{t-1}(-1)^i}{(-1)^k} \\ &= 2(-1)^{t(t-1)/2} + \sum_{k=0}^{t-1}\frac{(-1)^{[t(t-1)-k(k-1)]/2}}{(-1)^k} \end{aligned}$$

(b) $\{y_t\} = \{3, -2, -1, 2, 3\}$

(c) $y_5 = 3$

5.

$$Q_t = A + B\frac{Q_t - F}{G} + \theta Q_{t-1}$$

Rearranging gives

$$Q_{t+1} = \frac{\theta G}{G - B} Q_t + \frac{AG - BF}{G - B}$$

The solution is

$$Q_t = Q_0\left(\frac{\theta G}{G - B}\right)^t + \frac{AG - BF}{G(1-\theta) - B}\left[1 - \left(\frac{\theta G}{G - B}\right)^t\right]$$

$$\bar{Q} = \frac{AG - BF}{G(1-\theta) - B}$$

The solution converges monotonically to $\bar{Q}$ because $0 < G/(G - B) < 1$ and $0 < \theta < 1$.

7.

$$U_t = \left(U_0 - \frac{\alpha}{1-\beta}\right)\beta^t + \frac{\alpha}{1-\beta}$$

$\bar{U} = \alpha/(1-\beta)$ is the natural rate of unemployment. Since $\beta > 0$ is a given restriction, we add the restriction $\beta < 1$ to ensure that $\bar{U}$ exists and that $\bar{U} > 0$. This also ensures that U_t converges to $\bar{U}$.

(a) $U_t = U_0\beta^t + \sum_{k=0}^{t-1}(\alpha + e_k)\beta^{t-k-1}$

(b)

$$\begin{aligned} U_t &= U_0\beta^t + \alpha(\beta^{t-1} + \beta^{t-2} + \cdots + \beta + 1) \\ &\quad + \sum_{k=0}^{t-1} e_k\beta^{t-k-1} \\ &= U_0\beta^t + \frac{\alpha(1-\beta^t)}{1-\beta} + e_0\beta^{t-1} + e_1\beta^{t-2} + \cdots \\ &\quad + e_{t-2}\beta + e_{t-1} \end{aligned}$$

Shocks which occurred t periods ago are multiplied by β^{t-1}. Since $0 < \beta < 1$, β^{t-1} tends to 0 as t gets larger. Therefore, the influence of past shocks on the current value of U_t tends to 0 the further in the past the shock occurred. However, recent shocks prevent U_t from converging to $\alpha/(1-\beta)$ even as t goes to infinity.

(c) $\bar{U} = 6$, $U_0 = 6$, $U_1 = 9$, $U_2 = 7.5$, $U_3 = 6.75$, $U_4 = 6.38$, $U_5 = 5.19$, $U_6 = 5.59$.

(d) $\bar{U} = 6$, $U_0 = 6$, $U_1 = 9$, $U_2 = 8.4$, $U_3 = 7.92$, $U_4 = 7.54$, $U_5 = 6.23$, $U_6 = 6.18$.

CHAPTER 19

Section 19.1 (page 671)

1.

$$y_{t+1} = \frac{3}{16} + y_t^2$$

The steady-state points are the solution to

$$\bar{y}^2 - \bar{y} + \frac{3}{16} = 0$$

which gives $\bar{y}_1 = 1/4$ and $\bar{y}_2 = 3/4$. Using Theorem 19.1, we evaluate

$$\left.\frac{dy_{t+1}}{dy_t}\right|_{\bar{y}} = 2\bar{y}$$

which gives $1/2$ at $\bar{y}_1 = 1/4$ and $3/2$ at $\bar{y}_2 = 3/4$. Therefore, $\bar{y}_1 = 1/4$ is stable and at $\bar{y}_2 = 3/4$ is unstable locally. The phase diagram is in Figure 19.1.1.

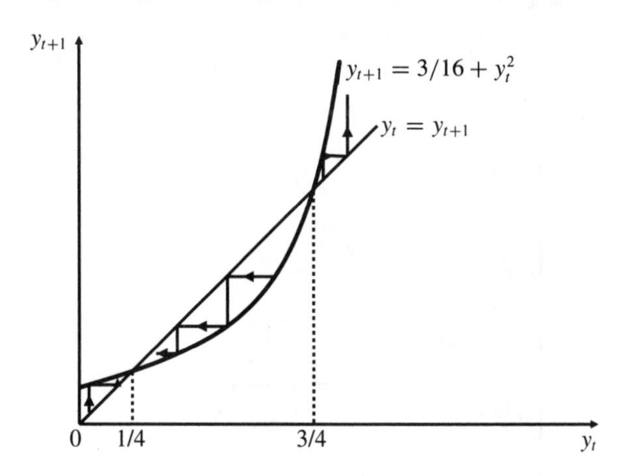

Figure 19.1.1

3. The steady-state points are the solution to

$$\bar{y} - \frac{9}{4\bar{y}} - 4 = 0$$

which becomes

$$\bar{y}^2 - 4\bar{y} - \frac{9}{4} = 0$$

This gives $\bar{y}_1 = 9/2$ and $\bar{y}_2 = -1/2$. Using Theorem 19.1,

$$\left.\frac{dy_{t+1}}{dy_t}\right|_{\bar{y}} = -\frac{9}{4}\bar{y}^{-2}$$

which equals $-1/9$ at $\bar{y}_1 = 9/2$. Therefore, this point is stable. The positive quadrant of the phase diagram is shown in Figure 19.1.3.

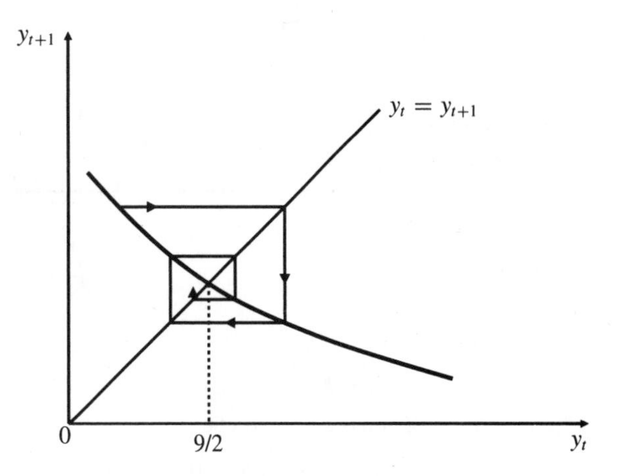

Figure 19.1.3

5. $\bar{k} = 100,000$, $k_1 = 95,019.59$, $k_2 = 95,039.11$, $k_3 = 95,058.55$, $k_4 = 95,077.92$. Since

$$\frac{dk_{t+1}}{dk_t} = 1 - \delta + \frac{s\alpha}{k_t^{1-\alpha}} > 0$$

for all $k > 0$ and $0 < \delta < 1$, then by Theorem 19.2, the path of y_t is monotonic. Further, since

$$\left.\frac{dk_{t+1}}{dk_t}\right|_{\bar{y}} = 1 - 0.02 + \frac{(0.2)(0.8)}{(100,000)^{0.2}} = 0.996$$

the path of k_t converges to $\bar{k}$.

Section 19.2 (page 677)

1. The steady-state points are the solution to

$$\bar{y}\left[1 - \frac{3}{2}(1 - \bar{y})\right] = 0$$

which gives $\bar{y} = 0$ and $\bar{y} = 1/3$. Using Theorem 19.1,

$$\frac{dy_{t+1}}{dy_t} = \frac{3}{2} - 3\bar{y}$$

which equals $3/2$ at $\bar{y} = 0$ and $1/2$ at $\bar{y} = 1/3$. Therefore, $\bar{y} = 0$ is unstable and $\bar{y} = 1/3$ is stable, locally. See Figure 19.2.1.

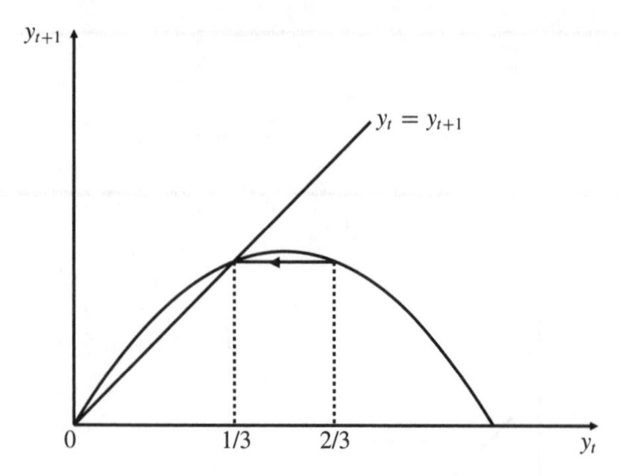

Figure 19.2.1

3.

$$E_{t+1} = b[aE_t(1 - E_t)]$$

The steady-state points are the solution to

$$\bar{E}[1 - ba(1 - \bar{E})] = 0$$

which gives $\bar{E} = 0$ and $\bar{E} = 1 - 1/(ba)$. Using Theorem 19.1,

$$\frac{dE_{t+1}}{dE_t} = ba - 2ba\bar{E}$$

at the steady-state points. The value of this derivative must lie between -1 and +1 when evaluated at $\bar{E}$ for convergence. For $\bar{E} = 1 - 1/(ba)$, the derivative is

$$ba - 2ba\left(1 - \frac{1}{ba}\right) = 2 - ba$$

We require $-1 < 2-ba < 1$ which simplifies to $1 < ba < 3$. The product of parameters b and a must lie between 1 and 3 for E_t to converge to the positive steady state. For $\bar{E} = 0$, the derivative is ba. For $\bar{E}$ to converge to 0, the product of these parameters (both positive) must be less than 1.

5. The steady-state points are the solution to

$$\bar{y} - 2\bar{y}(1 - \bar{y}) + H = 0$$

which becomes

$$\bar{y}^2 - \frac{\bar{y}}{2} + 0.04 = 0$$

This gives $\bar{y}_1 = 0.10$ and $\bar{y}_2 = 0.40$. Using Theorem 19.1, we determine that $\bar{y}_1$ is unstable and $\bar{y}_2$ is stable. $y_1 = 0.2950$, $y_2 = 0.3360$, $y_3 = 0.3662$, $y_4 = 0.3842$. See Figure 19.2.5.

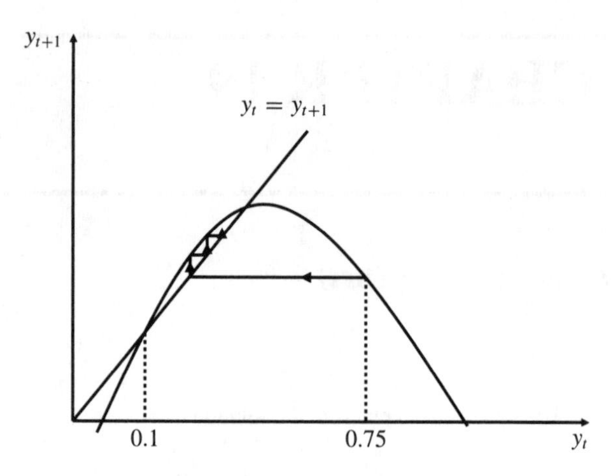

Figure 19.2.5

Review Exercises (page 679)

1. The steady-state points are the solution to

$$10\bar{y}^2 - 2\bar{y} + \bar{y} = 0$$

which simplifies to $\bar{y}(10\bar{y} - 1) = 0$, giving $\bar{y} = 0$ and $\bar{y} = 1/10$. Use Theorem 19.1 to determine that $\bar{y} = 0$ is unstable and $\bar{y} = 1/10$ is stable locally. See Figure 19.R.1.

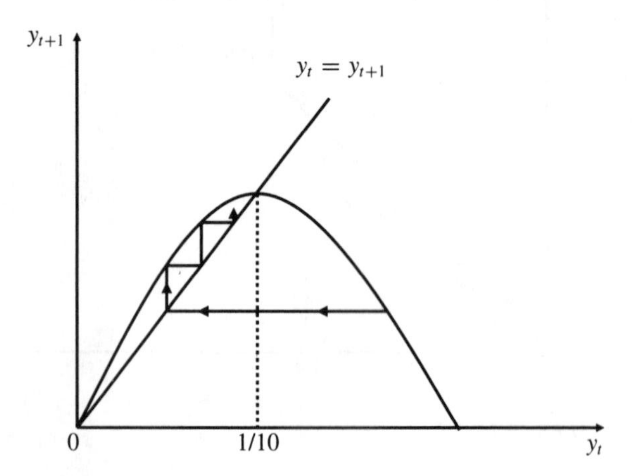

Figure 19.R.1

3. The steady-state points are the solution to

$$\bar{y}[1 - a + b\bar{y}] = 0$$

which gives $\bar{y} = 0$ and $\bar{y} = (a-1)/b$. Stability is determined using Theorem 19.1,

$$\frac{dy_{t+1}}{dy_t} = a - 2b\bar{y}$$

which gives a at $\bar{y} = 0$ and $2-a$ at $\bar{y} = (a-1)/b$. Therefore, the condition for stability is unaffected by the value of b.

5. The difference equation for capital is

$$k_{t+1} = k_t(1 - \delta) + sy_t$$

Substituting for y_t and rearranging gives

$$k_{t+1} = k_t(1 - \delta + sb) + sa$$

The steady-state is $\bar{k} = sa(\delta - sb)$. We therefore require $(\delta - sb) > 0$ to ensure $\bar{k} > 0$. Since this is a linear difference equation, we know that k_t converges if and only if the absolute value of $1 - \delta + sb$ is less than 1. That is, we require $-1 < 1 - (\delta - sb) < 1$. The restriction that $\delta - sb > 0$ ensures that the upper restriction is met. To ensure the lower restriction is met, we require $\delta - sb < 2$. The restriction $\delta - sb > 0$ ensures that the slope of the k_{t+1} line is less than 1 so that it intersects the 45° line. See Figure 19.R.5.

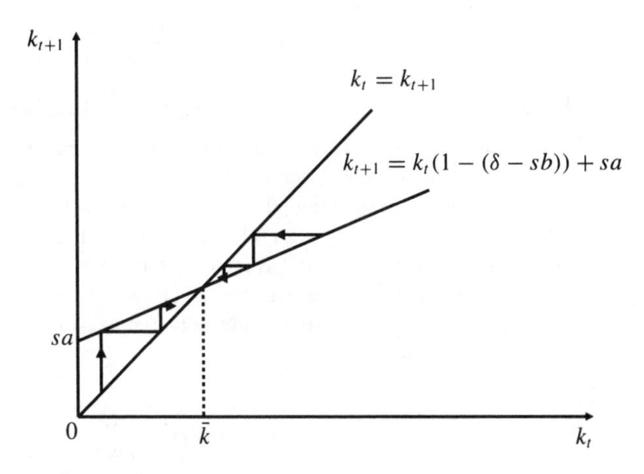

Figure 19.R.5

7. Since $Q_t^D = Q_t^S$ in equilibrium, we get

$$2 - P_t = P_{t-1}^{1/2}$$

Rearranging gives

$$P_{t+1} = 2 - P_t^{1/2}$$

The steady-state points are given by the solution to

$$\bar{P} + \bar{P}^{1/2} - 2 = 0$$

Define $x^2 = \bar{P}$ so that $x = \bar{P}^{1/2}$ or $-x = \bar{P}^{1/2}$. This gives

$$x^2 + x - 2 = 0$$

or

$$x^2 - x - 2 = 0$$

Solving gives $x = 1, -2$ or $x = 2, -1$. Since the price must be real-valued, we have $\bar{P} = \sqrt{1}$ or $\bar{P} = \sqrt{2}$. Trying each of these in the equation for $\bar{P}$ shows that $\bar{P} = 1$ is the only possible solution. To determine stability, take

$$\begin{aligned}\frac{dP_{t+1}}{dP_t} &= -\frac{1}{2}\bar{P}^{-1/2} \\ &= -\frac{1}{2} \quad \text{at} \quad \bar{P} = 1\end{aligned}$$

so the steady-state is locally stable. Price oscillates, however, as it converges. See Figure 19.R.7.

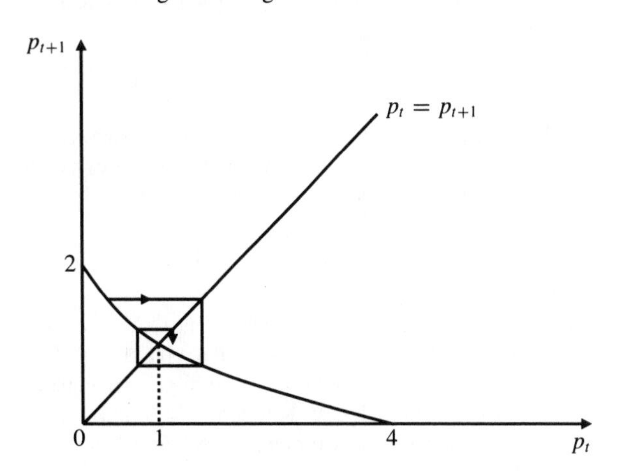

Figure 19.R.7

9. The steady-state points are the solution to

$$\bar{y}^2 - \frac{2}{3}\bar{y} + \frac{7}{144} = 0$$

which gives $\bar{y} = 1/12$ and $\bar{y} = 7/12$. Local stability is determined by evaluating

$$\frac{dy_{t+1}}{dy_t} = 3 - 6\bar{y}$$

which gives $5/2$ at $\bar{y} = 1/12$ and $-1/2$ at $\bar{y} = 7/12$. Therefore, $\bar{y} = 7/12$ is locally stable but $\bar{y} = 1/12$ is not.

CHAPTER 20

Section 20.1 (page 703)

1. (a) Here we have $a_1 = 0$, $a_2 = -1$ so $r_1, r_2 = (0 \pm \sqrt{0+4})/2 = \pm 1$. Since this is a homogeneous difference equation, the solution is

$$y_t = C_1(1)^t + C_2(-1)^t = C_1 + C_2(-1)^t$$

(b) Solve the homogeneous form first. The roots are $r_1, r_2 = 2, 1/2$. The steady-state solution, $\bar{y} = -4$, provides the particular solution. The complete solution is

$$y_t = C_1 2^t + C_2\left(\frac{1}{2}\right)^t - 4$$

(c) Solve the homogeneous form first. The roots are $r_1, r_2 = (-2 \pm \sqrt{4-4})/2 = -1$. Since the roots are equal, the homogeneous solution is of the form given in equation (20.4). The steady-state solution, $\bar{y} = 4$, provides the particular solution. The complete solution then is

$$y_t = C_1(-1)^t + C_2 t(-1)^t + 4$$

(d) Solve the homogeneous form first. The roots are complex-valued: $r_1, r_2 = 3 \pm 3i$. The homogeneous solution takes the form given in equation (20.5). The steady-state solution, $\bar{y} = 2$, provides the particular solution. The complete solution is

$$y_t = 18^{t/2}(C_1 \cos\theta t + C_2 \sin\theta t) + 2$$

where $\cos\theta = 6/(2\sqrt{18}) = 3/\sqrt{2 \times 9} = 1/\sqrt{2}$. Therefore, $\theta = \pi/4$.

3. Firm 1's profit in period $t+1$, assuming zero cost, is price times quantity:

$$\pi_{t+1} = [A - B(x_{t+1} + y_t)]x_{t+1}$$

To maximize π_{t+1}, set the derivative with respect to x_{t+1} equal to zero. After re-arranging this first-order condition, we get

$$x_{t+1} = \frac{A}{2B} - \frac{y_t}{2}$$

Following a similar procedure for firm 2 gives

$$y_{t+1} = \frac{A}{2B} - \frac{x_t}{2}$$

After using the reaction function for firm 2 to substitute for y_t in firm 1's reaction function, we get

$$x_{t+2} - 0.25x_t = \frac{A}{4B}$$

The roots of this difference equation are $r_1, r_2 = 1/2, -1/2$. The steady-state solution, $\bar{x} = A/(3B)$, provides the particular solution. The complete solution is

$$x_t = C_1\left(\frac{1}{2}\right)^t + C_2\left(-\frac{1}{2}\right)^t + \frac{A}{3B}$$

5. After making the substitutions and simplifying, we get

$$Y_{t+2} - \frac{\alpha}{1-m}Y_{t+1} + \frac{\alpha}{1-m}Y_t = \frac{\bar{G}}{1-m}$$

The roots are $r_1, r_2 = \frac{\alpha}{2(1-m)} \pm \frac{1}{2}\sqrt{\frac{\alpha}{1-m}(\frac{\alpha}{1-m} - 4)}$. The steady-state solution is $\bar{G}/(1-m)$. Since the roots may be complex-valued, we check the conditions for convergence before writing out the solution: 1) $1 + a_1 + a_2 = 1 > 0$ (satisfied); 2) $1 - a_1 + a_2 = 1 + 2\alpha/(1-m) > 0$ (satisfied); 3) $a_2 = \alpha/(1-m)$. Since $a_2 < 1$ is required for convergence, we would have to impose the restriction $\alpha/(1-m) < 1$ to ensure convergence. This also makes the roots complex-valued. The complete solution then is

$$Y_t = \left(\frac{\alpha}{1-m}\right)^{\frac{t}{2}}(C_1 \cos\theta t + C_2 \sin\theta t) + \frac{\bar{G}}{1-m}$$

where $\cos\theta = \sqrt{\alpha/(1-m)}/2$.

Section 20.2 (page 709)

1. (a) Start with the homogeneous form. The roots are $r_1, r_2 = 2, 1$. The homogeneous solution then is

$$y_t^h = C_1 2^t + C_2$$

The steady-state solution does not exist because $1 + a_1 + a_2 = 0$. Therefore, we use the method of undetermined coefficients. We note that $y_p = A$ will not work because this is a constant, and one of the terms in the homogeneous solution is a constant. Therefore, we try $y_p = At$ as the particular solution. Since it must satisfy the complete difference equation, we use this fact to find the value of A.

$$A(t+2) - 3A(t+1) + 2At = 12$$

Simplifying gives $A = -12$. Therefore, the complete solution is

$$y_t = C_1 2^t + C_2 - 12t$$

(b) The solution to the homogeneous form is

$$y_t^h = C_1 2^t + C_2\left(\frac{1}{2}\right)^t$$

To find the particular solution, we try $y_p = A3^t$ because the term in the difference equation is 3^t. In the complete difference equation, this gives

$$A3^{t+2} - \frac{5}{2}A3^{t+1} + A3^t = 3^t$$

Simplifying gives

$$A3^2 - \frac{15}{2}A + A = 1$$

which gives $A = 2/5$. Therefore, the complete solution is

$$y_t = C_1 2^t + C_2 \left(\frac{1}{2}\right)^t + \frac{2}{5}3^t$$

(c) The solution to the homogeneous form is

$$y_t^h = C_1(-1)^t + C_2 t(-1)^t$$

For a particular solution, we try $y_p = A_0 + A_1 t$. Substituting into the complete difference equation gives

$$A_0 + A_1(t+2) + 2A_0 + 2A_1(t+1) + A_0 + A_1 t = t$$

Collecting like terms gives

$$4A_0 + 4A_1 + t(4A_1 - 1) = 0$$

Solving gives $A_1 = 1/4$ and $A_0 = -1/4$. The complete solution then is

$$y_t = C_1(-1)^t + C_2 t(-1)^t - \frac{1}{4} + \frac{t}{4}$$

3. Substitute into the national income identity $Y_t = C_t + I_t + G_t$. After re-arranging, we get

$$Y_{t+2} - \frac{\alpha}{1-m}Y_{t+1} + \frac{\alpha}{1-m}Y_t = \frac{G_0}{1-m}(1+g)^t$$

The homogeneous solution is the same as the example solved in Section 20.2. For the particular solution, we try $y_p = A(1+g)^t$. Substitute this solution into the difference equation to get

$$A(1+g)^{t+2} - \frac{\alpha}{1-m}A(1+g)^{t+1} + \frac{\alpha}{1-m}A(1+g)^t$$
$$= \frac{G_0}{1-m}(1+g)^t$$

Simplifying and solving gives

$$A = \frac{G_0}{(1+g)^2(1-m) - \alpha g}$$

The complete solution then is

$$Y_t = C_1(r_1)^t + C_2(r_2)^t + \frac{G_0}{(1+g)^2(1-m) - \alpha g}(1+g)^t$$

where

$$r_1,\ r_2 = \frac{\alpha}{2(1-m)} \pm \frac{1}{2}\sqrt{\left(\frac{\alpha}{1-m}\right)^2 - \frac{4\alpha}{1-m}}$$

If the absolute values of the roots are less than 1, then the first two terms in the solution for Y_t tend towards 0 as t gets very large. Thus, Y_t converges in the limit to the third term, which produces exponential growth at the rate g.

Review Exercises (page 711)

1. Begin with the homogeneous solution. The roots are equal: $r_1, r_2 = 1/3$. Next, use the steady-state solution, $\bar{y} = 4$, for the particular solution. The complete solution then is

$$y_t = C_1\left(\frac{1}{3}\right)^t + C_2 t\left(\frac{1}{3}\right)^t + 4$$

3. The roots are $r_1, r_2 = 2, 1/2$. The particular solution is $\bar{y} = -10$. The complete solution is

$$y_t = C_1 2^t + C_2\left(\frac{1}{2}\right)^t - 10$$

5. The roots are equal: $r_1 = r_2 = 1$. The homogeneous solution is

$$y_t^h = C_1 + C_2 t$$

The steady-state solution does not exist because $1 + a_1 + a_2 = 0$. Therefore, we try $y_p = A$. However, that will not work because the homogeneous solution already contains a constant. Next, we try $y_p = At$ but that will not work either because the term $C_2 t$ appears in the homogeneous solution. Thus, we try $y_p = At^2$. Substituting this solution into the complete difference equation gives

$$A(t+2)^2 - 2A(t+1)^2 + At^2 = 2$$

Solving gives $A = 1$. The complete solution then is

$$y_t = C_1 + C_2 t + t^2$$

7. The roots are complex-valued: $r_1, r_2 = 1 \pm i/\sqrt{3}$. The homogeneous solution then is

$$y_t^h = \left(\frac{4}{3}\right)^{t/2}(C_1\cos\theta t + C_2\sin\theta t)$$

where $\cos\theta = 2/(2\sqrt{4/3}) = \sqrt{3}/2$. Therefore, $\theta = \pi/6$. The steady-state solution, $\bar{y} = 14$, provides the particular solution. The complete solution is

$$y_t^h = \left(\frac{4}{3}\right)^{t/2}\left(C_1\cos\frac{\pi}{6}t + C_2\sin\frac{\pi}{6}t\right) + 14$$

9. In Exercise 1, $C_1 = -3$ and $C_2 = -12$. In Exercise 3, $C_1 = 7/3$ and $C_2 = 26/3$. In Exercise 5, $C_1 = 1$ and $C_2 = -3$. In Exercise 7, we get $y_0 = 1 = C_1 + 14$. Therefore $C_1 = -13$. At $t = 1$ we get $y_1 = -1 = \sqrt{4/3}(C_1\cos\theta + C_2\sin\theta) + 14$. Using Theorem 20.1, $\cos\theta = \sqrt{3}/2$ and $\sin\theta = 1/2$. This gives $C_2 = -2\sqrt{3}$.

11. Substituting gives

$$x_{t+2} + \beta(b - \alpha x_{t-1}) = b$$

Simplifying gives

$$x_{t+2} - \beta\alpha x_t = b(1-\beta)$$

The roots are $r_1, r_2 = \pm\sqrt{\alpha\beta}$. The steady-state solution, $\bar{x} = b(1-\beta)/(1-\alpha\beta)$, provides the particular solution. The complete solution is

$$x_t = C_1(\alpha\beta)^{t/2} + C_2(-\alpha\beta)^{t/2} + \frac{b(1-\beta)}{1-\alpha\beta}$$

13. Substituting gives

$$y_t = \beta y_{t-1} + \rho e_{t-1} + u_t$$

But, from the given equation for y_t, we know that

$$e_{t-1} = y_{t-1} - \beta y_{t-2}$$

Substituting gives

$$y_t = \beta y_{t-1} + \rho(y_{t-1} - \beta y_{t-2}) + u_t$$

Simplifying and re-arranging gives

$$y_{t+2} - (\beta + \rho)y_{t+1} + \rho\beta y_t = u_{t+2}$$

15. For $y_{t+2} - (\beta + \rho)y_t + \rho\beta y_t = 0$, the solution is

$$y_t = C_1(r_1)^t + C_2(r_2)^t$$

where

$$r_1, \ r_2 = \frac{\beta + \rho}{2} \pm \sqrt{(\beta + \rho)^2 - 4\beta\rho/2}$$

Three conditions must be satisfied for convergence:

1. $1 + a_1 + a_2 = 1 - (\beta + \rho) + \beta\rho > 0$
2. $1 - a_1 + a_2 = 1 + (\beta + \rho) + \beta\rho > 0$
3. $a_2 = \beta\rho < 1$

These conditions are satisfied if the absolute values of β and ρ are less than 1.

CHAPTER 21

Section 21.1 (page 729)

1. **(a)** $y(t) = y_0 e^t = e^t$

(b) $y(t) = (y_0 - 12/3)e^{-3t} + 12/3 = 6e^{-3t} + 4$

(c) $y(t) = (y_0 - 6 \times 4)e^{-t/4} + 6 \times 4 = -14e^{-t/4} + 24$

(d) $y(t) = 5t + y_0 = 5t + 1$

(e) $y(t) = (y_0 - 6/6)e^{6t} + 6/6 = 2e^{6t} + 1$

3. We are given $\dot{p}/p = 0.05$. Therefore $\ln p(t) = 0.05t + c$ or $p(t) = Ce^{0.05t}$. We are also given $p(0) = 100$. Therefore

$$p(t) = 100e^{0.05t}$$

5. We are given $\dot{y}/y = -\alpha$. Therefore, $y(t) = y_0 e^{-\alpha t} = 500e^{-\alpha t} \times 10^6$.

7. Equilibrium occurs when $p^D = p^S$. This condition gives equilibrium quantity as $-(a-g)/(b-h)$. The differential equation in this model is

$$\dot{q} - \alpha(b-h)q = \alpha(a-g)$$

The solution is

$$q(t) = \left(q_0 + \frac{a-g}{b-h}\right)e^{\alpha(b-h)t} - \frac{a-g}{b-h}$$

For stability, we require that $q(t)$ converge to the steady-state quantity, which is also the equilibrium quantity. Convergence occurs if and only if $b - h < 0$ assuming $\alpha > 0$.

Section 21.2 (page 734)

1. The integrating factor is e^{-t}. Multiply both sides of the differential equation by e^{-t}, then we can write it as

$$\frac{d}{dt}(ye^{-t}) = e^{2t}e^{-t}$$

Integrating both sides gives

$$ye^{-t} = \int e^t + C = e^t + C$$

Therefore, $y(t) = e^{2t} + Ce^t$. Since $y(0) = 1$, we set $1 = 1 + C$ or $C = 0$. This gives $y(t) = e^{2t}$

3. The integrating factor is $e^{3t^{-1}}$. Multiply both sides of the differential equation by this factor and rewrite it as

$$\frac{d}{dt}(ye^{3t^{-1}}) = t^{-2}e^{3t^{-1}}$$

Integrate both sides to obtain

$$\begin{aligned} ye^{3t^{-1}} &= \int t^{-2}e^{3t^{-1}}dt \\ &= -\frac{e^{3t^{-1}}}{3} + C \end{aligned}$$

Rewrite as

$$y(t) = -1/3 + Ce^{-3t^{-1}}$$

5. Evaluating the solution at $t = 4$ gives $m(4) = 18$. Similarly, $m(6) = 28$.

7. The differential equation now becomes

$$\dot{k} = y - \beta t = \alpha t k - \beta t$$

which we write as

$$\dot{k} - \alpha t k = -\beta t$$

The integrating factor is $e^{-\alpha t^2/2}$. Multiply both sides by this factor and write as

$$\frac{d}{dt}(ke^{-\alpha t^2/2}) = -\beta t e^{-\alpha t^2/2}$$

Integrating yields

$$ke^{-\alpha t^2/2} = \frac{\beta}{\alpha}e^{-\alpha t^2/2} + C$$

and re-arranging gives

$$k(t) = \frac{\beta}{\alpha} + Ce^{\alpha t^2/2}$$

Evaluating the solution at $t = 0$ gives $C = k_0 - \beta/\alpha$. The solution becomes

$$k(t) = \frac{\beta}{\alpha} + \left(k_0 - \frac{\beta}{\alpha}\right)e^{\alpha t^2/2}$$

The steady state occurs at $\dot{k} = 0$ which gives $\bar{k} = \beta/\alpha$. However, since $\alpha > 0$, $k(t)$ does not converge to $\bar{k}$.

Review Exercises (page 735)

1. We are given $\dot{E}/E = 0.02$ and $E(t_0) = 2$. Solving gives $E(t) = Ce^{0.02t}$. At t_0, we have $2 = Ce^{0.02t_0}$. Solving for C and substituting it into the solution gives

$$E(t) = 2e^{0.02(t-t_0)}$$

3. Given $\dot{K} = I - \delta K$ and the values for I, δ, and K_0, we have

$$\dot{K} + 0.05K = 100$$

$$\begin{aligned} K(t) &= \left(500 - \frac{100}{0.05}\right)e^{-0.05t} + \frac{100}{0.05} \\ &= -1500e^{-0.05t} + 2000 \end{aligned}$$

$K(t)$ converges to the steady state, $\bar{K} = 2000$, because the exponential term goes to 0 in the limit.

5. The money market clears instantly and the equilibrium interest rate is $R = kY/h - \bar{M}/h$. The goods market adjusts according to

$$\dot{Y} = \alpha[a + bY - lR + \bar{I}(1-R) + \bar{G} - Y]$$

Substituting for R and simplifying gives

$$\dot{Y} + AY = B$$

where

$$\begin{aligned} A &= \alpha[1 - b + (l + \bar{I})k/h] \\ B &= \alpha[a + (l + \bar{I})\bar{M}/h + \bar{I} + \bar{G}] \end{aligned}$$

We require $A > 0$ for stability. The solution is

$$Y(t) = \left(Y_0 - \frac{B}{A}\right) e^{-At} + \frac{B}{A}$$

7. Since $\dot{Y} = gY$, we have $Y(t) = Y_0 e^{gt}$. Substituting this into the differential equation for debt gives

$$\dot{D} = bY_0 e^{gt} + rD$$

Re-arranging gives

$$\dot{D} - rD = bY_0 e^{gt}$$

The integrating factor is e^{-rt}. Multiply both sides by this factor and rewrite the equation as

$$\frac{d}{dt}(De^{-rt}) = bY_0 e^{(g-r)t}$$

Integrating gives

$$De^{-rt} = \frac{bY_0 e^{(g-r)t}}{g - r} + C$$

which, after re-arranging, becomes

$$D(t) = \frac{bY_0 e^{gt}}{g - r} + Ce^{rt}$$

At $t = 0$, $D(0) = D_0$. This gives $C = D_0 - bY_0/(g - r)$. The solution for debt becomes

$$D(t) = \frac{bY_0 e^{gt}}{g - r} + \left(D_0 - \frac{bY_0}{g - r}\right) e^{rt}$$

The ratio of interest payments, $rD(t)$, to national income is

$$\frac{rD(t)}{Y(t)} = \frac{rb}{g - r} + e^{(r-g)t}\left(\frac{rD_0}{Y_0} - \frac{rb}{g - r}\right)$$

If $g > r$, the exponential term goes to 0 in the limit and the solution converges to the finite limit $rb/(g-r)$. If $g < r$, the ratio of interest payments to national income grows without limit. To see this, it helps to rewrite our expression as

$$\frac{rD(t)}{Y(t)} = \frac{rb}{r - g}\left(e^{(r-g)t} - 1\right) + e^{(r-g)t}\frac{rD_0}{Y_0}$$

With $r - g > 0$, both terms on the right-hand side are positive and grow without limit.

9. We are given $\dot{I}/I = g$ so that $I(t) = I_0 e^{gt}$. The differential equation becomes

$$\dot{K} + \delta K = I_0 e^{gt}$$

The integrating factor is $e^{\delta t}$. The differential equation can be re-rewritten as

$$\frac{d}{dt}(Ke^{\delta t}) = I_0 e^{(g+\delta)t}$$

Integrating gives

$$Ke^{\delta t} = \frac{I_0 e^{(g+\delta)t}}{g + \delta} + C$$

Solving and using initial condition $K(0) = K_0$ gives

$$K(t) = \frac{I_0 e^{gt}}{g + \delta} + \left(K_0 - \frac{I_0}{g + \delta}\right) e^{-\delta t}$$

Because $g > 0$, $K(t)$ grows without limit.

11. After making the substitutions for q^D and q^S, the differential equation for price becomes

$$\dot{p} - \alpha(B - G)p = \alpha[A - F - H(1 - e^{-\mu t})]$$

To simplify, let $a = -\alpha(B - G)$ and $b = \alpha(A - F - H)$. Then

$$\dot{p} + ap = b + \alpha He^{-\mu t}$$

The integrating factor is e^{at}. Use it to rewrite the differential equation as

$$\frac{d}{dt}(pe^{at}) = be^{at} + \alpha He^{(a-\mu)t}$$

Integrating gives

$$pe^{at} = \frac{b}{a}e^{at} + \frac{\alpha H}{a - \mu}e^{(a-\mu)} + C$$

and then

$$p(t) = \frac{b}{a} + \frac{\alpha H}{a - \mu}e^{-\mu t} + Ce^{-at}$$

Using $p(0) = p_0$ to evaluate C gives

$$p(t) = \frac{b}{a} + \frac{\alpha H}{a - \mu}e^{-\mu t} + \left(p_0 - \frac{b}{a} - \frac{\alpha H}{a - \mu}\right) e^{-at}$$

Using our definitions for a and b now, this becomes

$$\begin{aligned} p(t) &= \frac{A - F - H}{G - B} + \frac{H}{G - B - \frac{\mu}{\alpha}}e^{-\mu t} \\ &\quad + \left(p_0 - \frac{A - F - H}{G - B} - \frac{H}{G - B - \frac{\mu}{\alpha}}\right) e^{\alpha(B-G)t} \end{aligned}$$

Note that if $G - B > 0$ as it is if the supply curve is positively sloped ($G > 0$) and the demand curve is negatively sloped ($B < 0$), price converges to $(A - F - H)/(G - B)$.

CHAPTER 22

Section 22.1 (page 745)

1.

$$\dot{y} = -y + y^2 + \frac{3}{16}$$

Find the steady-state points. Setting $\dot{y} = 0$ gives $y^2 - y + 3/16 = 0$ to which the solutions are $1/4$ and $3/4$. Next, determine the shape of the curve from the first and second derivative:

$$\frac{d\dot{y}}{dy} = -1 + 2y$$

This equals 0 at $y = 1/2$ and is positive for $y > 1/2$ and negative for $y < 1/2$. This suggests a U-shaped curve, a suspicion confirmed by the positive second derivative.

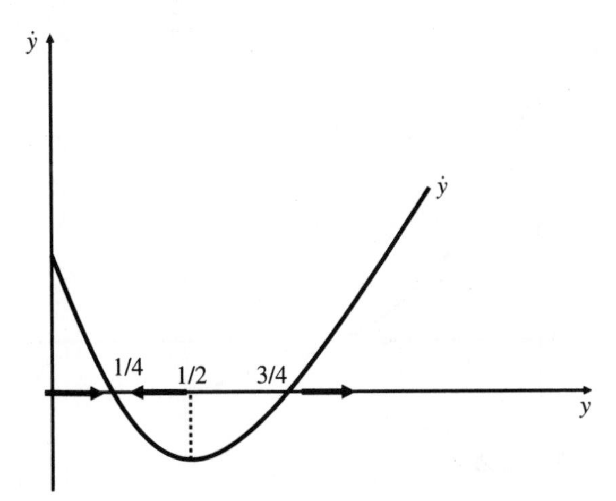

Figure 22.1.1

The $\dot{y}$ curve shown above indicates the motion of y is towards the point $\bar{y} = 1/4$ but away from $\bar{y} = 3/4$. This is confirmed by using Theorem 22.2. We see that $d\dot{y}/dy = -1/2$ at $\bar{y} = 1/4$ indicating a stable point, but $d\dot{y}/dy = 1/2$ at $\bar{y} = 3/4$ indicating instability.

3.

$$\dot{y} = y - y^{1/2}$$

The steady-state points are $\bar{y}_1 = 0$ and $\bar{y}_2 = 1$. The slope of the curve is $d\dot{y}/dy = 1 - y^{-1/2}/2$. The second derivative is positive. At $\bar{y} = 1$, $d\dot{y}/dy = 1/2$ so this point is unstable. At $\bar{y} = 0$, we cannot evaluate $d\dot{y}/dy$ because it would mean dividing by zero. However, we see that as $\bar{y} \Longrightarrow 0$, the derivative is negative, indicating that the direction of motion is negative so that y converges to 0.

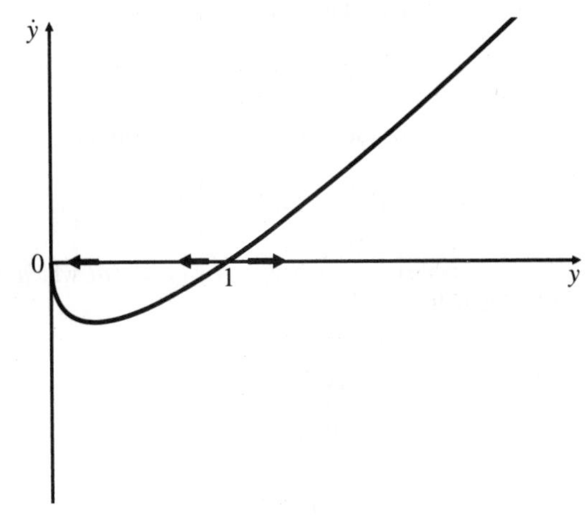

Figure 22.1.3

5. Substituting for q^d and q^s gives $\dot{p} = \alpha(p^{-2} - 8p)$. We see that as $p \Longrightarrow 0$, then $\dot{p} \Longrightarrow \infty$ indicating the curve is asymptotic to the vertical axis. We set $\dot{p} = 0$ to find the steady state. This gives $p^{-2} - 8p = 0$ or $p^3 = 1/8$. The solution is $\bar{p} = 1/2$. We use $d\dot{p}/dp = \alpha(-2p^{-3} - 8)$ to see that the slope is always negative and that the second derivative is positive. This information tells us the $\dot{p}$ curve must cut the horizontal axis as shown in Figure 22.1.5. The phase diagram indicates that the equilibrium is stable. This is confirmed by evaluating $d\dot{p}/dp$ at $\bar{p} = 1/2$. This gives a value of -24α. Since $\alpha > 0$, the derivative is negative, indicating stability.

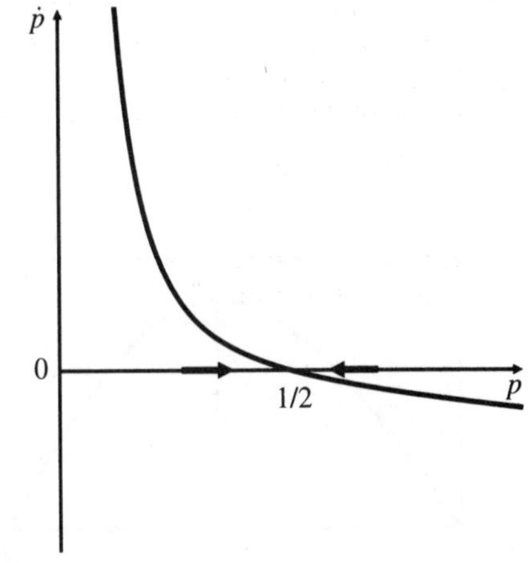

Figure 22.1.5

Section 22.2 (page 750)

1. Multiply through by y to obtain $\dot{y}y + 2y^2 = 3$. Now define $x = y^2$ so that $\dot{x} = 2y\dot{y}$. Making these substitutions transforms the differential equation into

$$\dot{x} + 4x = 6$$

The solution to this linear equation is

$$x(t) = Ce^{-4t} + \frac{3}{2}$$

Substituting now for y gives the general solution

$$y(t) = \left(Ce^{-4t} + \frac{3}{2}\right)^{1/2}$$

3. This separable equation becomes $ydy = t^2dt$ which can be integrated directly to obtain

$$\frac{y^2}{2} = \frac{t^3}{3} + C$$

Solving for y explicitly gives

$$y(t) = \left(\frac{2t^3}{3} + 2C\right)^{1/2}$$

5. This separable equation becomes $y^2dy = -tdt$ which integrates to

$$\frac{y^3}{3} = -\frac{t^2}{2} + C$$

Solving gives

$$y(t) = \left(-\frac{3}{2}t^2 + 3C\right)^{1/3}$$

Review Exercises (page 751)

1. To draw the phase diagram, note that $\dot{y} = 0$ at $\bar{y}_1 = 0$ and $\bar{y}_2 = 1/3$. Also note that the curve reaches a maximum (second derivative negative) at $y = 1/6$.

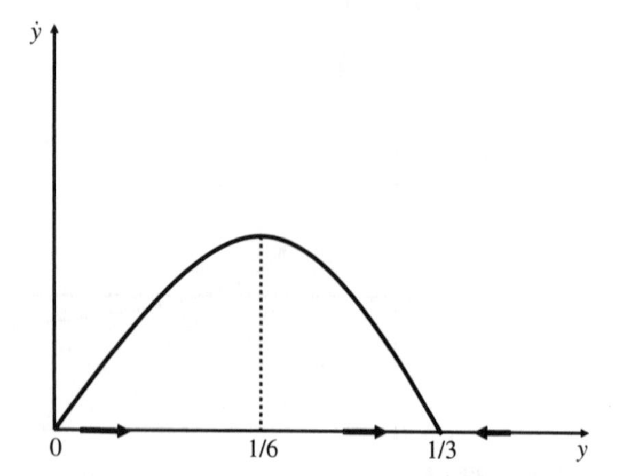

Figure 22.R.1

The phase diagram shows that $\bar{y}_2 = 1/3$ is a stable point. To solve the differential equation, multiply through by y^{-2}. This gives $\dot{y}y^{-2} - 2y^{-1} = -6$. Define $x = y^{-1}$ so that $\dot{x} = -\dot{y}y^{-2}$. This transforms the differential equation into

$$\dot{x} + 2x = 6$$

for which the solution is $x(t) = Ce^{-2t} + 3$. Use $y = x^{-1}$ to transform the solution back. This gives

$$y(t) = \frac{1}{Ce^{-2t} + 3}$$

3. This is equivalent to $ydy = tdt$ which integrates to

$$\frac{y^2}{2} = \frac{t^2}{2} + C$$

Solving gives

$$y(t) = (t^2 + 2C)^{1/2}$$

5. Set $\dot{q} = 0$ to find $\bar{q} = (p/a)^{1/2} > 0$ as the steady state. The slope of the $\dot{q}$ curve is

$$\frac{d\dot{q}}{dq} = -2\alpha aq < 0$$

This gives the phase diagram in Figure 22.R.5 which shows that $\bar{q}$ is a stable point. This is confirmed with Theorem 22.2 because $d\dot{q}/dq < 0$ at any $q > 0$ including $\bar{q}$.

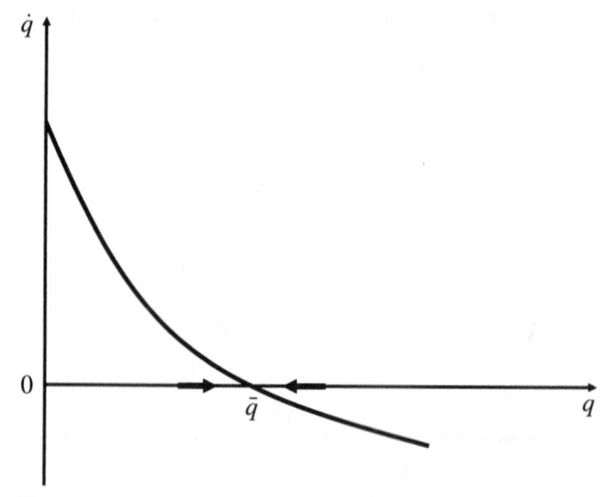

Figure 22.R.5

7. We are given $\dot{k} = sk^{1/2} - nk$. Set $\dot{k} = 0$ to find the steady state. For $k \neq 0$, this gives $\bar{k} = (s/n)^2$. Now write the differential equation as $\dot{k} + nk = sk^{1/2}$ and multiply through by $k^{-1/2}$. This gives $\dot{k}k^{-1/2} + nk^{1/2} = s$. Define $x = k^{1/2}$ so that $\dot{x} = k^{-1/2}\dot{k}/2$. Substituting transforms the differential equation into

$$\dot{x} + \frac{nx}{2} = \frac{s}{2}$$

The solution is $x(t) = Ce^{-nt/2} + s/n$. Since $k = x^2$, this becomes

$$k(t) = \left(Ce^{-nt/2} + \frac{s}{n}\right)^2$$

If $n > 0$, then $k(t)$ converges to $(s/n)^2$ in the limit.

CHAPTER 23

Section 23.1 (page 771)

1. For all of the differential equations in Exercise 1, begin by solving the homogeneous form. Then use the steady-state solution as the particular solution. Add them together to obtain the complete solution. Finally, evaluate the complete solution and its derivative at $t = 0$ and set $y(0) = 10$ and $\dot{y}(0) = 8$ to obtain the values of the constants of integration.

   (a) The roots are $r_1,\ r_2 = 1,\ -3$. The particular solution is $\bar{y} = 2$.

   $$y(t) = C_1e^t + C_2e^{-3t} + 2$$

   with $C_1 = 8$ and $C_2 = 0$.

   (b) $r_1,\ r_2 = -1/2$. $\bar{y} = 8$.

   $$y(t) = C_1e^{-t/2} + C_2te^{-t/2} + 8$$

   with $C_1 = 2$ and $C_2 = 9$.

   (c) $r_1,\ r_2 = 1/2,\ -1/3$; $\bar{y} = -30$.

   $$y(t) = C_1e^{t/2} + C_2e^{-t/3} - 30$$

   with $C_1 = 25.6$ and $C_2 = 14.4$.

   (d) $r_1,\ r_2 = -1$; $\bar{y} = 3$.

   $$y(t) = C_1e^{-t} + C_2te^{-t} + 3$$

   with $C_1 = 7$ and $C_2 = 15$.

3. The restriction is $G - B > 0$. With $G = 10$, we require $B < 10$.

5. $r_1,\ r_2 = -1/2 \pm i$; $\bar{p} = 10$.

   $$p(t) = e^{-t/2}[A_1 \cos t + A_2 \sin t] + 10$$

Section 23.2 (page 776)

1. The solution to the homogeneous form is

   $$y^h(t) = C_1e^t + C_2e^{-3t}$$

   Use the method of undetermined coefficients to find the particular solution. Try $y_p = Ae^{2t}$. This means $\dot{y}_p = 2Ae^{2t}$ and $\ddot{y}_p = 4Ae^{2t}$. Substitute into the differential equation to get

   $$3(4Ae^{2t}) + 6(2Ae^{2t}) - 9(Ae^{2t}) = -18e^{2t}$$

   Solving gives $A = -18/15$. The complete solution then is

   $$y(t) = C_1e^t + C_2e^{-3t} - \frac{18}{15}e^{2t}$$

3. The homogeneous solution is

   $$y^h(t) = C_1e^{3t} + C_2e^{2t}$$

   Try $y_p = Ae^{-t/2}$. Substituting this solution and its derivatives into the differential equation gives

   $$\frac{A}{4}e^{-t/2} - 5\left(-\frac{A}{2}e^{-t/2}\right) + 6Ae^{-t/2} = e^{-t/2}$$

   Solving gives $A = 4/35$. The complete solution becomes

   $$y(t) = C_1e^{3t} + C_2e^{2t} + \frac{4}{35}e^{-t/2}$$

5. The homogeneous solution is

   $$y^h(t) = C_1 + C_2e^{-t/2}$$

   For the particular solution, the steady-state solution does not exist because $a_2 = 0$ and $y_p = A$ will not work because a constant already appears in the homogeneous solution. Try $y_p = At$. This gives $\dot{y}_p = A$ and $\ddot{y}_p = 0$. Substituting into the differential equation gives $0 + A/2 = 4$. Therefore, $A = 8$ and the complete solution becomes

   $$y(t) = C_1 + C_2e^{-t/2} + 8t$$

Review Exercises (page 777)

1. In the following problems, solve the homogeneous forms first, then find the particular solution.

   (a) $r_1,\ r_2 = 2,\ -1$; $\bar{y} = -5$.

   $$y(t) = C_1e^{2t} + C_2e^{-t} - 5$$

   (b) $r_1 = r_2 = -3$; $\bar{y} = 3$.

   $$y(t) = C_1e^{-3t} + C_2te^{-3t} + 3$$

   (c) $r_1,\ r_2 = -2 \pm i$; $\bar{y} = 2$.

   $$y(t) = e^{-2t}(A_1 \cos t + A_2 \sin t) + 2$$

   At $t = 0$, we have

   $$y(0) = A_1 + 2$$

   Given $y(0) = 2$, we set $A_1 = 0$. The derivative of $y(t)$ is

   $$\begin{aligned}\dot{y}(t) &= -2e^{-2t}(A_1 \cos t + A_2 \sin t) + \\ &\quad e^{-2t}(-A_1 \sin t + A_2 \cos t)\end{aligned}$$

   Evaluated at $t = 0$ with $A_1 = 0$ and $\dot{y}(0) = 1$ gives $A_2 = 1$. The complete solution is

   $$y(t) = e^{-2t} \sin t + 2$$

3. Differentiate the first equation and substitute to get

   $$\ddot{y} = a_{11}\dot{y} + a_{12}(a_{21}y)$$

   Re-arranging gives

   $$\ddot{y} - a_{11}\dot{y} - a_{12}a_{21}y = 0$$

   The solution is $y(t) = C_1e^{r_1t} + C_2e^{r_2t}$ where

   $$r_1,\ r_2 = \frac{a_{11}}{2} \pm \sqrt{a_{11}^2 + 4a_{12}a_{21}}/2$$

   To ensure the roots are both negative, we need $a_{11} < 0$ and $a_{12}a_{21} < 0$. If the roots are complex-valued, the real part will be negative.

5. The solution is

$$y(t) = C_1 e^{r_1 t} + C_2 e^{r_2 t} + \frac{A - F}{G - B}$$

where $r_1,\ r_2 = -a_1/2 \pm (1/2)\sqrt{a_1^2 - 4a_2}$.

(a) We get $a_1 = 9,\ a_2 = 18$. This gives $r_1,\ r_2 = -3,\ -6$. The solution is

$$y(t) = C_1 e^{-3t} + C_2 e^{-6t} + \frac{A - F}{45}$$

(b) We get $a_1 = 6,\ a_2 = 18$. This gives $r_1,\ r_2 = -3 \pm 3i$. The solution is

$$y(t) = e^{-3t}[A_1 \cos(3t) + A_2 \sin(3t)] + \frac{A - F}{45}$$

CHAPTER 24

Section 24.1 (page 797)

1. (a)

$$2\ddot{y} - 5\dot{y} + y - 10 = 0$$

Let $x = \dot{y}$. Then $\dot{x} = \ddot{y}$. Making the substitutions and re-arranging gives

$$\begin{aligned} \dot{y} &= x \\ \dot{x} &= -\frac{y}{2} + \frac{5}{2}x + 10 \end{aligned}$$

The characteristic roots are

$$r_1,\ r_2 = \frac{5}{4} \pm \frac{1}{4}\sqrt{17}$$

Because the roots are real-valued and different, the solutions, using Theorem 24.2, are

$$y(t) = C_1 e^{r_1 t} + C_2 e^{r_2 t} + 20$$

$$x(t) = r_1 C_1 e^{r_1 t} + r_2 C_2 e^{r_2 t} + 0$$

(b)

$$\ddot{y} - 2y = 1$$

Let $x = \dot{y}$. Then $\dot{x} = \ddot{y}$. Making the substitutions and re-arranging gives

$$\begin{aligned} \dot{y} &= x \\ \dot{x} &= 2y + 1 \end{aligned}$$

The roots are $r_1,\ r_2 = \pm\sqrt{2}$. The solutions are

$$y(t) = C_1 e^{r_1 t} + C_2 e^{r_2 t} - \frac{1}{2}$$

$$x(t) = r_1 C_1 e^{r_1 t} + r_2 C_2 e^{r_2 t} + 0$$

(c)

$$\ddot{y} + 10\dot{y} + y = 1$$

Letting $x = \dot{y}$ gives

$$\begin{aligned} \dot{y} &= x \\ \dot{x} &= -y - 10x + 1 \end{aligned}$$

The roots are $r_1,\ r_2 = -5 \pm 2\sqrt{6}$. The solutions are

$$y(t) = C_1 e^{r_1 t} + C_2 e^{r_2 t} + 1$$

$$x(t) = r_1 C_1 e^{r_1 t} + r_2 C_2 e^{r_2 t} + 0$$

3. (i)

$$\begin{aligned} \dot{y}_1 &= -2y_1 + 2y_2 + 12 \\ \dot{y}_2 &= y_1 - 3y_2 - 12 \end{aligned}$$

(a) *Substitution Method*

Begin by solving the homogeneous form. Differentiate the homogeneous form of the first equation. After using the homogeneous form of the second equation to substitute for y_2, this becomes

$$\ddot{y}_1 = -2\dot{y}_1 + 2(y_1 - 3y_2)$$

Using the first equation to substitute for y_2, and re-arranging, gives

$$\ddot{y}_1 + 5\dot{y}_1 + 4y_1 = 0$$

The characteristic roots of this second-order, linear, homogeneous differential equation are $r_1,\ r_2 = -1,\ -4$. The solution, using Theorem 23.2 is

$$y_1^h(t) = C_1 e^{r_1 t} + C_2 e^{r_2 t}$$

To obtain $y_2(t)$, differentiate the solution for $y_1(t)$ and use it and the solution for $y_1(t)$ in the expression for $y_2(t)$ obtained from the first differential equation. This gives

$$y_2(t) = \frac{1}{2}(r_1 C_1 e^{r_1 t} + r_2 C_2 e^{r_2 t}) + C_1 e^{r_1 t} + C_2 e^{r_2 t}$$

After simplifying and substituting for $r_1,\ r_2$, this becomes

$$y_2^h(t) = \frac{C_1}{2} e^{r_1 t} - C_2 e^{r_2 t}$$

The particular solutions are given by the steady-state solutions. Setting $\dot{y}_1 = \dot{y}_2 = 0$ in the complete equations gives

$$\begin{aligned} -2\bar{y}_1 + 2\bar{y}_2 + 12 &= 0 \\ \bar{y}_1 - 3\bar{y}_2 - 12 &= 0 \end{aligned}$$

Solving gives $\bar{y}_1 = 3$ and $\bar{y}_2 = -3$. The complete solutions are the sum of the homogeneous and particular solutions; that is

$$\begin{aligned} y_1(t) &= C_1 e^{r_1 t} + C_2 e^{r_2 t} + 3 \\ y_2(t) &= \frac{C_1}{2} e^{r_1 t} - C_2 e^{r_2 t} - 3 \end{aligned}$$

The solutions must also satisfy the initial conditions. Evaluating the solutions at $t = 0$ and using the given initial conditions, gives

$$\begin{aligned} -2 &= C_1 + C_2 + 3 \\ 5 &= \frac{C_1}{2} - C_2 - 3 \end{aligned}$$

Solving gives $C_1 = 2$ and $C_2 = -7$.

(b) *Direct Method*

The characteristic equation is

$$\begin{vmatrix} -2 - r & 2 \\ 1 & -3 - r \end{vmatrix} = 0$$

which gives $r_1,\ r_2 = -1,\ -4$. For $r_1 = -1$, the eigenvector is the solution to

$$\begin{bmatrix} -2+1 & 2 \\ 1 & -3+1 \end{bmatrix}\begin{bmatrix} k_1 \\ k_2 \end{bmatrix} = 0$$

This gives $-k_1 + 2k_2 = 0$. Setting $k_1 = 1$ gives $k_2 = 1/2$. For $r_2 = -4$, the eigenvector is the solution to

$$\begin{bmatrix} -2+4 & 2 \\ 1 & -3+4 \end{bmatrix}\begin{bmatrix} k_1 \\ k_2 \end{bmatrix} = 0$$

This gives $2k_1 + 2k_2 = 0$. Setting $k_1 = 1$ gives $k_2 = -1$. The homogeneous solutions therefore are

$$\begin{bmatrix} y_1^h(t) \\ y_2^h(t) \end{bmatrix} = \begin{bmatrix} 1 \\ \frac{1}{2} \end{bmatrix} C_1 e^{-t} + \begin{bmatrix} 1 \\ -1 \end{bmatrix} C_2 e^{-4t}$$

The particular solutions and initial conditions are found as in the substitution method. The final complete solutions then are

$$\begin{bmatrix} y_1(t) \\ y_2(t) \end{bmatrix} = \begin{bmatrix} 2 \\ 1 \end{bmatrix} e^{-t} + \begin{bmatrix} -7 \\ 7 \end{bmatrix} e^{-4t} + \begin{bmatrix} 3 \\ -3 \end{bmatrix}$$

(ii)

$$\begin{aligned} \dot{y}_1 &= -y_1 - \frac{9}{4}y_2 + 2 \\ \dot{y}_2 &= -3y_1 + 2y_2 - 1 \end{aligned}$$

(a) *Substitution Method*

Begin by solving the homogeneous form. The second-order differential equation we obtain after following the procedure used in part (a) is

$$\ddot{y}_1 - \dot{y}_1 - \frac{35}{4}y_1 = 0$$

The characteristic roots are $r_1 = r_2 = 7/2,\ -5/2$. The solution given in Theorem 23.2 is

$$y_1^h(t) = C_1 e^{7t/2} + C_2 e^{-5t/2}$$

After substituting this solution and its differential into the expression for y_2, we get

$$y_2^h(t) = -2C_1 e^{7t/2} + \frac{2}{3}C_2 e^{-5t/2}$$

The particular solutions are the steady-state solutions. These are $\bar{y}_1 = 0.2$ and $\bar{y}_2 = 0.8$. The complete solutions therefore are

$$\begin{aligned} y_1(t) &= C_1 e^{7t/2} + C_2 e^{-5t/2} + 0.2 \\ y_2(t) &= -2C_1 e^{7t/2} + \frac{2}{3}C_2 e^{-5t/2} + 0.8 \end{aligned}$$

To satisfy the given initial conditions, we must set $C_1 = 4.5$ and $C_2 = 15.3$.

(b) *Direct Method*

The characteristic equation is

$$\begin{vmatrix} -1-r & -\frac{9}{4} \\ -3 & 2-r \end{vmatrix} = 0$$

which gives $r_1,\ r_2 = 7/2,\ -5/2$. For $r_1 = 7/2$, the eigenvector is

$$\begin{bmatrix} -1-\frac{7}{2} & -\frac{9}{4} \\ -3 & 2-\frac{7}{2} \end{bmatrix}\begin{bmatrix} k_1 \\ k_2 \end{bmatrix} = 0$$

This gives $-9k_1/2 - 9k_2/4 = 0$. Setting $k_1 = 1$ gives $k_2 = -2$. For $r_2 = -5/2$, the eigenvector is

$$\begin{bmatrix} -1+\frac{5}{2} & -\frac{9}{4} \\ -3 & 2+\frac{5}{2} \end{bmatrix}\begin{bmatrix} k_1 \\ k_2 \end{bmatrix} = 0$$

This gives $3k_1/2 - 9k_2/4 = 0$. Setting $k_1 = 1$ gives $k_2 = 2/3$. Using the particular solutions found above, the complete solutions are

$$\begin{bmatrix} y_1(t) \\ y_2(t) \end{bmatrix} = \begin{bmatrix} 1 \\ -2 \end{bmatrix} C_1 e^{7t/2} + \begin{bmatrix} 1 \\ \frac{2}{3} \end{bmatrix} C_2 e^{-5t/2} + \begin{bmatrix} 0.2 \\ 0.8 \end{bmatrix}$$

The constants are found as above.

(iii)

$$\begin{aligned} \dot{y}_1 &= 2y_1 - 2y_2 + 5 \\ \dot{y}_2 &= 2y_1 + 2y_2 + 1 \end{aligned}$$

(a) *Substitution Method*

Begin with the homogeneous form. The second-order differential equation we obtain after following the procedure used above is

$$\ddot{y}_1 - 4\dot{y}_1 + 8y_1 = 0$$

The roots are complex-valued: $r_1,\ r_2 = 2 \pm 2i$. The homogeneous solution, using Theorem 23.3, is

$$y_1^h(t) = A_1 e^{2t}\cos(2t) + A_2 e^{2t}\sin(2t)$$

The derivative of this solution is

$$\begin{aligned} \dot{y}_1^h(t) &= 2e^{2t}[A_1\cos(2t) + A_2\sin(2t)] \\ &\quad + e^{2t}[-2A_1\sin(2t) + 2A_2\cos(2t)] \end{aligned}$$

Substitute this and the solution, $y_1^h(t)$, into the expression for y_2 obtained from the first differential equation. After simplifying, this becomes

$$y_2^h(t) = e^{2t}[A_1\sin(2t) - A_2\cos(2t)]$$

The particular solutions are given by the steady-state solutions: $\bar{y}_1 = -3/2$ and $\bar{y}_2 = 1$. Therefore, the complete solutions are

$$\begin{aligned} y_1(t) &= e^{2t}[A_1\cos(2t) + A_2\sin(2t)] - 1.5 \\ y_2(t) &= e^{2t}[A_1\sin(2t) - A_2\cos(2t)] + 1 \end{aligned}$$

To satisfy the initial conditions, evaluate the solutions at $t = 0$ and use the given initial conditions.

$$\begin{aligned} 2.5 &= A_1 - 1.5 \\ -1 &= -A_2 + 1 \end{aligned}$$

This gives $A_1 = 4$ and $A_2 = 2$.

(b) *Direct Method*

The characteristic equation is

$$\begin{vmatrix} 2-r & -2 \\ 2 & 2-r \end{vmatrix} = 0$$

which gives $r_1, r_2 = 2 \pm 2i$. For $r_1 = 2+2i$, the eigenvector is the solution to

$$\begin{bmatrix} 2-2-2i & -2 \\ 2 & 2-2-2i \end{bmatrix} \begin{bmatrix} k_1 \\ k_2 \end{bmatrix} = 0$$

This gives $-2ik_1 - 2k_2 = 0$. Setting $k_1 = 1$ gives $k_2 = -i$. For $r_2 = 2-2i$, we get $k_1 = 1$ and $k_2 = i$. Therefore, the homogeneous solutions are

$$\begin{bmatrix} y_1^h(t) \\ y_2^h(t) \end{bmatrix} = \begin{bmatrix} 1 \\ -i \end{bmatrix} C_1 e^{(2+2i)t} + \begin{bmatrix} 1 \\ i \end{bmatrix} C_2 e^{(2-2i)t}$$

Using Euler's formula, this becomes

$$\begin{aligned} \begin{bmatrix} y_1^h(t) \\ y_2^h(t) \end{bmatrix} &= \begin{bmatrix} 1 \\ -i \end{bmatrix} C_1 e^{2t}[\cos(2t) + i\sin(2t)] \\ &+ \begin{bmatrix} 1 \\ i \end{bmatrix} C_2 e^{2t}[\cos(2t) - i\sin(2t)] \end{aligned}$$

Simplifying gives

$$\begin{bmatrix} y_1^h(t) \\ y_2^h(t) \end{bmatrix} = \left(A_1 \begin{bmatrix} \cos(2t) \\ \sin(2t) \end{bmatrix} + A_2 \begin{bmatrix} \sin(2t) \\ -\cos(2t) \end{bmatrix} \right) e^{2t}$$

where $A_1 = C_1 + C_2$ and $A_2 = (C_1 - C_2)i$.

The particular solutions are found and the initial conditions are satisfied as for the substitution method. The final complete solutions then are

$$\begin{aligned} \begin{bmatrix} y_1(t) \\ y_2(t) \end{bmatrix} &= 4\begin{bmatrix} \cos(2t) \\ \sin(2t) \end{bmatrix} e^{2t} + 2\begin{bmatrix} \sin(2t) \\ -\cos(2t) \end{bmatrix} e^{2t} \\ &+ \begin{bmatrix} -1.5 \\ 1 \end{bmatrix} \end{aligned}$$

Section 24.2 (page 819)

1. (a) Both roots are positive. ($r_1, r_2 \doteq 8.85, 2.15$) Unstable node.

(b) Complex roots; real part positive ($r_1, r_2 = 1 \pm 2.45i$). Unstable focus.

(c) Roots of opposite sign ($r_1, r_2 = 8, -1$). Saddlepoint.

(d) Both roots negative ($r_1, r_2 = -1, -10$). Stable node.

3. The determinant of the coefficient matrix is negative, indicating a saddlepoint. The roots are $r_1, r_2 = 5, -7$. The solution to the homogeneous forms are

$$\begin{aligned} y_1^h(t) &= C_1 e^{5t} + C_2 e^{-7t} \\ y_2^h(t) &= -\frac{C_1}{3} e^{5t} + C_2 e^{-7t} \end{aligned}$$

The steady-state solutions provide the particular solutions. These are $\bar{y}_1 = 14$, $\bar{y}_2 = 7$. Adding these to the homogeneous solutions gives the complete solutions:

$$\begin{aligned} y_1(t) &= C_1 e^{5t} + C_2 e^{-7t} + 14 \\ y_2(t) &= -\frac{C_1}{3} e^{5t} + C_2 e^{-7t} + 7 \end{aligned}$$

The equation for the saddlepath is found by setting $C_1 = 0$. This gives

$$C_2 = y_1(t)e^{7t} - 14e^{7t}$$

and

$$C_2 = y_2(t)e^{7t} - 7e^{7t}$$

Therefore, the saddlepath is given by the line

$$y_2 = y_1 - 7$$

To draw the phase diagram, begin by finding the y_1 isocline which is defined by $\dot{y}_1 = 0$. This gives the positively sloped line $y_2 = 2y_1/9 + 35/9$. Since $\partial\dot{y}_1/\partial y_1 = 2 > 0$, then y_1 is increasing to the right and decreasing to the left of the isocline. Now draw the y_2 isocline. Setting $\dot{y}_2 = 0$ gives the negatively sloped line $y_2 = -3y_1/4 + 70/4$. Since $\partial\dot{y}_2/\partial y_2 = -4 < 0$, then y_2 is decreasing to the right and increasing to the left of the isocline.

5. The solutions are $\bar{p} = 2$, $\bar{e} = 1.4$ and

$$\begin{aligned} p(t) &= C_1 e^{-3t/2} + C_2 e^{t/2} + 2 \\ e(t) &= -\frac{C_1}{2} e^{-3t/2} + \frac{3}{2} C_2 e^{t/2} + 1.4 \end{aligned}$$

The saddlepath is defined by $C_2 = 0$. This gives the line

$$e = \frac{-p}{2} + 2.4$$

Therefore, if $p(0) = 4$, then $e(0)$ must jump to the value 0.4 to put the economy on the saddlepath.

Section 24.3 (page 833)

1.

$$\begin{aligned} y_{t+1} &= y_t + 5x_t - 10 \\ x_{t+1} &= \frac{1}{4} y_t - x_t + 10 \end{aligned}$$

Begin with the homogeneous forms. The determinant of the coefficient matrix is $-9/4$. Its trace is 0. Therefore, the roots are

$$r_1, r_2 = 0 \pm \frac{1}{2}\sqrt{0 + 4(\frac{9}{4})} = \pm\frac{3}{2}$$

The homogeneous solutions are

$$\begin{aligned} y_t^h &= C_1 \left(\frac{3}{2}\right)^t + C_2 \left(-\frac{3}{2}\right)^t \\ x_t^h &= \frac{C_1}{10}\left(\frac{3}{2}\right)^t - \frac{C_2}{2}\left(-\frac{3}{2}\right)^t \end{aligned}$$

The particular solutions are the steady-state solutions. These are $\bar{y} = -24$ and $\bar{x} = 2$. The complete solutions are the sum of the homogeneous and particular solutions:

$$\begin{aligned} y_t &= C_1 \left(\frac{3}{2}\right)^t + C_2 \left(-\frac{3}{2}\right)^t - 24 \\ x_t &= \frac{C_1}{10}\left(\frac{3}{2}\right)^t - \frac{C_2}{2}\left(-\frac{3}{2}\right)^t + 2 \end{aligned}$$

To satisfy the initial conditions given, evaluate these solutions at $t = 0$. This gives $C_1 = 20$ and $C_2 = 10$.

3.

$$y_{t+1} = -y_t + \frac{3}{4}x_t$$
$$x_{t+1} = -3y_t + 2x_t$$

The determinant of the coefficient matrix is $1/4$; the trace is 1. Therefore, the roots are

$$r_1, r_2 = \frac{1}{2} \pm \frac{1}{2}\sqrt{1 - 4(\frac{1}{4})} = \frac{1}{2}$$

The solution, for the case of equal roots, is

$$y_t = (C_1 + C_2 t)\left(\frac{1}{2}\right)^t$$
$$x_t = (2C_1 + \frac{2}{3}C_2 + 2C_2 t)\left(\frac{1}{2}\right)^t$$

Evaluate this solution at $t = 0$ and use the given initial conditions to get $C_1 = 5$ and $C_2 = -3$.

5.

$$y_{t+1} = 60 - \frac{1}{4}x_t$$
$$x_{t+1} = 60 - \frac{1}{4}y_t$$

The determinant of coefficient matrix is $-1/16$; the trace is zero. Therefore, the roots are

$$r_1, r_2 = 0 \pm \frac{1}{2}\sqrt{0 - 4\left(\frac{-1}{16}\right)} = \pm\frac{1}{4}$$

The steady states are $\bar{y} = 48$ and $\bar{x} = 48$. Because the absolute values of both roots are less than 1, the system converges to the steady state. Therefore, the steady state is stable. The complete solution is

$$y_t = C_1\left(\frac{1}{4}\right)^t + C_2\left(-\frac{1}{4}\right)^t + 48$$
$$x_t = -C_1\left(\frac{1}{4}\right)^t + C_2\left(-\frac{1}{4}\right)^t + 48$$

7.

$$y_{t+1} = 0.9x_t$$
$$x_{t+1} = 0.9y_t$$

The roots are $r_1, r_2 = \pm 0.9$. The solution is

$$y_t = C_1(0.9)^t + C_2(-0.9)^t$$
$$x_t = C_1(0.9)^t - C_2(-0.9)^t$$

If $y_0 = x_0$, then $C_2 = 0$ and $C_1 = y_0$. The solution becomes

$$y_t = y_0(0.9)^t$$
$$x_t = y_0(0.9)^t$$

Both solutions converge to 0.

Review Exercises (page 835)

1. (a)

$$\dot{y}_1 = \frac{1}{2}y_1 + \frac{1}{4}y_2 + 3$$
$$\dot{y}_2 = 3y_1 + \frac{1}{2}y_2 + 2$$

Begin with the homogeneous forms. After differentiating the first equation and using the homogeneous form of the second to substitute for $\dot{y}_2$, we get

$$\ddot{y}_1 = \frac{1}{2}\dot{y}_1 + \frac{1}{4}\left(3y_1 + \frac{1}{2}y_2\right)$$

Using the homogeneous form of the first equation to substitute for y_2, this becomes

$$\ddot{y}_1 - \dot{y}_1 - \frac{1}{2}y_1 = 0$$

The characteristic roots are

$$r_1,\ r_2 = \frac{1}{2} \pm \frac{\sqrt{3}}{2}$$

The homogeneous solution is

$$y_1^h(t) = C_1 e^{r_1 t} + C_2 e^{r_2 t}$$

Differentiate this solution and use it in the expression for y_2 obtained from the homogeneous form of the first equation. This gives

$$y_2^h(t) = 2\sqrt{3}C_1 e^{r_1 t} - 2\sqrt{3}C_2 e^{r_2 t}$$

The particular solutions are $\bar{y}_1 = 2$ and $\bar{y}_2 = -16$. The complete solutions then are

$$y_1(t) = C_1 e^{r_1 t} + C_2 e^{r_2 t} + 2$$
$$y_2(t) = 2\sqrt{3}C_1 e^{r_1 t} - 2\sqrt{3}C_2 e^{r_2 t} - 16$$

The steady-state point $(2, -16)$ is a saddle point because the roots are of opposite sign.

(b)

$$\dot{y}_1 = -y_1 + \frac{3}{4}y_2 - 4$$
$$\dot{y}_2 = -3y_1 + 2y_2 - 1$$

Begin with the homogeneous forms and follow the procedure of differentiation and substitution used above to get the following second-order differential equation:

$$\ddot{y}_1 - \dot{y}_1 + \frac{1}{4}y_1 = 0$$

The roots are real and equal: $r_1,\ r_2 = 1/2$. The homogeneous solution then is

$$y_1^h(t) = (C_1 + C_2 t)e^{t/2}$$

Substitute this solution and its derivative into the homogeneous form of the first equation to obtain

$$y_2^h(t) = 2(C_1 + C_2 t)e^{t/2} + \frac{4}{3}C_2 e^{t/2}$$

The steady-state solutions serve as the particular solutions. These are $\bar{y}_1 = 29$ and $\bar{y}_2 = 44$. The complete solutions then are

$$y_1(t) = (C_1 + C_2 t)e^{t/2} + 29$$
$$y_2(t) = 2(C_1 + C_2 t)e^{t/2} + \frac{4}{3}C_2 e^{t/2} + 44$$

Because the roots are equal and positive, the steady state is an unstable improper node.

3. The matrix of coefficients for this problem is

$$\begin{bmatrix} -\alpha & 1 \\ -\beta & 0 \end{bmatrix}$$

The determinant is $\beta > 0$ and the trace is $-\alpha < 0$. Therefore, both roots are negative: $r_1,\ r_2 = -\alpha/2 \pm (1/2)\sqrt{\alpha^2 - 4\beta}$. If $\alpha^2 - 4\beta < 0$ the roots are complex-valued with negative real parts. The steady-state solutions provide the particular solutions. These are $\bar{y} = a/\beta$ and $\bar{x} = \alpha a/\beta$. If the roots are real-valued, the solutions are

$$y(t) = C_1 e^{r_1 t} + C_2 e^{r_2 t} + \frac{a}{\beta}$$
$$x(t) = (r_1 + \alpha)C_1 e^{r_1 t} + (r_2 + \alpha)C_2 e^{r_2 t} + \frac{\alpha a}{\beta}$$

Because the real parts of both roots are negative, the steady state is asymptotically stable (a stable node if roots are real-valued; stable focus if roots are complex-valued). The model implies that the stock of pollution converges to the finite amount $\bar{y} = a/\beta$ even though industrial emissions converge to the positive amount $\bar{x} = \alpha a/\beta$.

5.

$$\dot{I} = \delta I - \frac{\alpha K^{\alpha-1}}{2}$$
$$\dot{K} = I - \delta K$$

The steady-state capital stock is found by setting $\dot{I} = 0$ and $\dot{K} = 0$. This gives

$$\bar{K} = \left(\frac{2\delta^2}{\alpha}\right)^{\frac{1}{\alpha - 2}}$$

The steady-state investment is $\bar{I} = \delta\bar{K}$. To show that the steady state is a saddle point, find the determinant of the coefficient matrix of the linearized system. The coefficient matrix is

$$\begin{bmatrix} \delta & \frac{-(\alpha-1)\alpha}{2}\bar{K}^{\alpha-2} \\ 1 & -\delta \end{bmatrix}$$

which has determinant

$$-\delta^2 + \frac{(\alpha-1)\alpha}{2}\bar{K}^{\alpha-2} < 0$$

It is negative because $\delta > 0$ and $0 < \alpha < 1$ and $\bar{K} > 0$. The negative determinant means the roots are of opposite signs; the steady state is therefore a saddle point.

The $\dot{I} = 0$ isocline is the curve given by

$$I = \frac{\alpha}{2\delta}K^{\alpha-1}$$

This curve is asymptotic to both axes and has a negative slope because $0 < \alpha < 1$. To find the motion of I in the two regions divided by this isocline, take $\partial\dot{I}/\partial I = \delta > 0$. This indicates I is increasing above and decreasing below the isocline, as shown in the Figure 24.R.5.

The $\dot{K} = 0$ isocline is the line given by

$$I = \delta K$$

This is a straight line emanating from the origin with slope δ. K is decreasing to the right and increasing to the left of the isocline.

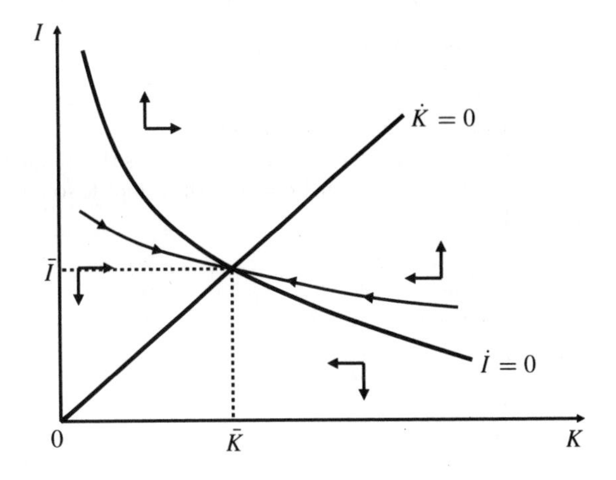

Figure 24.R.5

7. (a)

$$y_{t+1} = y_t + x_t + 1$$
$$x_{t+1} = y_t + x_t + 2$$

Begin with the homogeneous forms. The first equation becomes

$$y_{t+2} = y_{t+1} + (y_t + x_t)$$

after using the second equation to substitute for x_{t+1}. Now use the first equation to substitute for x_t. After re-arranging, we get

$$y_{t+2} - 2y_{t+1} = 0$$

The roots are $r_1,\ r_2 = 2/2 \pm \sqrt{4}/2 = 0,\ 2$. The homogeneous solution then is

$$y_t = C_2(2)^t$$

Substitute this into the expression for x_t obtained from the first equation. This gives

$$x_t = C_2(2)^t$$

The particular solutions are the steady-state solutions. These are $\bar{y} = -2$ and $\bar{x} = -1$. The complete solutions then are

$$y_t = C_2(2)^t - 2$$
$$x_t = C_2(2)^t - 1$$

(b)

$$y_{t+1} = -2y_t + 2x_t + 1$$
$$x_{t+1} = y_t - 3x_t + 2$$

Begin with the homogeneous forms. After making the appropriate substitutions as outlined above, the first equation becomes

$$y_{t+2} + 5y_{t+1} + 4y_t = 0$$

The roots are $r_1,\ r_2 = -5/2 \pm \sqrt{25-16}/2 = -4,\ -1$. The homogeneous solution then is

$$y_t = C_1(-4)^t + C_2(-1)^t$$

Substitute into the expression for x_t to get

$$x_t = -C_1(-4)^t + \frac{C_2}{2}(-1)^t$$

The particular solutions are the steady-state solutions, which are $\bar{y} = 0.8$ and $\bar{x} = 0.7$. The complete solutions then are

$$\begin{aligned} y_t &= C_1(-4)^t + C_2(-1)^t + 0.8 \\ x_t &= -C_1(-4)^t + \frac{C_2}{2}(-1)^t + 0.7 \end{aligned}$$

9. Let p_t be the probability of high sales and q_t be the probability of low sales.

$$\begin{aligned} p_{t+1} &= 0.75p_t + 0.5q_t \\ q_{t+1} &= 0.25p_t + 0.5q_t \end{aligned}$$

The roots of this system are $r_1,\ r_2 = 1,\ 0.25$. The solutions then are

$$\begin{aligned} p_t &= C_1 + C_2(0.25)^t \\ q_t &= \frac{C_1}{2} - C_2(0.25)^t \end{aligned}$$

At $t = 0$, we have

$$\begin{aligned} p_0 &= C_1 + C_2 \\ q_0 &= \frac{C_1}{2} - C_2 \end{aligned}$$

Solving gives

$$\begin{aligned} C_2 &= \frac{p_0 - 2q_0}{3} \\ C_1 &= \frac{2}{3}(p_0 + q_0) \end{aligned}$$

However, since $p_0 + q_0 = 1$, we have $C_1 = 2/3$. The solutions become

$$\begin{aligned} p_t &= \frac{2}{3} + \frac{p_0 - 2q_0}{3}(0.25)^t \\ q_t &= \frac{1}{3} - \frac{p_0 - 2q_0}{3}(0.25)^t \end{aligned}$$

The limiting value of p_t as t goes to infinity is $2/3$; the limit of q_t is $1/3$.

CHAPTER 25

Section 25.1 (page 854)

1. Form the Hamiltonian:

$$H = -(ay + by^2) + \lambda(x - y)$$

Maximize H with respect to y. Since H is concave, we get

$$\frac{\partial H}{\partial y} = -a - 2by - \lambda = 0$$

This gives $y = (-a - \lambda)/(2b)$. Now form the system of differential equations:

$$\begin{aligned} \dot{\lambda} &= -\lambda \\ \dot{x} &= x + \frac{a + \lambda}{2b} \end{aligned}$$

The boundary conditions are $x(0) = x_0$ and $\lambda(T) = 0$. Since the first equation depends only on λ, we solve it directly. This gives $\lambda(t) = Ce^{-t}$. Using $\lambda(T) = 0$, we conclude that $C = 0$ and therefore $\lambda(t) = 0$. This implies $y(t) = -a/2b$. Now solve the second equation. It has become

$$\dot{x} - x = \frac{a}{2b}$$

which has the following solution, (using $x(0) = x_0$):

$$x(t) = \left(x_0 + \frac{a}{2b}\right)e^t - \frac{a}{2b}$$

3.

$$H = -(ay + by^2 + cx) + \lambda(\alpha x + \beta y)$$

Maximize the Hamiltonian with respect to the control variable.

$$\frac{\partial H}{\partial y} = -a - 2by + \beta\lambda = 0$$

This gives $y = (-a + \beta\lambda)/(2b)$. Next, obtain the differential equations.

$$\begin{aligned} \dot{\lambda} &= -(-c + \alpha\lambda) = c - \alpha\lambda \\ \dot{x} &= \alpha x + \frac{\beta(-a + \beta\lambda)}{2b} \end{aligned}$$

with boundary conditions $x(0) = x_0$ and $\lambda(T) = 0$. Solve the first equation directly. This gives

$$\lambda(t) = C_1 e^{-\alpha t} + \frac{c}{\alpha}$$

Using $\lambda(T) = 0$ means $C_1 = -ce^{\alpha T}/\alpha$. Therefore

$$\lambda(t) = \frac{c}{\alpha}(1 - e^{\alpha(T-t)})$$

The second differential equation now becomes

$$\dot{x} - \alpha x = -\frac{a\beta}{2b} + \frac{\beta^2 c}{2b\alpha}(1 - e^{\alpha(T-t)})$$

To solve, multiply both sides by the integrating factor (Chapter 21) $e^{-\alpha t}$ and rewrite the equation as

$$\frac{d}{dt}(xe^{-\alpha t}) = -\frac{a\beta}{2b}e^{-\alpha t} + \frac{\beta^2 c}{2b\alpha}(e^{-\alpha t} - e^{\alpha(T-2t)})$$

Integrate

$$xe^{-\alpha t} = \frac{a\beta}{2b\alpha}e^{-\alpha t} + \frac{\beta^2 c}{2b\alpha}\left(\frac{-e^{-\alpha t}}{\alpha} + \frac{e^{\alpha(T-2t)}}{2\alpha}\right) + C_2$$

Now use $x(0) = x_0$ to obtain

$$C_2 = x_0 - \frac{a\beta}{2b\alpha} - \frac{\beta^2 c}{2b\alpha}\left(-\frac{1}{\alpha} + \frac{e^{\alpha T}}{2\alpha}\right)$$

Now simplify to express $x(t)$ as

$$x(t) = x_0 e^{\alpha t} + \frac{\beta^2 c}{4b\alpha^2}e^{-\alpha t} + \frac{\alpha\beta + \beta^2 c}{4b\alpha^2}(e^{\alpha T} - 2)(1 - e^{\alpha t})$$

5.

$$H = y - y^2 + 4x - 3x^2 + \lambda(x + y)$$

Maximizing with respect to y gives $y = (1 + \lambda)/2$. Next, we obtain

$$\begin{aligned} \dot{\lambda} &= -(-4 - 6x + \lambda) = 4 + 6x - \lambda \\ \dot{x} &= x + \frac{1 + \lambda}{2} \end{aligned}$$

The system can be written as

$$\begin{bmatrix} \dot{\lambda} \\ \dot{x} \end{bmatrix} = \begin{bmatrix} -1 & 6 \\ \frac{1}{2} & 1 \end{bmatrix}\begin{bmatrix} \lambda \\ x \end{bmatrix} + \begin{bmatrix} 4 \\ \frac{1}{2} \end{bmatrix}$$

The roots are $r_1 = 2$ and $r_2 = -2$. The complete solutions are

$$\begin{aligned} \lambda(t) &= C_1 e^{2t} + C_2 e^{-2t} + \bar{\lambda} \\ x(t) &= \frac{C_1}{2}e^{2t} - \frac{C_2}{6}e^{-2t} + \bar{x} \end{aligned}$$

where $\bar{\lambda} = 1/4$ and $\bar{x} = -5/8$. Use $\lambda(T) = 0$ and $x(0) = x_0$ to obtain the values for C_1 and C_2. After simplification, we obtain

$$\begin{aligned} C_1 &= \frac{6x_0 e^{-4T} + 3.5e^{-2T}}{1 + 3e^{-4T}} \\ C_2 &= \frac{-(0.75e^{-2T} + 6x_0 + 15/4)}{1 + 3e^{-4T}} \end{aligned}$$

7.

$$H = p(K - aK^2) - bI^2 + \lambda(I - \delta K)$$

Maximizing with respect to I yields $I = \lambda/(2b)$. Next, we obtain the differential equations.

$$\begin{aligned} \dot{\lambda} &= -(p - 2paK - \lambda\delta) = -p + 2paK + \lambda\delta \\ \dot{K} &= \frac{\lambda}{2b} - \delta K \end{aligned}$$

These can be written as

$$\begin{bmatrix} \dot{\lambda} \\ \dot{K} \end{bmatrix} = \begin{bmatrix} \delta & 2pa \\ \frac{1}{2b} & -\delta \end{bmatrix}\begin{bmatrix} \lambda \\ K \end{bmatrix} + \begin{bmatrix} -p \\ 0 \end{bmatrix}$$

The roots are $r_1,\ r_2 = \pm\sqrt{\delta^2 + pa/b}$. The complete solutions are

$$\begin{aligned}\lambda(t) &= C_1e^{r_1t} + C_2e^{r_2t} + \bar{\lambda}\\ K(t) &= \frac{r_1-\delta}{2pa}C_1e^{r_1t} + \frac{r_2-\delta}{2pa}C_2e^{r_2t} + \bar{K}\end{aligned}$$

where

$$\begin{aligned}\bar{\lambda} &= \frac{b\delta p}{b\delta^2 + pa}\\ \bar{K} &= \frac{p}{2(b\delta^2+pa)}\end{aligned}$$

The values of C_1 and C_2 are determined by using the boundary conditions $\lambda(T) = 0$ and $K(0) = K_0$. After simplifying, we obtain

$$\begin{aligned}C_1 &= \frac{-2pa(K_0-\bar{K})e^{(r_2-r_1)T} - (r_2-\delta)\bar{\lambda}e^{-r_1T}}{r_2-\delta-(r_1-\delta)e^{(r_2-r_1)T}}\\ C_2 &= \frac{2pa(K_0-\bar{K}) + (r_1-\delta)\bar{\lambda}e^{-r_1T}}{r_2-\delta-(r_1-\delta)e^{(r_2-r_1)T}}\end{aligned}$$

Section 25.2 (page 863)

1. The current-valued Hamiltonian is

$$\mathcal{H} = -y^2 + \mu(x+y)$$

Maximizing with respect to y gives $y = \mu/2$. Next, we obtain

$$\dot{\mu} - \rho\mu = -\mu$$

The system of differential equations then is

$$\begin{aligned}\dot{\mu} &= \mu(\rho-1)\\ \dot{x} &= x + \frac{\mu}{2}\end{aligned}$$

Solving the first directly gives $\mu(t) = C_1e^{(\rho-1)t}$. Using the boundary condition $\mu(T) = 0$ (since T is finite) gives $C_1 = 0$ and, hence, $\mu(T) = 0$. Therefore, $y(t) = 0$. Next, we have $\dot{x} = x$ after substituting for μ. The solution is $x(t) = C_2e^t$. Using the boundary condition $x(0) = x_0$, we get $C_2 = x_0$. Therefore, the final solution is $x(t) = x_0e^t$.

3. The current-valued Hamiltonian is

$$\mathcal{H} = yx - y^2 - x^2 + \mu(x+y)$$

Maximizing with respect to y gives $y = (x+\mu)/2$. The differential equations we obtain are

$$\begin{bmatrix}\dot{\mu}\\ \dot{x}\end{bmatrix} = \begin{bmatrix}\rho - \frac{3}{2} & \frac{3}{2}\\ \frac{1}{2} & \frac{3}{2}\end{bmatrix}\begin{bmatrix}\mu\\ x\end{bmatrix}$$

The roots are $r_1,\ r_2 = \rho/2 \pm (1/2)\sqrt{\rho^2 - 6\rho + 12}$. The solutions are

$$\begin{aligned}\mu(t) &= C_1e^{r_1t} + C_2e^{r_2t}\\ x(t) &= \frac{r_1-\rho+3/2}{3/2}C_1e^{r_1t} + \frac{r_2-\rho+3/2}{3/2}C_2e^{r_2t}\end{aligned}$$

because the steady-state solutions are $\bar{\mu} = \bar{x} = 0$. The values for C_1 and C_2 are found by using $\mu(T) = 0$ and $x(0) = x_0$.

After simplifying, we get

$$\begin{aligned}C_1 &= \frac{-x_0e^{(r_2-r_1)T}}{k_2 - k_1e^{(r_2-r_1)T}}\\ C_2 &= \frac{x_0}{k_2 - k_1e^{(r_2-r_1)T}}\end{aligned}$$

where

$$k_i = \frac{r_i - \rho + 3/2}{3/2}$$

5.

$$\mathcal{H} = p(L - aL^2) - wL - qH^2 + \mu(H - \delta L)$$

Maximizing with respect to the control variable, H, gives $H = \mu/(2q)$. The differential equation system then is

$$\begin{bmatrix}\dot{\mu}\\ \dot{L}\end{bmatrix} = \begin{bmatrix}\rho+\delta & 2pa\\ \frac{1}{2q} & -\delta\end{bmatrix}\begin{bmatrix}\mu\\ L\end{bmatrix} + \begin{bmatrix}-(p-w)\\ 0\end{bmatrix}$$

The roots are

$$r_1,\ r_2 = \rho/2 \pm (1/2)\sqrt{\rho^2 + 4\delta(\rho+\delta) + 4pa/q}$$

The solutions are

$$\begin{aligned}\mu(t) &= C_1e^{r_1t} + C_2e^{r_2t} + \bar{\mu}\\ L(t) &= \frac{r_1-\rho-\delta}{2pa}C_1e^{r_1t}\\ &\quad + \frac{r_2-\rho-\delta}{2pa}C_2e^{r_2t} + \bar{L}\end{aligned}$$

where

$$\begin{aligned}\bar{\mu} &= \frac{q\delta(p-w)}{q\delta(\rho+\delta)+pa}\\ \bar{L} &= \frac{p-w}{2[q\delta(\rho+\delta)+pa]}\end{aligned}$$

To find C_1 and C_2, use the boundary conditions, $\mu(T) = 0$ and $L(0) = L_0$. This gives

$$\begin{aligned}C_1 &= \frac{-(L_0-\bar{L})e^{(r_2-r_1)T} - k_2\bar{\mu}e^{-r_1T}}{k_2 - k_1e^{(r_2-r_1)T}}\\ C_2 &= \frac{L_0 - \bar{L} + k_1\bar{\mu}e^{-r_1T}}{k_2 - k_1e^{(r_2-r_1)T}}\end{aligned}$$

where

$$k_i = \frac{r_i - \rho - \delta}{2pa}$$

To get the optimal path of hiring, use $H^*(t) = \mu(t)/2$.

7. Differentiate the first-order condition for I, $c'(I) = \mu$, to get

$$c''(I)\dot{I} = \dot{\mu}$$

Substitute for $\dot{\mu}$ and we have

$$\dot{I} = \frac{c'(I)(\rho+\delta) + 2aK - 1}{c''(I)}$$

For $\dot{K}$, we have $\dot{K} = I - \delta K$. The isocline for I is the curve defined by $\dot{I} = 0$. This gives

$$c'(I) = \frac{1-2aK}{\rho+\delta}$$

Its slope is

$$\frac{dI}{dK} = \frac{-2a}{c''(I)(\rho+\delta)} < 0$$

The motion of I is given by taking $\partial \dot{I}/\partial K = 2a/c'' > 0$ which means I is increasing to the right and decreasing to the left of the isocline.

The isocline for K is the line $I = \delta K$; K is decreasing to the right and increasing to the left of the isocline.

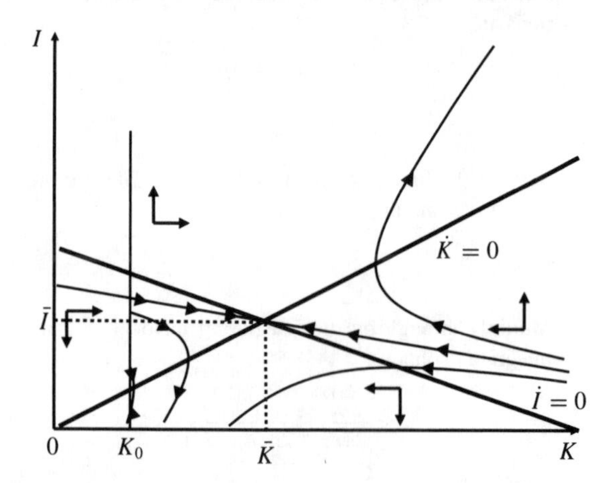

Figure 25.2.7

We know the steady state is a saddle point because the determinant of the linearized coefficient matrix is negative. If $K_0 < \bar{K}$ in Figure 25.2.7, the solution requires choosing the $I(0)$ that will reach, in the given time T, the boundary condition $c'(I(T)) = \mu(T) = 0$. If $c'(0) = 0$, this means $I(T) = 0$.

Section 25.3 (page 876)

1. The free-endpoint version of this problem was solved in Exercise 3 of Section 25.2. The general solutions are

$$\mu(t) = C_1 e^{r_1 t} + C_2 e^{r_2 t}$$
$$x(t) = k_1 C_1 e^{r_1 t} + k_2 C_2 e^{r_2 t}$$

where

$$k_i = \frac{r_i - \rho + 3/2}{3/2}$$
$$r_1,\ r_2 = \frac{\rho}{2} \pm \frac{\sqrt{\rho^2 - 6\rho + 12}}{2}$$

We now use the boundary conditions $x(0) = x_0$ and $x(T) = x_T$ to solve for C_1 and C_2. After simplifying, we get

$$C_1 = \frac{-x_0 e^{(r_2 - r_1)T} + x_T e^{-r_1 T}}{k_1[1 - e^{(r_2 - r_1)T}]}$$
$$C_2 = \frac{x_0 - x_T e^{-r_1 T}}{k_2[1 - e^{(r_2 - r_1)T}]}$$

3.

$$\mathcal{H} = \frac{c^{1-\alpha}}{1-\alpha} + \mu(rx - c)$$

Maximizing with respect to c gives $c = \mu^{-1/\alpha}$. The differential equations are

$$\dot{\mu} = \mu(\rho - r)$$
$$\dot{x} = rx - \mu^{-1/\alpha}$$

Solving the first equation directly gives $\mu(t) = C_1 e^{(\rho - r)t}$. Putting this in the second equation gives

$$\dot{x} - rx = -\left[C_1 e^{(\rho - r)t}\right]^{-1/\alpha} = -C_1^{-1/\alpha} e^{-(\rho - r)t/\alpha}$$

Multiply both sides by the integrating factor, e^{-rt}, to get

$$\frac{d}{dt}(xe^{-rt}) = -C_1^{-1/\alpha} e^{-(\rho - r)t/\alpha} e^{-rt}$$

Integrating gives

$$xe^{-rt} = -\frac{C_1^{-1/\alpha} e^{\left(\frac{r-\rho}{\alpha} - r\right)t}}{\frac{r-\rho}{\alpha} - r} + C_2$$

$$x(t) = -\frac{C_1^{-1/\alpha} e^{\frac{(r-\rho)t}{\alpha}}}{\frac{r-\rho}{\alpha} - r} + C_2 e^{rt}$$

Using $x(0) = x_0$ gives

$$x_0 = -\frac{C_1^{-1/\alpha}}{\frac{r-\rho}{\alpha} - r} + C_2$$

which gives

$$C_2 = x_0 + \frac{\alpha C_1^{-1/\alpha}}{r - \rho - r\alpha}$$

Next, use $x(T) = x_T$ to get

$$x(T) = -\frac{\alpha C_1^{-1/\alpha} e^{\frac{(r-\rho)T}{\alpha}}}{r - \rho - r\alpha} + C_2 e^{rT}$$

which gives

$$C_1^{-1/\alpha} = \frac{r - \rho - r\alpha}{\alpha} e^{-\frac{(r-\rho)T}{\alpha}} (C_2 e^{rT} - x_T)$$

Using the expression for C_2 to substitute gives, after simplifying

$$C_1^{-1/\alpha} = \frac{(r - \rho - r\alpha)(x_0 e^{rT} - x_T)}{\alpha(e^{(r-\rho)T/\alpha} - e^{rT})}$$

Because we are asked to solve for the optimal consumption path, substitute this expression into $c(t) = \mu(t)^{-1/\alpha} = C_1^{-1/\alpha} e^{-(\rho - r)t/\alpha}$. This gives

$$c^*(t) = \frac{(r - \rho - r\alpha)(x_0 e^{rT} - x_T)}{\alpha(e^{(r-\rho)T/\alpha} - e^{rT})} e^{-(\rho - r)t/\alpha}$$

5. The differential equations are

$$\dot{\mu} = \mu(\rho + \delta) + 2x$$
$$\dot{x} = \frac{1 + \mu}{2} - \delta x$$

Set $\dot{\mu} = 0$ to obtain the μ isocline. This gives

$$\mu = \frac{-2x}{\rho + \delta}$$

which has a negative slope and goes through the origin. In addition, $\partial \dot{\mu}/\partial \mu > 0$ so μ is increasing above and decreasing below the isocline.

Set $\dot{x} = 0$ to obtain the x isocline. This gives

$$\mu = 2\delta x - 1$$

which has a positive slope and intercept -1. In addition, $\partial\dot{x}/\partial x < 0$ so x is decreasing to the right and increasing to the left of the isocline.

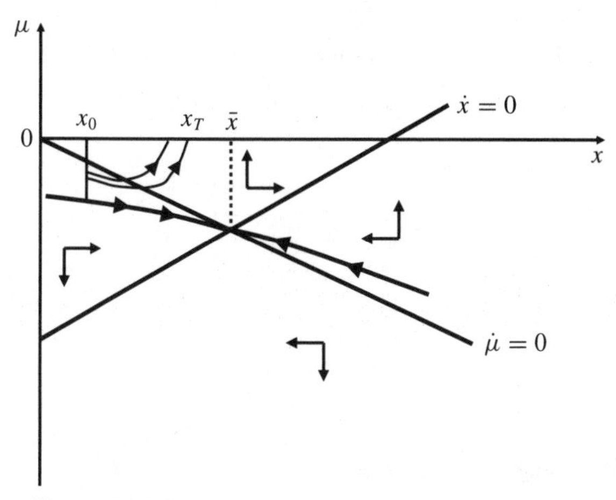

Figure 25.3.5

Suppose x_0 and x_T are as shown in Figure 25.3.5. Clearly, not all trajectories beginning with $x(0) = x_0$ reach x_T. For example, trajectories that also start with $\mu(0)$ close to 0 return to $\mu = 0$ before x_T is reached. Thus, if T is too small, we can only select a trajectory which reaches $\mu = 0$ before x_T is reached—there is no solution. Only if T is large enough does there exist a solution trajectory. As T gets large, the solution trajectory approaches the saddlepath.

7.

$$\mathcal{H} = \ln y - \mu y$$

Maximizing gives $y = 1/\mu$. The differential equations are

$$\begin{aligned} \dot{\mu} &= \rho\mu \\ \dot{x} &= -\frac{1}{\mu} \end{aligned}$$

Solving the first equation gives $\mu(t) = C_1 e^{\rho t}$. If we used the transversality condition, $\mu(T) = 0$, as the boundary condition, it would imply $C_1 = 0$ and then $\mu(t) = 0$ which in turn implies $y(t) \longrightarrow \infty$ which will clearly cause the boundary condition $x(T) \geq 0$ to be violated. We therefore shall use $x(T) = 0$ instead of $\mu(T) = 0$. Substituting the solution for μ into the second equation gives

$$\dot{x} = -\frac{e^{-\rho t}}{C_1}$$

Solving by direct integration gives

$$x(t) = \frac{e^{-\rho t}}{\rho C_1} + C_2$$

Using $x(0) = x_0$ and $x(T) = 0$ gives

$$\begin{aligned} C_1 &= \frac{1 - e^{-\rho T}}{\rho x_0} \\ C_2 &= \frac{-x_0 e^{-\rho T}}{1 - e^{-\rho T}} \end{aligned}$$

Substituting C_1 into $\mu(t)$ and then into $y(t)$ gives the optimal extraction path:

$$y^*(t) = \frac{\rho x_0}{1 - e^{-\rho T}} e^{-\rho t}$$

Section 25.4 (page 887)

1.

$$\mathcal{H} = c - ac^2 + \mu(rx - c)$$

Maximizing gives $c = (1 - \mu)/(2a)$. The differential equations are

$$\begin{aligned} \dot{\mu} &= (\rho - r)\mu \\ \dot{x} &= rx - \frac{1 - \mu}{2a} \end{aligned}$$

Solving the first gives $\mu(t) = C_1 e^{(\rho - r)t}$. Substituting this into the second gives

$$\dot{x} - rx = -\frac{1}{2a} + \frac{C_1 e^{(\rho - r)t}}{2a}$$

Multiply through by the integrating factor, e^{-rt}, and then integrate to obtain

$$xe^{-rt} = \frac{e^{-rt}}{2ar} + \frac{C_1 e^{(\rho - 2r)t}}{2a(\rho - 2r)} + C_2$$

Simplifying gives

$$x(t) = \frac{1}{2ar} + \frac{C_1 e^{(\rho - r)t}}{2a(\rho - 2r)} + C_2 e^{rt}$$

Before applying boundary conditions, we confirm that the steady state is a saddle point. The determinant of the coefficient matrix is $(\rho - r)r < 0$ given the assumption that $\rho - r < 0$. Therefore, we use the steady-state value of x as the boundary condition. To find this value, set $\dot{x} = 0$ and $\dot{\mu} = 0$. This gives $\bar{\mu} = 0$ and $\bar{x} = 1/(2ar)$.

We use

$$\lim_{t \longrightarrow \infty} x(t) = \frac{1}{2ar}$$

to get $C_2 = 0$. Next, we use

$$x_0 = \frac{1}{2ar} + \frac{C_1}{2a(\rho - 2r)}$$

to get $C_1 = (x_0 - \bar{x})2a(\rho - 2r)$. Substituting these values for C_1 and C_2 into the expression for $x(t)$ gives

$$x(t) = \bar{x} + (x_0 - \bar{x})e^{(\rho - r)t}$$

Substituting C_1 into the expression for consumption gives the optimal consumption path:

$$c^*(t) = r\bar{x} + (x_0 - \bar{x})(\rho - 2r)e^{(\rho - r)t}$$

3. Differentiate the first-order condition $c = \mu^{-1}$ to obtain

$$\dot{c} = -\mu^{-2}\dot{\mu}$$

Substitute for $\dot{\mu}$ to obtain

$$\dot{c} = -\mu^{-2}(\rho - k^{-\alpha} + \delta)\mu$$

Substitute for μ to get the differential equation for c:

$$\dot{c} = -c(\rho - k^{-\alpha} + \delta)$$

The differential equation for k is

$$\dot{k} = \frac{k^{1-\alpha}}{1 - \alpha} - c - \delta k$$

Analyze the motion of c first. Set $\dot{c} = 0$ to obtain the isocline $k = (\rho + \delta)^{-1/\alpha}$ which is a vertical line at what turns out to be the steady-state capital stock level. Use

$\partial \dot{c}/\partial k = -\alpha c k^{-\alpha-1} < 0$ to determine that c is decreasing to the right and increasing to the left of the isocline.

Next, set $\dot{k} = 0$ to obtain the isocline defined by

$$c = \frac{k^{1-\alpha}}{1-\alpha} - \delta k$$

Its slope is

$$\frac{dc}{dk} = k^{-\alpha} - \delta$$

which equals 0 at $\hat{k} = \delta^{-1/\alpha}$ which exceeds $\bar{k}$. The second derivative is negative so the shape is as drawn in Figure 25.4.3.

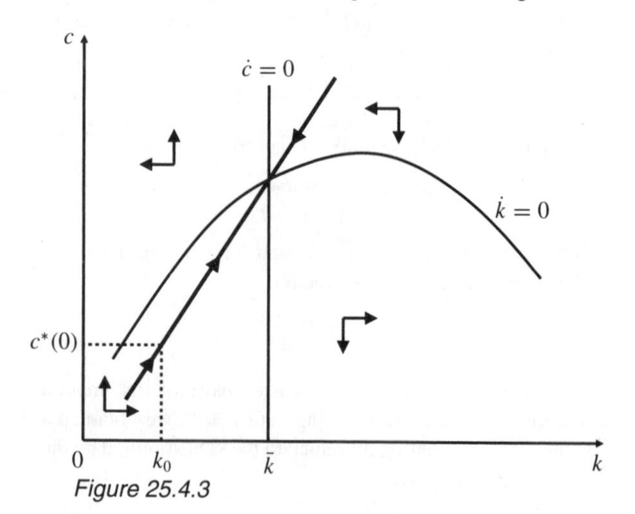

Figure 25.4.3

If $k_0 < \bar{k}$, the solution requires choosing a low value for consumption, $c^*(0)$, initially, which implies a high level of saving. As a result, the capital stock grows as we follow the saddlepath, with $c(t)$ and $k(t)$ rising as they approach the steady state.

5. The finite-time version of this problem is solved in Exercise 7 of Section 25.3. The general solutions obtained are

$$\mu(t) = C_1 e^{\rho t}$$

$$x(t) = \frac{e^{-\rho t}}{\rho C_1} + C_2$$

If we tried to find the steady state in this problem by setting $\dot{\mu} = 0$, we would find that implies $\mu = 0$ which in turn implies $y \longrightarrow \infty$. Since this would violate the constraint on the resource stock, $x(t) \geq 0$, we can safely assume that it is optimal to deplete the stock. We therefore use

$$\lim_{t \longrightarrow \infty} x(t) = 0$$

which gives $C_2 = 0$ because $\rho > 0$. Next, use

$$x_0 = \frac{1}{\rho C_1}$$

to obtain $C_1 = 1/(\rho x_0)$. The solutions become

$$\mu(t) = \frac{e^{\rho t}}{\rho x_0}$$

$$x(t) = x_0 e^{-\rho t}$$

Therefore, $y^*(t) = \rho x_0 e^{-\rho t}$.

Section 25.5 (page 895)

1.

$$\mathcal{H} = c + \mu(rx - c)$$

Form the Lagrangean

$$\mathcal{L} = c + \mu(rx - c) + \theta_1 c + \theta_2(c_{\max} - c)$$

Maximizing with respect to c gives

$$1 - \mu + \theta_1 - \theta_2 = 0$$

with

$$\theta_1 \geq 0 \qquad \theta_1 c = 0$$

$$\theta_2 \geq 0 \qquad \theta_2(c_{\max} - c) = 0$$

These imply that whenever $c > 0$, we have $\theta_1 = 0$ (and $c = 0$ implies $\theta_1 \geq 0$). Whenever $c < c_{\max}$, $\theta_2 = 0$ (and $c = c_{\max}$ implies $\theta_2 \geq 0$).

The differential equation for μ is $\dot{\mu} = (\rho - r)\mu$ and for x is $\dot{x} = rx - c$. Analyze the three cases to construct the solution.

Case 1: $c = 0$ so that $\theta_1 \geq 0$ and $\theta_2 = 0$. Therefore, we have $\mu = 1 + \theta_1$ so $\mu \geq 1$ in this case.

Case 2: $c = c_{\max}$ so that $\theta_1 = 0$ and $\theta_2 \geq 0$. Therefore, we have $\mu = 1 - \theta_2$ so $\mu \leq 1$ in this case.

Case 3: $0 < c < c_{\max}$ so that $\theta_1 = \theta_2 = 0$. Therefore, $\mu = 1$.

These three cases define three critical regions for μ. They tell us that for $\mu > 1$, we set $c = 0$ so that $\dot{x} = rx > 0$. That is, consumption is zero and k is rising in this region. Further, for $\mu < 1$, we set $c = c_{\max}$ so that $\dot{x} = rx - c_{\max}$ which we assume is negative.

To construct the phase diagram in Figure 25.5.1, note that $\dot{\mu} = 0$ implies $\mu = 0$. Thus, as long as $\mu > 0$, it is declining. Next, we do not graph the $\dot{x} = 0$ isocline in the region $\mu < 1$ since we assume it occurs at a larger-than-relevant size.

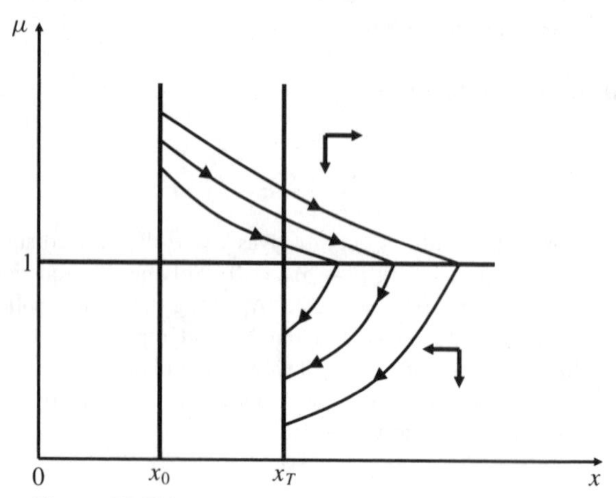

Figure 25.5.1

Suppose x_0 and x_T are as shown in Figure 25.5.1. Then, the solution is to choose a $\mu(0)$ that puts us on a trajectory that first moves to the right ($c = 0$ and $\dot{x} > 0$) until it reaches $\mu = 1$, then switches by moving to the left ($c = c_{\max}$ and $\dot{x} < 0$) until it reaches x_T. Many trajectories do this but we must pick the one that takes exactly the given amount of time, T, to complete its journey.

3.

$$\mathcal{H} = pK^\alpha L^\beta - wL - qI + \mu(I - \delta K)$$

$$\mathcal{L} = pK^\alpha L^\beta - wL - qI + \mu(I - \delta K) + \theta_1 I + \theta_2(b - I)$$

Maximizing with respect to L gives

$$\frac{\partial \mathcal{L}}{\partial L} = \beta p K^\alpha L^{\beta-1} - w = 0$$

which implies

$$L = \left(\frac{w}{\beta p}\right)^{\frac{1}{\beta - 1}} K^{\frac{-\alpha}{\beta - 1}}$$

Maximizing with respect to I gives

$$-q + \mu + \theta_1 - \theta_2 = 0$$

which implies the following three cases:

Case 1: $I = 0$ so that $\theta_1 \geq 0$ and $\theta_2 = 0$. Then $\mu \leq q$. In this region, $\dot{K} = -\delta K < 0$.

Case 2: $I = b$ so that $\theta_1 = 0$ and $\theta_2 \geq 0$. Then $\mu \geq q$. In this region, $\dot{K} = b - \delta K < 0$ by assumption.

Case 3: $0 < I < b$ so that $\theta_1 = \theta_2 = 0$. Then $\mu = q$.

The differential equation for μ is

$$\dot{\mu} - \rho\mu = -[p\alpha K^{\alpha-1} L^\beta - \mu\delta]$$

To analyze the motion of μ, set $\dot{\mu} = 0$ to find its isocline. This gives

$$\mu = \frac{p\alpha A}{\rho + \delta} K^{\frac{1 - \alpha - \beta}{\beta - 1}}$$

which has a negative slope because $\alpha + \beta < 1$ and $\alpha > 0$, $\beta > 0$. Use $\partial\dot{\mu}/\partial\mu = \rho + \delta > 0$ to determine that μ is increasing above and decreasing below the isocline.

The phase diagram has the same appearance and properties as Figure 25.13 which is used to analyze the linear investment problem in the chapter. The optimal solution is to follow the saddlepath, reaching the steady state in finite time and staying there forever.

Section 25.6 (page 902)

1.

$$\mathcal{H} = \frac{y^{1-\alpha}}{1 - \alpha} - c - \mu y$$

Maximizing with respect to y gives $y = \mu^{-1/\alpha}$. The co-state variable must satisfy $\dot{\mu} - \rho\mu = 0$. Solving this equation directly gives $\mu(t) = C_1 e^{\rho t}$. Applying the transversality condition $\mu(T) = 0$ (if T is finite) would imply $y \longrightarrow \infty$ so that the constraint $x(T) \geq 0$ would be violated. Therefore, if T is finite, we use $x(T) = 0$ as the boundary condition.

Substitute into $\dot{x} = -y$ to get

$$\dot{x} = -C_1^{-1/\alpha} e^{-\rho t/\alpha}$$

Solving this by direct integration gives

$$x(t) = \frac{\alpha C_1^{-1/\alpha}}{\rho} e^{-\rho t/\alpha} + C_2$$

Before we know what boundary condition to use, we must determine whether T is finite or not. Setting $e^{-\rho T}\mathcal{H}(T) = 0$ gives

$$e^{-\rho T}\left[\frac{y^{1-\alpha}}{1 - \alpha} - c - \mu y\right] = 0$$

at time T. Substitute for μ and simplify to get

$$e^{-\rho T} y^{1-\alpha}\left(\frac{\alpha}{1 - \alpha}\right) = c e^{-\rho T}$$

Therefore $y(T) = [c(1-\alpha)/\alpha]^{1/(1-\alpha)} > 0$. Since $y(T) > 0$, then $y(t) > 0$ for all t which means the resource will be used up in finite time. Therefore the optimal T is finite. We use $x(T) = 0$ then to get

$$0 = \frac{\alpha C_1^{-1/\alpha}}{\rho} e^{-\rho T/\alpha} + C_2$$

which gives $C_2 = -\alpha/\rho C_1^{-1/\alpha} e^{-\rho T/\alpha}$. Next use $x(0) = x_0$ to get

$$x_0 = \frac{\alpha}{\rho} C_1^{-1/\alpha} + C_2$$

Using these two expressions for C_2 gives

$$C_1^{-1/\alpha} = \frac{\rho x_0/\alpha}{1 - e^{-\rho T/\alpha}}$$

Substituting this into the expression for $\mu(t)$ and into $y(t)$ gives the optimal path of extraction.

$$y^*(t) = \frac{\rho x_0/\alpha}{1 - e^{-\rho T/\alpha}} e^{-\rho t/\alpha}$$

T has not yet been determined but we have the information required to do so. From setting $\mathcal{H}(T) = 0$, we obtained a value for $y(T)$. Setting this equal to the value implied by the above expression gives

$$\left[\frac{c(1 - \alpha)}{\alpha}\right]^{1/(1-\alpha)} = \frac{\rho x_0/\alpha}{1 - e^{-\rho T/\alpha}} e^{-\rho T/\alpha}$$

Solving for T gives

$$T^* = \frac{\alpha}{\rho} \ln\left[1 + \frac{\rho x_0}{\alpha}\left(\frac{c(1 - \alpha)}{\alpha}\right)^{-1/(1-\alpha)}\right]$$

3.

$$\mathcal{H} = \frac{y^{1-\alpha}}{1 - \alpha} - \mu y$$

Maximizing gives $y = \mu^{-1/\alpha}$. The co-state variable must satisfy $\dot{\mu} - \rho\mu = 0$ which gives $\mu(t) = C_1 e^{\rho t}$. Again, we do not set $\mu(T) = 0$ as this would cause the constraint $x(T) \geq 0$ to be violated. Therefore, we use $x(T) = 0$ if T is finite or the limit of $x(t) = 0$ as $t \longrightarrow \infty$ if T is infinite as the boundary condition.

For x, we have $\dot{x} = -y$ which becomes

$$\dot{x} = -C_1^{-1/\alpha} e^{-\rho t/\alpha}$$

Solving gives

$$x(t) = \frac{\alpha C_1^{-1/\alpha}}{\rho} e^{-\rho t/\alpha} + C_2$$

Using $x(0) = x_0$ gives $C_2 = x_0 - \alpha C_1^{-1/\alpha}/\rho$. To find T, set $e^{-\rho T}\mathcal{H}(T) = 0$. This gives, after substituting for $\mu(T)$,

$$e^{-\rho T}\left(\frac{y^{1-\alpha}}{1 - \alpha} - y^{1-\alpha}\right) = 0$$

This is solved by $y(T) = 0$ or $T = \infty$. If $y(T) = 0$ in finite time, this implies $\mu(T) = \infty$ which implies $\mu(t) = \infty$ for all t and therefore $y(t) = 0$ for all t. Thus, $y(T) = 0$ cannot be the solution. Therefore, $T^* = \infty$.

We therefore use

$$\lim_{t \longrightarrow \infty} x(t) = 0$$

which implies $C_2 = 0$. Our other expression for C_2 then implies

$$C_1^{-1/\alpha} = \frac{\rho x_0}{\alpha}$$

The solution then is

$$y^*(t) = \frac{\rho x_0}{\alpha} e^{-\rho t/\alpha}$$

Review Exercises (page 903)

1.

$$H = F(x, y) + \lambda G(x, y)$$

Maximizing H with respect to y leads to

$$F_y + \lambda G_y = 0$$

where F_y is notation for $\partial F(x, y)/\partial y$ and so on. Take the derivative of H to obtain

$$\dot{H} = F_x \dot{x} + F_y \dot{y} + \dot{\lambda} G(x, y) + \lambda(G_x \dot{x} + G_y \dot{y})$$

Collect like terms to get

$$\dot{H} = (F_x + \lambda G_x)\dot{x} + (F_y + \lambda G_y)\dot{y} + \dot{\lambda} G(x, y)$$

Along the optimal path, the first-order conditions for y hold, $F_y + \lambda G_y = 0$. Also, the costate variable obeys the equation $\dot{\lambda} = -(F_x + \lambda G_x)$. Making these substitutions gives

$$\dot{H} = (F_x + \lambda G_x)\dot{x} - (F_x + \lambda G_x)G(x, y)$$

but since $\dot{x} = G(x, y)$, we have shown that $\dot{H} = 0$ along the optimal path.

3.

$$\mathcal{H} = K^\alpha - I^2 + \mu(I - \delta K)$$

Maximizing with respect to I gives $I = \mu/2$. The costate variable must satisfy

$$\dot{\mu} - \rho\mu = -[\alpha K^{\alpha-1} - \mu\delta]$$

so

$$\dot{\mu} = (\rho + \delta)\mu - \alpha K^{\alpha-1}$$

and $\dot{K}$ is given by

$$\dot{K} = \frac{\mu}{2} - \delta K$$

To construct the phase diagram, begin with the motion of μ. Set $\dot{\mu} = 0$ to obtain the isocline

$$\mu = \frac{\alpha K^{\alpha-1}}{\rho + \delta}$$

which has slope $d\mu/dK = (\alpha - 1)\alpha K^{\alpha-2}/(\rho+\delta) < 0$ (since $0 < \alpha < 1$). Also note that as $K \longrightarrow 0$, $\mu \longrightarrow \infty$. This gives the shape drawn in Figure 25.R.3. Use $\partial\dot{\mu}/\partial\mu = \rho + \delta > 0$ to determine that μ is increasing above and decreasing below the isocline.

Next, set $\dot{K} = 0$ to obtain the isocline $\mu = 2\delta K$ which is a line with a positive slope emanating from the origin. Use $\partial\dot{K}/\partial K = -\delta$ to determine that K is decreasing to the right and increasing to the left of the isocline. This is a free-endpoint problem ($K(T)$ is unspecified). If T is fixed, the boundary conditions are $\mu(T) = 0$ and $K(0) = K_0$. For example, if K_0 is as shown in the diagram, the optimal trajectory starts along the vertical line at K_0 and ends with $\mu(T) = 0$. Many trajectories do this but the optimal trajectory is the one that takes exactly T to do this.

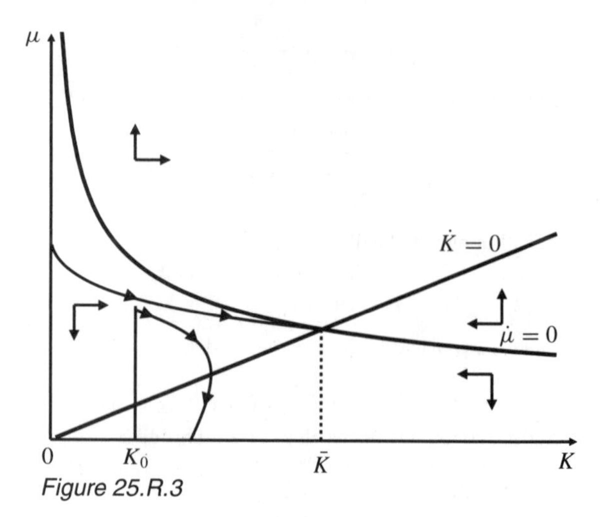

Figure 25.R.3

5.

$$\mathcal{H} = \frac{y^{1-\alpha}}{1-\alpha} - \mu(rx - y)$$

Maximizing with respect to y gives $y = \mu^{-1/\alpha}$. The costate variable must satisfy $\dot{\mu} = \mu(\rho - r)$. Solving gives $\mu(t) = C_1 e^{(\rho-r)t}$. Since $x(T)$ is not fixed, we could use $\mu(T) = 0$ as a boundary condition. However, this implies $C_1 = 0$ and therefore $\mu(t) = 0$ which in turn implies $y(t) \longrightarrow \infty$. This would cause the constraint $x(T) \geq 0$ to be violated. Therefore, we leave $\mu(T)$ free and use $x(T) = 0$ as the boundary condition.

The differential equation for x becomes

$$\dot{x} - rx = -C_1^{-1/\alpha} e^{-(\rho-r)t/\alpha}$$

Multiply through by the integrating factor and rewrite it as

$$\frac{d}{dt}(xe^{-rt}) = -C_1^{-1/\alpha} e^{-[(\rho-r)/\alpha + r]t}$$

Integrating gives

$$xe^{-rt} = \frac{\alpha C_1^{-1/\alpha}}{\rho - r + r\alpha} e^{-[(\rho-r)/\alpha+r]t} + C_2$$

which becomes

$$x(t) = \frac{\alpha C_1^{-1/\alpha}}{\rho - r + r\alpha} e^{-[(\rho-r)/\alpha]t} + C_2 e^{rt}$$

Using $x(0) = x_0$ gives

$$C_2 = x_0 - \frac{\alpha C_1^{-1/\alpha}}{\rho - r + r\alpha}$$

Using $x(T) = 0$ gives

$$C_2 = -\frac{\alpha C_1^{-1/\alpha}}{\rho - r + r\alpha} e^{-[(\rho-r)/\alpha]T} e^{-rT}$$

Using these two expressions to solve for C_1 gives

$$C_1^{-1/\alpha} = \frac{(\rho - r + r\alpha)x_0}{\alpha[1 - e^{-(\rho-r)T/\alpha} e^{-rT}]}$$

Using this gives the solution for the optimal consumption path

$$y^*(t) = \frac{(\rho - r + r\alpha)x_0}{\alpha[1 - e^{-(\rho-r)T/\alpha}e^{-rT}]}e^{-(\rho-r)t/\alpha}$$

7.

$$\mathcal{H} = py - \frac{y^2}{2(R+\alpha)} - \mu y$$

Maximizing with respect to y gives $y = (p-\mu)(R+\alpha)$. The costate variable must satisfy

$$\dot{\mu} - \rho\mu = -\left[\frac{y^2}{2(R+\alpha)^2}\right]$$

Substituting for y and re-arranging gives

$$\dot{\mu} = \rho\mu - \frac{(p-\mu)^2}{2}$$

and we also have

$$\dot{R} = -(p-\mu)(R+\alpha)$$

Set $\dot{\mu} = 0$ to draw the isocline for μ. This gives

$$2\rho\mu - (p-\mu)^2 = 0$$

which has slope $d\mu/dR = 0$. The isocline is therefore a horizontal line; however, because the equation is quadratic, there are two values of μ at which $\dot{\mu} = 0$. It is possible to prove that one of the values is smaller than p and the other is larger. Only the smaller one is relevant because if $\mu > p$, then $y < 0$ which is impossible. To prove this, solve the quadratic. Then, since $\rho - \sqrt{\rho^2 + 2p\rho} < 0$ and $\rho + \sqrt{\rho^2 + 2p\rho} > 0$, you can prove that $\bar{\mu}_1 < p$ and $\bar{\mu}_2 > p$ in Figure 25.R.7.

Now, use $\partial\dot{\mu}/\partial\mu = \rho + (p-\mu)$ to find the motion of μ. Because we are interested only in the range of μ values satisfying $p - \mu > 0$, we see that μ is increasing above and decreasing below the isocline.

Set $\dot{R} = 0$ to find the isocline $(p-\mu)(R+\alpha) = 0$. This is satisfied when $p = \mu$ or $R = -\alpha$. The motion of R is always negative for $\mu < p$.

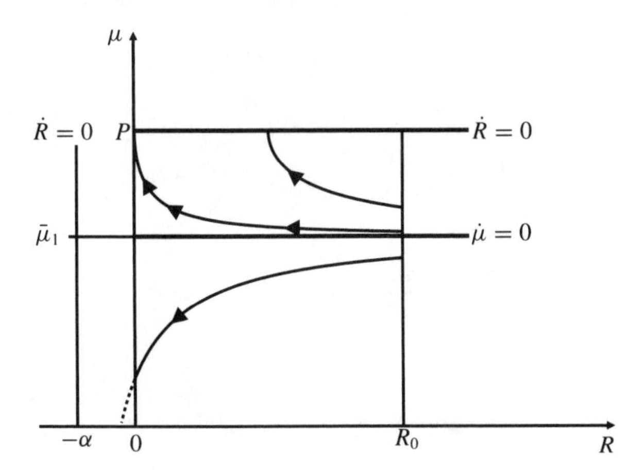

Figure 25.R.7

The phase diagram in Figure 25.R.7 shows the steady state occurs at $\bar{R} = -\alpha$ and $\mu = \bar{\mu}_1$. The saddlepath is the horizontal line along $\dot{\mu} = 0$. It is clearly not possible to reach the steady state because it violates the constraint $R(T) \geq 0$.

What are the correct boundary conditions then? We first look at the condition determining the optimal T.

$$e^{-\rho T}\mathcal{H}(T) = e^{-\rho T}\left[py - \frac{y^2}{2(R+\alpha)} - \left(p - \frac{y}{R+\alpha}\right)y\right] = 0$$

This is solved with $y(T) = 0$ for a finite T. If T were infinite, we would still need $y(T) \longrightarrow 0$ as $T \longrightarrow \infty$ so as to not violate the constraint $R(T) \geq 0$. Thus, $y(T) \longrightarrow 0$ implies $\mu(T) \longrightarrow p$ from the solution for y. This tells us that we cannot use $\mu(T) = 0$ as a boundary condition. Thus, we leave $\mu(T)$ free and use $R(T) = 0$ as the relevant boundary condition.

The phase diagram helps to understand the solution. The optimal trajectory must begin at $R(0) = R_0$; it must finish at $R(T) = 0$ to satisfy the boundary condition; and it must reach $\mu(T) = p$ to satisfy the condition $y(T) = 0$ given by the condition for choosing the optimal T. There is only one trajectory satisfying these three conditions. Moreover, since the point it reaches $(R = 0,\ \mu = p)$ is not a steady state, it is reached in a finite amount of time. Therefore, T^* is finite.

9.

$$\mathcal{H} = \left(p - \frac{c}{x}\right)h + \mu[rx(1-x) - h]$$

$$\mathcal{L} = \left(p - \frac{c}{x}\right)h + \mu[rx(1-x) - h] + \theta_1 h + \theta_2[h_{max} - h]$$

$$\frac{\partial\mathcal{L}}{\partial h} = p - \frac{c}{x} - \mu + \theta_1 - \theta_2 = 0$$

We have three cases to consider.

Case 1: $h = 0$ implies $\theta_2 = 0$ and $\theta_1 \geq 0$ so $\mu \geq p - c/x$

$$\dot{\mu} - \rho\mu = -\mu(r - 2rx)$$
$$\dot{x} = rx(1-x); \qquad \dot{x} > 0 \qquad \text{for } 0 < x < 1$$

Case 2: $h = h_{max}$ implies $\theta_1 = 0$ and $\theta_2 \geq 0$ so $\mu \leq p - c/x$

$$\dot{\mu} - \rho\mu = -\left[\frac{ch_{max}}{x^2} + \mu(r - 2rx)\right]$$
$$\dot{x} = rx(1-x) - h_{max} < 0 \qquad \text{by assumption}$$

Case 3: $0 < h < h_{max}$ implies $\mu = p - c/x$.

$$\dot{\mu} - \rho\mu = -\left[\frac{ch}{x^2} + \mu(r - 2rx)\right]$$
$$\dot{x} = rx(1-x) - h < 0$$

The relevant boundary conditions are $x(0) = x_0$ and

$$\lim_{t\longrightarrow\infty} x(t) = \bar{x}$$

where $\bar{x}$ is the steady-state value of x, as long as the steady state is a saddle point.

The steady state is found by setting $\dot{\mu} = \dot{x} = 0$. Since $\dot{x} = 0$ implies $h = rx(1-x)$, this can occur only in Case 3 when $\mu = p - c/x$. Setting $\dot{\mu} = 0$ in that case gives

$$\rho\mu = \frac{ch}{x^2} + \mu(r - 2rx)$$

Substituting $h = rx(1-x)$ and $\mu = p - c/x$ and solving for x gives

$$x^2 + \frac{p(\rho - r) - cr}{2rp}x - \frac{\rho c}{2rp} = 0$$

Because $-\rho c/(2rp) < 0$, the two solutions for x are of opposite sign. We are interested in the one which is positive. However, writing out the expression for x offers no insights to the solution but is helpful only if we want to calculate the actual numerical value for $\bar{x}$ given values for all the parameters on which it depends. Since we do not wish to do that, we shall just write $\bar{x} > 0$ as the solution. Equation (25.64) in the text provides a clue to the economic insights to this solution.

The optimal solution then depends on x_0. If $x_0 < \bar{x}$, we set $h = 0$ and Case 1 applies. In this case, μ falls and x rises. When $\mu = p - c/x$, which occurs in finite time, we set $h = r\bar{x}(1 - \bar{x})$ and remain there forever.

If $x_0 > \bar{x}$, we set $h = h_{max}$ and Case 2 applies. In this case, μ rises and x falls. When $\mu = p - c/x$, which again occurs in finite time, we set $h = r\bar{x}(1 - \bar{x})$ and remain there forever.